V&R Academic

John P. A. Lamers / Asia Khamzina /
Inna Rudenko / Paul L. G. Vlek (eds.)

Restructuring land allocation, water use and agricultural value chains

Technologies, policies and practices for the lower Amudarya region

With numerous figures

V&R unipress

Bonn University Press

Bibliographic information published by the Deutsche Nationalbibliothek

The Deutsche Nationalbibliothek lists this publication in the Deutsche Nationalbibliografie;
detailed bibliographic data are available on the Internet at http://dnb.d-nb.de.

ISBN 978-3-8471-0297-7
ISBN 978-3-8470-0297-0 (E-Book)

**Publications of Bonn University Press
are published by V&R unipress GmbH.**

© Copyright 2014 by V&R unipress GmbH, 37079 Goettingen, Germany
All rights reserved, including those of translation into foreign languages. No part of this work may
be reproduced or utilized in any form or by any means, electronic or mechanical, including
photocopying, microfilm and recording, or by any information storage and retrieval system,
without permission in writing from the publisher.
Printed in Germany.
Cover image: © Dr. Kirsten Kienzler
Printing and binding: CPI buchbuecher.de GmbH, Birkach

Inhalt

Foreword (Dr. Georg Schütte) 9

Foreword (Ruzumbay Eshchanov) 11

Section 1: Introduction

Paul L. G. Vlek, John P. A. Lamers, Asia Khamzina, Inna Rudenko,
Christopher Martius, Bernhard Tischbein, Ruzumbay Eshchanov
Restructuring land allocation, water use and agricultural value chains.
Technologies, policies and practices for the lower Amudarya region . . . 15

Section 2: Production Systems

Nodir Djanibekov, Ihtiyor Bobojonov, Kristof Van Assche,
Inna Rudenko, Kudrat Nurmetov, John P. A. Lamers
2.1 Farm restructuring in Uzbekistan through fragmentation to
consolidation 33

Alexandra Conliffe
2.2 Geography Matters: Understanding Smallholder Livelihoods in Rural
Khorezm 45

Krishna P. Devkota, Ahmad M. Manschadi, John P. A. Lamers,
Erkin Ruzibaev, Mina K. Devkota, Oybek Egamberdiev,
Raj K. Gupta, Paul L. G. Vlek
2.3 Exploring innovations to sustain rice production in Central Asia: A
case study from the Khorezm region of Uzbekistan 63

Inna Rudenko, John P. A. Lamers, Utkur Djanibekov,
Sanjar Davletov
2.4 Virtual water along the Uzbek cotton value chain 77

Section 3: Natural Resource Management

Bernhard Tischbein, Usman Khalid Awan, Fazlullah Akhtar,
Pulatbay Kamalov, Ahmad M. Manschadi
3.1 Improving irrigation efficiency in the lower reaches of the Amu
Darya River 91

Mirzakhayot Ibrakhimov, Bernhard Tischbein, Usman Khalid Awan
3.2 Knowledge on groundwater – A prerequisite for water management in
Khorezm 109

Akmal Akramkhanov, Bernhard Tischbein, Usman Khalid Awan
3.3 Effective management of soil salinity – revising leaching norms ... 121

Alexander Tupitsa, John P. A. Lamers, Asia Khamzina, Evgeniy Botman,
Martin Worbes, Christopher Martius, Paul L. G. Vlek
3.4 Adaptation of photogrammetry for tree hedgerow and windbreak
assessment in the irrigated croplands of the Khorezm region 135

Dilfuza Djumaeva, John P. A. Lamers, Asia Khamzina,
Shirin Babajanova, Ruzumbay Eshchanov, Paul L. G. Vlek
3.5 Nitrogen fixation by trees with P-fertilization enhances growth and
carbon sequestration in degraded irrigated croplands 153

Sebastian Fritsch, Christopher Conrad, Teresa Dürbeck,
Gunther Schorcht
3.6 Mapping marginal land in Khorezm using GIS and remote sensing
techniques 167

Elena N. Ginatullina, Laurel Saito, Lisa Atwell, Diana B. Shermetova,
Dilorom Fayzieva, John P. A. Lamers, Sudeep Chandra,
Margaret Shanafield
3.7 Water chemistry and zooplankton communities in drainage lakes in
downstream Amu Darya, Central Asia 179

Section 4: Production and Resource Economics

Maksud Bekchanov, John P. A. Lamers, Christopher Martius
4.1 Coping with water scarcity in the irrigated lowlands of the lower
Amudarya basin, Central Asia 199

Anik Bhaduri, Nodir Djanibekov
4.2 Potential Water Price Flexibility, Tenure Uncertainty and Cotton Restrictions on Adoption of Efficient Irrigation Technology in Uzbekistan 217

Aziz A. Karimov, Miguel Niño-Zarazúa
4.3 Assessing Efficiency of input Utilization in Wheat Production in Uzbekistan 231

V. S. Saravanan, Mehmood Ul-Hassan, Benjamin Schraven
4.4 Irrigation water management in Uzbekistan: analyzing the capacity of households to improve water use profitability 253

Yadira Mori Clement, Anik Bhaduri, Nodir Djanibekov
4.5 Food price fluctuations in Uzbekistan: Evidences from local markets in 2002–2010 275

Section 5: Society, Policy and Institutions

Bekchanov Maksud, John P. A. Lamers, Kudrat Nurmetov
5.1 Economic incentives for adopting irrigation innovations in arid environments 297

Anastasiya Shtaltovna, Anna-Katharina Hornidge, Peter P. Mollinga
5.2 Caught in a Web – Travails of a Machine Tractor Park in Khorezm, Uzbekistan 317

Anna-Katharina Hornidge, Mehmood Ul-Hassan,
Laurens van Veldhuizen
5.3 Follow the innovation: transdisciplinary innovation research in Khorezm, Uzbekistan 341

Section 6: Conclusions and Options for Action

John P. A. Lamers, Paul L. G. Vlek, Asia Khamzina,
Bernhard Tischbein, Inna Rudenko
Conclusions, recommendations and outlook 365

List of abbreviations and acronyms 379

Glossary of Latin, Russian and Uzbek words 381

List of authors/co-authors and postal addresses 383

Foreword (Dr. Georg Schütte)

Central Asia is one of the largest irrigated zones worldwide. For centuries, irrigated agriculture has been the most important economic basis in the Khorezm Province in northwestern Uzbekistan. Unsustainable land management practices introduced during the soviet era combined with outdated and inefficient irrigation and drainage infrastructures inevitably led to wide-spread soil degradation and low-productivity systems. The overuse of water resources has been exacerbating the desiccation of the Aral Sea. For that reason the German engagement in the Aral Sea region in research and capacity development aimed to make an important contribution to economic and social development while simultaneously safeguarding water and land resources in the region.

Under the funding initiative "Integrated Water Resources Management (IWRM) – From Research to Implementation" of the German Federal Ministry of Education and Research (BMBF), new approaches and concepts have been developed and adapted to the different natural and socio-economic conditions in cooperation with emerging and developing countries. The strategic framework for this IWRM funding initiative was set in the BMBF funding programme entitled "Research for Sustainable Development – FONA", which includes the funding priority "Sustainable Water Management – NaWaM". With these measures, German research contributes innovatively to the design and implementation of solutions for better water resources management worldwide. Precondition for these research projects was the active participation and cooperation of industrial partners and stakeholders from Germany and the respective countries. Besides the development and testing of technological solutions to improve water use efficiency, all projects had to consider the enabling institutional and policy environment. An important aspect of all project's activities was local capacity development to strengthen participants and stakeholders to secure sustainable project continuation in the future. The Khorezm project in Uzbekistan was one of seventeen research projects that developed and implemented IWRM concepts.

The aforementioned key elements of IWRM guided the activities of the German-Uzbek project entitled "Economic and ecological restructuring of land and water use in the Khorezm region of Uzbekistan – A pilot project in development research". It has been led by the Center for Development Research (ZEF) on the German side, the State University of Urgench (UrDU) on the Uzbek side and UNESCO at the international level. Activities started in 2001 and project results were delivered to the local institutions until 2011. In the process, a sustainable structure of local institutions and the UNESCO has been established to take over the project after its lifetime and to disseminate project findings in and beyond the model region Khorezm. The results achieved within one decade of interdisciplinary research to improve land and water management in agriculture demonstrate the enormous efforts of the bilateral project team.

The project recommended a range of technical innovations to improve irrigation water use efficiency supported by economic tools and improved enabling institutional and policy environments. A number of these innovations are being adopted by farmers and found their way into national agricultural policies. Moreover, a modern GIS and soil laboratory at UrDU funded by BMBF supports the state-of-the-art research and education in natural and social sciences at UrDU for future generations. The partnership between ZEF, UrDU and UNESCO resulted in the establishment of a UNESCO Chair on Education for Sustainable Development at UrDU, which ensures the transfer of scientific methods and research findings from the project into teaching. More than 50 PhD students and more than 100 MSc students, most of them from Uzbekistan, have become experts and decision makers assuring a broad communication of research findings to regions with similar agro-ecological conditions.

For Germany, the improvement of the environment, particularly of water resources is of highest importance. That is why BMBF engages with local partners in research projects worldwide. Thus it contributes to the development of sustainable solutions in water management for a better future for all. In the geographically strategic region of Central Asia, Germany promotes security and stability, supports economic development and sustains water and land resources. This book aims to share the experiences gained in the Khorezm project with you and is testimony of the BMBF commitment to integrated water management.

In addition to scholarly and scientific insight, this book may also raise your interest in our BMBF research initiatives. I certainly hope that it triggers also new ideas for future research activities on all levels. Thus I wish you a pleasant time while reading this publication.

Dr. Georg Schütte, State Secretary
German Federal Ministry of Education and Research

Foreword (Ruzumbay Eshchanov)

Uzbekistan needs cooperation for research and higher education development.

Seven decades of integration in the Soviet Union have brought our country not only laureates. No doubt, the expansion of irrigated agriculture to produce cotton throughout much of Central Asia has been of paramount social and economic importance. Cotton revenues still account for a considerable share in foreign exchange revenues and national income, and employment and income security for rural families. However, the (ab)use of the natural resources for irrigated crop production has also had ecological and social consequences: the desiccation of the Aral Sea, land degradation and desertification arising from soil erosion, salinization, overgrazing of pastures, unsustainable agricultural practices, sand encroachment, seasonal drought, and more. The implementation of the ZEF/UNESCO project in the Khorezm region aiming at good governance of natural resources in general but especially of land and water management in the irrigated areas of the Aral Sea Basin came, therefore, just at the right time.

We knew some facts. Between 1950 and 1990, the irrigated cropland areas in Central Asia grew from 2 to ca. 8 million ha, and between 1950 and 2000 the population increased by about 300 %. Irrigated agriculture became the keystone for the welfare of the region and its rural population. Modern means of production replaced traditional crop rotations. Only later were we confronted with widespread land and ecological degradation of the irrigated dryland ecosystems. We were convinced that fertilizers, seeds, machinery, pesticides, water schemes, etc., were all that was needed. However, the consideration given to the environment and to our farmers did not match these efforts; a high price is being paid for this.

The texts of the Avesta, the sacred book of the Zoroastrians, teach us that one should be committed to a life based on good thoughts, good words and good deeds. In this context, the ZEF/UNESCO project was exemplary for us in various ways. It combined the best knowledge, experience and approaches from both worlds, the east and the west, together with local and international views. In light of our immense need for trained and educated people in sustainable development, the initiative by the German Federal Ministry of Education and Research (BMBF) and ZEF/Bonn boosted education in the region. The project served as an example of educational structures spreading their outputs for sustainable development throughout Uzbekistan. The reputation of our regional university, the number of internationally accepted publications, the number of international collaborations, all increased dramatically. The spread of innovations and insights benefited both the environment and the population of the region already during the lifetime of the project.

This unique partnership in higher education and research and science development resulted in extensive media coverage and parliamentary support. It also contributed to the development of options for the use of our natural resources and markets that will be sustainable for a long time, and be acceptable to farmers, decision makers, and those worrying about the environment. This book provides an overview of these important aspects.

Ruzumbay Eshchanov
Rector State University of Urgench,
Chairman Khorezm Mamun Academy

Section 1: Introduction

Paul L. G. Vlek, John P. A. Lamers, Asia Khamzina, Inna Rudenko, Christopher Martius, Bernhard Tischbein, Ruzumbay Eshchanov

Restructuring land allocation, water use and agricultural value chains. Technologies, policies and practices for the lower Amudarya region

Abstract

During the seven decades of the Soviet Union (SU), the irrigated farming areas in Central Asia became some of the largest in the world. Though highly lucrative due to the cultivation of cotton, the introduced agricultural practices ended up being the cause of severe environmental degradation. The loss of the natural resources base during the SU period could not be arrested despite post-independence reforms. The Khorezm region, covering part of the lower reaches of the Amudarya River, is a typical example of this phenomenon, with land users and managers struggling to find efficient and environmentally friendly options for sustainable management of land and water resources. These "sustainable intensification" options must, on the one hand, help reverse land degradation and, on the other hand, ensure sufficient production to meet increasing demands and secure livelihoods. In 2000, the Center for Development Research (ZEF) at the University of Bonn accepted the offer from the German Federal Ministry of Education and Research (BMBF) to develop and implement a research and education program to serve as an example of international cooperation in the field of higher education in Uzbekistan. The aim was to address the necessary restructuring of the economic and ecological management of this Aral Sea region. The development of governance and incentive structures as well as sustainable intensification measures that spare the natural resources base, yet provide for sustainable livelihoods, were the overarching goals pursued through an interdisciplinary program. This approach implied realigning the land and water resources in such a way as to render farming profitable, sustainable, and acceptable to the local land users and managers. Policy makers are facing tough decisions in forging this transition and accommodating the concerns of producers and environmentalists alike. This chapter summarizes the overarching findings of the last phase of this project, and concludes with the lessons learned over a decade of research and educational activities.

1.1 Introduction

The development of irrigated agriculture in Central Asia has contributed to the reign of the various khanates in this region, which today is known as the Aral Sea Basin (ASB). The ASB used to be a vital and vibrant part of the famous Great Silk Road that brought Chinese silk, bronze ware, cosmetics, paint, rice, and tea to the West, whilst glassware, dried fruits, vegetables, cotton, horses, and semi-precious stones were brought to the East (Tolstov 1948). But since the days of the Great Silk Road, the ASB has served as a crossroad not only for trade, but also for cultures, ideas and agricultural development with impressive achievements as exemplified by irrigation and drainage networks, which were created along with the establishment of the khanates such as Kokand, Khiva and Bukhara.

In the arid environment of the ASB, only irrigated agriculture is possible. During the SU era (ca. 1925–1991), but particularly since the 1960s, about 8 million ha of land, including natural forest and desert biomes, were put under irrigation in the ASB. This required an annual supply of 96 km^3 of irrigation water conveyed through 323,000 km of channels (Orlowsky et al. 2000). To date, the irrigation and drainage infrastructure established during the SU period is used for the cultivation of cotton, wheat, rice and some other crops (Bobojonov et al. 2012). However, the present management of irrigated cropland is becoming increasingly unsustainable, as evidenced by a widespread land degradation that threatens the ecological and economic sustainability of the agricultural sector and the livelihoods of the local population. In the Khorezm region of Uzbekistan, with an average annual rainfall of ca. 100 mm, unsustainable land and water management practices have caused widespread waterlogging and soil salinization and a consequent decline in agricultural productivity.

During 2000–2012, the Center for Development Research (ZEF) at Bonn University, Germany, in collaboration with the Science Sector of UNESCO, the Urgench State University (UrDU) in Khorezm, Uzbekistan, and the German Space Agency (DLR) with the University of Würzburg conducted a research and education project in the ASB, funded by the German Federal Ministry of Research and Education (BMBF). The study region Khorezm covers 6,800 km^2 located in northwestern Uzbekistan, approximately 250 km upstream of the present shores of the Aral Sea. Roughly 270,000 ha are under irrigated agriculture. The Khorezm region is an over-seeable unit of management owing to its well delineated borders, which facilitates the calculation of regional balances for water, salinity, produce and the economy. Of the 1.7 million people that live in Khorezm, about 70 % are rural dwellers. The project focused on Khorezm located in the lower Amudarya Basin rather than on the entire ASB, which comprises a string of similar oases. It is widely recognized that reducing or shutting down the irrigation water supplies to re-fill the Aral Sea would create immense human suffering, potential unrest and conflicts, and is therefore only theoretically an option. The only viable option is to render irrigated agriculture sustainable, thus making it possible for millions of farm families to make a living while having access to sufficient drinking water and healthy food. The project aimed, therefore, to define and test sustainable options for land and water use, develop ecologically and economically sound practices to increase the resource use efficiency, combat land degradation, mitigate greenhouse gas emissions, and increase rural incomes. The innovative concept and approach was based on four cornerstones: (i) development of science-based options for improving land and water use; (ii) building human and institutional capacity in the intervention area and creating an educational center of excellence; (iii) integrating research and education at national and international levels; and (iv) long-term commitment and policy outreach.

This book is the fourth contribution of the project's research findings focusing on options and policies for sustainable agricultural intensification in the lower Amudarya region. It is a companion to the documentation of the social and economic findings (mainly Phase I of the project) covered by Wehrheim et al. (2008), and to the overview by Martius et al. (2012) on cropping and irrigation management and land-use systems in the given biophysical, economic and anthropological context (covering Phase I and II). The third book by Ul-Hassan et al. (2011) offers a guide for researchers and practitioners for conducting participatory testing and adaptation of agricultural innovations (part of Phase III).

1.2 The ZEF/UNESCO project in Khorezm

In Phase III (2007–2011), the ZEF/UNESCO project aimed at providing comprehensive, science-based options for *restructuring land and water use and governance* at three nested levels: policies, institutions, and technologies. The project intended to provide decision support for effective agricultural policies at the regional and national levels.

Only few youngsters in Central Asia are attracted to an education in science (Mukhammadiev 2010). The growing lack of qualified people in the agricultural and environmental sectors motivated the project to help change this situation. This called for a long-term collaboration between local and international (German) institutions aiming at the education of a larger group of Uzbekistan/Khorezmian researchers. The demand for education was fulfilled by creating a unique and conducive learning atmosphere that prepared young, talented people to get embedded in the local and international science community. Linking national and international institutions for addressing the intractable problems of natural resources degradation in the Khorezm region was thus used as a means to lure young talents from the region and expose them to international scientific standards.

The education of a new generation of scientists was integrated into all levels of project activities. During the decade of collaboration, 53 Ph.D. candidates, of whom more than half came from Uzbekistan, participated. By the end of 2012, 33 students (14 females), of whom 15 came from Uzbekistan (7 females), completed their doctoral degree. The Ph.D. candidates, who conducted their core research under the supervision of local and foreign experts, in turn supervised a large number of M.Sc. and B.Sc. students. For example, 105 M.Sc. students, of whom 31 came from a wide range of countries[1], completed their higher education in this intercultural and international cooperation program. The alumni also represent a future network of international collaboration.

ZEF successfully conducted an inter-disciplinary research and education program that, with active participation of the students, pioneered a series of innovations for increasing the sustainability of land and water resource use. An international partnership in research and education was fostered with members of the Consultative Group on International Agricultural Research (CGIAR) for which the project became a co-recipient of "The King Baudouin Award" for Sustainable Agricultural Development in Central Asia and the Caucasus in 2008. In 2013, the project was awarded the national Energy Globe Award for Uzbekistan by the Energy Globe Award[2].

A suitable and necessary infrastructure was set up to support advanced research and education. The capacity building was not limited to individual people, but also involved institutions such as educational organizations by supporting the development and dissemination of training and teaching materials. Also senior local scholars were involved in collaborative research and joint scientific publications, and offered opportunities for further academic promotion. Special training courses previously uncommon in the higher education system included Geographic Information Systems (GIS) and remote sensing methodologies, advanced statistical analyses, and linear programming.

---

1 These included Germany, Finland, Italy, Czech Republic, The Netherlands, New Zealand, Vietnam, Hong Kong, Iran, Afghanistan, Brazil, and Columbia.
2 The goal of the Globe Energy Award is to present successful sustainable projects to a broad audience, guided by the conviction that many of the environmental problems already have good, feasible solutions. Projects which conserve and protect resources or that employ renewable energy can participate (see www.energyglobe.info).

1.3 Overview of this book

This book is grouped into five sections covering the main themes policies, institutions, and technologies, and a final summary section. The five sections cover production systems (part 2), natural resources management (part 3), production and resource economics (part 4), society, policy and institutions (part 5) and the conclusions and options for actions section (part 6). The summaries of the contributions to each of these sections are presented in the following.

1.3.1 Part 2: Production systems

The agricultural production systems during the SU era relied on large farms that were managed by well-trained agronomists who put a large number of laborers to work largely according to their individual skills. After independence, the inherited production units were supposed to have been broken up, but the government dithered until 2001, and then the land reform was poorly conceived and even worse in its implementation and management. The lack of farming skills among the many new small farmers threatened the cotton crop and food security and, according to **Djanibekov et al. (Chapter 2.1)**, the Government of Uzbekistan (GoU) thus had to reverse various land reforms and upscale the individual farm size. The land fragmentation that started after independence was reversed in 2008. The authors concluded that farmers presently often do not know how to manage their newly gained land. The authors also argued that as long as the benefits of overall economic growth remain out of reach of the rural population, the optimization of farm sizes for improving production efficiency will be stifled.

After dismantling the collective farms in 1996–2002, irrigated crop production in Khorezm as in most of Uzbekistan, was mandated to two types of production units: rural households (*dehqons*) and private farmers (*fermers*). The *dehqons* account for the largest share in horticultural and animal husbandry production (i.e., dairy, eggs, etc.) whereas the *fermers* mainly fulfil the state-ordered production of cotton and wheat. **Conliffe (Chapter 2.2)** differentiates and analyzes irrigation-based agricultural livelihoods in rural Khorezm. She groups smallholder (*dehqon*) households around cropping, entrepreneurship, and migration as rural livelihood types. She argues that spatial location even within one relatively small agro-climatic zone such as the Khorezm region is an important but often overlooked factor influencing livelihood opportunities. She furthermore illustrates convincingly that livelihood strategies vary between upstream and downstream locations and are associated with the location along irrigation channels. Also, household cropping decisions are influenced by access to natural resources and by the availability of male labor, propensity to take risks, and access to niche markets. Upstream households are able to double-crop the staple crop wheat with the region's most lucrative and water-intensive crop, rice. In contrast, downstream households have to engage in seasonal migration because they are doubly hard-hit, i.e., by inferior agricultural potential due to a poorer access to natural resources such as water, and by their remoteness, which reduces market access. Spatial differentiation and stratification thus impacts the households' abilities to respond to shocks that adversely affect livelihoods. Although **Conliffe (Chapter 2.2)** cautions against the use of geography in a rigid manner, she underlines that when broadly interpreting the findings in a geographical context a better understanding emerges about the abilities of rural households to respond to political and environmental change.

Flooded rice production in the Khorezm region has been controversial: "income generation but water waste". For many outsiders, paddy rice production in an arid region suffering from regular irrigation water shortages is incomprehensible given the huge water demands. Despite the notoriously unreliable regional statistics about rice production, paddy rice seems to occupy annually ca. 10 % of the arable land but consumes up to 30 % of the total water resources (Bekchanov et al., 2012). But in some regards, this is putting a magnifying glass on a small part of the picture thereby ignoring the whole: paddy rice is the most remunerative crop in the entire crop portfolio of the Khorezm farming population (Bobojonov et al., 2012). Since rice cultivation can therefore hardly be reasoned away, **Devkota et al. (Chapter 2.3)** examined a series of innovations to meet present rice yields with water-saving approaches including water-saving irrigation methods combined with conservation agriculture practices. The latter reduced water demand but also yields, which is thus unlikely to be acceptable to rice producers. Additional approaches dealt with seeding/planting methods (transplanted versus wet-direct seeding), rice varieties (short- and middle-duration rather than long-duration varieties) and optimal seeding date taking into account the type of rice and climatic conditions. Some of these innovations are best recommended as stand-alone measures (e.g., rice transplanting), some in combination with others (e.g., short-duration varieties and transplanting date), some only when accompanied by additional changes in cultivation practices (e.g., yield reductions under conservation agricultural practices can be reduced when flood irrigation is applied till the grain-setting phase). But since all measures bear the potential to reduce irrigation water input for rice production, they merit further attention.

Improved water management options were also assessed through value chain analysis (VCA) combined with a water footprint analysis (WFA) for identifying options to improve water use efficiency **(Rudenko et al. (Chapter 2.4))**. This analysis was completed for different management levels and different sectors of the economy, with a special focus on the most prominent agro-industrial cotton sector. The approach considered simultaneously financial and ecological aspects of regional development, and thus accounted for the present socio-ecological challenges. The combined findings indicate two paths for reducing water use and coping with water scarcity with lowest possible detriment to the regional economy. One is to reduce agricultural water use (because agricultural production of raw cotton uses the most water along the cotton value chain) through upgrading irrigation and drainage networks, and to introduce innovations that have a high potential for increasing water use efficiency on the field. The other is to shift water use from the high water-demanding agricultural sector (such as raw cotton) to the less water consuming industrial sector, such as the cotton processing industry. The latter in particular suggests the production and subsequent export of cotton yarn, which would allow the reduction in water use but could maintain the same or even increase export revenues over those presently gained with raw cotton. This shift should, however, be made with care to keep the untreated waste water (grey water component) from the cotton processing (textile) industry at a minimum level.

1.3.2 Part 3: Natural resources management

Irrigation water use efficiency elsewhere is better than that in Uzbekistan, and this by large margins (WWF 2002). The low irrigation water use efficiency both at on-farm and perimeter level remains the prism through which outsiders view Uzbekistan's irrigated agriculture. In **Chapter 3.1, Tischbein et al.** present numerous options for improving this low efficiency by employing alternative irrigation practices. The focus is on technological improvements, which is the language most easily understood by the national administration in Uzbekistan. The authors show that with relatively simple technological improvement, the efficiency of water use can be drastically increased to the benefit of the farmers and society. They argue that the present low level of water use efficiency should be considered a floor rather than a ceiling, which thus leaves ample room for improvement.

In downstream and lowland regions, adequate management of shallow groundwater is essential in the irrigated agriculture context. The ancient khanate system had mastered the management of land and water resources with a variety of land use and water management systems to suit the lay of the land. They understood that the depth of the groundwater table affects not only crop growth but also soil salinity. The oasis was finely tuned to maximize the returns on the land with a minimum of salt damage (Rakhmatullaev et al., 2003; O'Hara 1997; Kats 1976). This traditional practice was largely ignored and then lost as the SU reformed and extended the irrigation systems in the region, which led to large-scale cropping, inefficient water use and, consequently, rising groundwater table and soil salinization of considerable portions of the land (O'Hara 1997). **Ibrakhimov et al. (Chapter 3.2)** concluded from an analysis of multi-locational monitoring of groundwater salinity over a period of 16 years that knowledge of the spatial variability of groundwater salinity, which can vary from 1.3 to 15 g l^{-1} (average is not more than 1.75 g l^{-1}), can help identify the areas where conjunctive use of ground and surface water is an effective way of meeting crop water demand. Although this information is being collected by regional water managers, it is not made use of by farmers and irrigation channel managers to increase overall efficiency of irrigation water, as traditional systems of learning and knowledge sharing have disappeared. To this end, an advisory institution needs to be put in place that can serve the community in improving the use of its water resources.

Khorezm is fighting a continuous battle to avoid an environmental crisis caused by water insecurity and soil degradation. **Akramkhanov et al. (Chapter 3.3)** illustrate that the dominating practice of pre-season leaching for soil salinity removal is far less effective and efficient than might be expected when considering the high amounts of fresh water applied (400–500 mm). The present leaching rarely results in the intended salt removal but rather shifts the salts from upper (20 cm) to lower soil layers. Much higher efficiencies were obtained when using an accurate, high-resolution, pre-leaching map of soil salinity at field level and a consequent timely targeting of site-specific leaching. Even though the necessary infrastructure is not available yet for implementing this combination of measures, the modest costs per hectare should be a stimulus to farmers and governmental officials alike to improve leaching effectiveness and avoid further salt accumulation during the vegetation periods.

Even during the SU era, the importance of protecting natural resources was recognized. This led to such practical measures as the establishment of tree hedgerow systems to combat wind erosion and improve the microclimate of the protected fields. Yet the lack of funds after independence limited the regular monitoring and evaluation of such hedges resulting in a knowledge gap about the condition and functioning of these protective systems. To fill this gap, **Tupitsa et al. (Chapter 3.4)** developed a cost-effective methodology combining photogrammetry with GIS technology and field surveys to assess the state of hedgerows. The findings illustrate that the hedgerow structure in the study region Khorezm needs to be improved by better design and maintenance as well as by introducing specific harvesting techniques to support the windbreak structure and function. The developed method could be transferred to other regions with similar agro-ecology.

The traditional multi-storey cultivation of crops and trees in the Khorezm oasis has been abandoned over the past century. In fact, laws exist that limit the coverage of trees. However, it was proven by Khamzina et al. (2012) that highly salinized cropping sites characterized by low yields of common crops could be made productive through the re-introduction of trees and with minimal irrigation input, as the trees rapidly reached shallow groundwater. In **Chapter 3.5, Djumaeva et al.** provide evidence that afforestation of degraded croplands could benefit from localized additions of phosphorus, thus enhancing N_2 fixation rates and tree growth. Consequently, the nitrogen and carbon stocks of the agroecosystems can be increased. Afforesting saline, degraded croplands is thus an option for (a) rehabilitation of nutrient-poor soil and carbon sequestration; (b) provision of benefits to the land owners and households due to increased land value and aesthetics; and (c) generation of alternative income and livelihood security for the land users (farmers) engaged in forest harvesting activities. Uzbekistan, a signatory to the desertification convention, is however slow when it comes to implementing sustainable land-use measures. But by adopting afforestation as a rehabilitation measure and land-use option for degraded croplands, Uzbekistan could set an example for Central Asia, where many similar agro-ecological landscapes exist.

If measures are to be taken to ameliorate the areas of low productivity, decision makers need to know where such hot-spots or unproductive areas are located. To enhance awareness of the spatial distribution of marginal croplands, **Fritsch et al. (Chapter 3.6)** elaborated a methodology based on GIS and remote-sensing data for detecting degraded areas through a weighted, multi-criteria analysis. The resulting maps can be used as land-use planning tools and strategic priority setting. The approach developed thus supports land users and land-use planners alike by not only targeting afforestation activities (Khamzina et al, 2012; **Chapter 3.4, 3.5**), but also by implementing measures to improve the distribution of irrigation water (**Chapter 3.1**) and innovative practices to increase water use efficiencies (**Chapter 2.3, 4.1, 4.3**). However, for a widespread use of this technology, investments in human capital are needed.

The potential of lakes, another underused natural resource in the water-scarce study region of Khorezm, was analyzed by **Ginatullina et al. (Chapter 3.7)** concerning its potential for aquaculture development. This could not only support families in their quest to meet nutritional needs but also could increase their economic viability. Therefore, as a first step in the exploration of this potential, the food supply by lakes and its availability to fishes was assessed. The findings show the presence of a large number of zooplankton taxa. Their density/biomass, diversity and composition were impacted by the salinity level of the lake water, but subject to temporal and spatial variability. The lake temperatures affected the seasonal cycles of zooplankton in abundance and community composition. The salinity level of the examined lakes periodically appeared to override temperature as the dominant factor when salinity became high. The authors concluded that, due to the unpredictable fluctuations in salinity and its potential influence on zooplankton biomass and seasonal declines in the zooplankton communities, fisheries in the lakes of the Khorezm region may be more likely to succeed when cultivating fishes that do not rely directly on zooplankton.

1.3.3 Part 4: Production and resource economics

Regional statistics on water supply are hardly suitable for reflecting water use and water use efficiency as they do not account for conveyance losses. This gap was filled by the analyses of **Bekchanov et al. (Chapter 4.1)**. Although water use usually lags behind water supply, the evidence indicates that in years of water scarcity, water use efficiencies turned out to be higher. Since the present state-order strategy makes the GoU the main body accountable for the water management institutions, the authors argue that it has a major role to play in supporting farmers to shift to more water-wise technologies and ensuring that water management institutions become more accountable to farmers.

The need for introducing and applying efficient irrigation technologies, especially at the field level, has been clear from various studies included in this book (e.g., **Chapter 3.1, 3.3**). This finding is further confirmed by farmers' perceptions as well as by the rate of acceptance and adoption of irrigation technologies by the farming population. **Bhaduri and Djanibekov (Chapter 4.2)** investigated different institutional and economic factors that may induce farmers to adopt water efficient technologies in the irrigated agriculture of Uzbekistan. The authors analyzed several scenarios with respect to water price flexibility, cotton policy restrictions, land tenure security and stability of the water supply. Their findings underline that a more flexible mechanism of water pricing holds the potential for increasing adoption rates of efficient irrigation technologies by 20 % compared to the fixed water price levels presently applied and foreseen in future policy measures. On the other hand, variability in water supply, restriction in decision making with regard to cotton production and land use slow down technology adoption and are inversely proportional to the adoption rate of efficient irrigation technologies by farmers. These findings thus confirm earlier suggestions that water pricing in Uzbekistan can be a viable tool for increasing the efficiency of water use. However, the absence of strong institutional arrangements and insufficient farm capital remain a constraint to the implementation of such water-pricing instruments. Hence, water-pricing measures must match the agricultural policy set-up, infrastructural capacities and the capital availability of farming entrepreneurs.

Not only irrigation water use efficiency on field and network level has been assessed as low (**Chapter 3.1**). **Karimov and Zarazúa (Chapter 4.3)** see room also for output gains when increasing the overall technical efficiency of the current production technologies. However, options for implementation differed according to region, crop growing area, location and soil fertility. Based on surveys conducted before land consolidation in 2008 (Djanibekov et al. 2012, **Chapter 2.1**), the findings show that gaps between present and attainable yields do not stem from scale differences, although increasing farm size has been an important argument by the GoU for introducing land consolidation in 2008. Furthermore, as long as farm-level management does not improve, overall technical efficiencies are likely to remain low. This supports the conclusion by the authors that the on-going land consolidation must be accompanied by measures such as farmer education. Noteworthy is the great difference in technical efficiency on land with different fertility levels, which indicates that farms operating on nutrient-poor land are relatively more efficient in the use of resources should they need to use them. Furthermore, differences in technical efficiency between regions and between different locations within regions indicate that advice to farmers must become differentiated. Two decades of farming have created a diversity of farms that consequently have different advisory needs and conditions.

The findings of a number of studies (e.g., **Chapter 4.1, 4.2, 4.3**) are based on production functions that are rooted in bio-physical and economic relationships used to estimate water productivity. Yet production functions alone are insufficient to determine water productivity, as they do not account for the interplay between the endowment and contextual factors that influence a production function. Therefore, water productivity assessments must be complemented by water profitability estimates as argued by **Saravanan et al. (Chapter 4.4)**, who defined this as the *net value of products per unit of consumed water*. This approach allows including and thereby understanding a farmer's choice of water management practices to maximize profit given the physical and bio-physical settings. The authors argue that water profitability could be estimated through the Bayesian Network analysis, which helps identify factors other than bio-physical and economic ones, since annual farm profit over a longer period is only a part of a farmer's business objective. While taking endowment and contextual factors into consideration, the authors concluded that agricultural water profitability cannot be explained by the optimization of a single objective, but rather by a trade-off by the farmers between multiple objectives. The combination of endowment, contextual and production factors thus determines the space within which a farmer operates.

Clement et al. (Chapter 4.5) analyzed the seasonal and inter-annual price movements to pinpoint the factors influencing price fluctuation. In transition countries like Uzbekistan that pursue food self-sufficiency policies (indicated by the proportion of food import to total consumption), the effects of local factors on price formation are expected to be dominant. Various studies have looked at food security concerns in Uzbekistan, but very few have addressed food price variability and its determinants at the local level, which is necessary to understand household welfare impacts. As the income of households to a large extent depends on agricultural production, and since the largest share of the budget is spent on food consumption (WFP, 2008; Musaev et al., 2010), the fluctuations in prices will have a considerable effect on the level of both production and consumption, and thus on the overall welfare of the population in the region. The authors used price behavior of ten agricultural commodities, collected weekly from local markets in the Khorezm region during 2002–2010. The results show two general patterns of agricultural price fluctuations: (i) price fluctuations of rice, meat and wheat are more sensitive to external factors such as their respective international prices, market exchange rate and oil prices; (ii) the price movements of apple, onion, potato, and tomato are more locally determined, and particularly affected by seasonal patterns where the minimum price occurs during the harvest season and the maximum during off-season. To reduce the fluctuation of food prices, the authors argue for the creation of storage and processing facilities. These have deteriorated following independence, and should be given more attention in national policies.

1.3.4 Part 5: Society, policy and institutions

After a decade of interdisciplinary research, it has become clear that the expansion of land and water use undertaken during the SU era has been fraught with miscalculations, bureaucracy and lack of commitment. Furthermore, the many layers of decision making combined with the institutional distance between the decision makers and the farming population have resulted in an inability to reply quickly to the farming reality. The present irrigation water distribution system and network in the Khorezm region still has the capacity to convey adequate water supplies, but it is not a sustainable system yet. Steps to eliminate the present inefficiencies of natural resource use are needed. Science and technology research has indicated various pathways that could be followed to increase the efficiencies of natural resources use, but their adoption needs to be facilitated by laws and institutions. These institutions should provide the farming population with incentives to implement water-wise technologies as is argued by **Bekchanov et al. (Chapter 5.1)**. This would also benefit the state and thus create a win-win situation. Institutional measures such as guaranteeing timely water availability, introducing adequate water pricing measures, and creating an environment for capital investments were seen to be less promising in the short term, as they do not address the high conveyance losses, which require costly infrastructural improvements. Instead, various water-saving measures that increase water use efficiencies (e.g., crop diversification), and technical innovations (e.g., drip irrigation for selected crops such as vegetables) could be implemented by the farmers and reduce conveyance needs, thus benefiting regional income as a whole.

The overall findings illustrate that a region-wide implementation of innovations necessitates investments and institutional changes, which would be substantial, but so would be the consequent pay-offs. Based on this view, **Shtaltovna et al. (Chapter 5.2)** point out that, of the nationwide changes so far, the machine tractor parks are among the most prominent in reaching the agricultural goal of modernization. Such parks had been mandated following independence and agricultural reforms, but the remaining state controls actually prevent these from becoming financially independent service providers. According to the authors, loosening this tight grip would demand a substantial shift in the mind-set of administrators, but the potential gains could be enormous.

The overarching challenge in the study region is not a straightforward issue of water scarcity even though it is perceived as such by farmers and others (Oberkircher et al., 2012). As seductively simple as this might sound, it is not helpful when searching for solutions to overcome the water crises in the region. The generation of farmers that emerged after independence and land consolidation has been seen to lack experience as private farm entrepreneurs. The same farmers are also overwhelmed in the face of the recurrent economic water scarcity and soil salinity. In fact, the entire region is wrestling with the rising demands for water and a generally declining supply caused by humans and nature. The older (farming) generations of Uzbekistan grew up during a time when many, if not all, innovation choices were made by the state or by the working place. This situation has in many ways not changed in Uzbekistan. **Hornidge et al. (Chapter 5.3)** report on the experiences of a Follow-the-Innovation (FTI)-approach developed and implemented over three years with four FTI-teams. The key of these experiments was to illustrate the gains possible when introducing project-based water and land-use innovations. It was demonstrated that innovation must be more than merely creating an environment in which such innovations are perceived as benefitting the adopter. Participation and democratic principles form a pillar in the FTI approach, conditions that the post SU period have yet to create.

1.3.5 Part 6: Conclusions

The approach selected by the project for the development of sustainable agricultural intensification options in Uzbekistan has been innovative in various ways. For example, the establishment of an extended research and educational infrastructure in the region, including a well-equipped GIS laboratory with skilled staff, which serves as a centerpiece for offering services and products. The project benefited from the trans-disciplinary view and approach used. The combined use of GIS, mathematical modelling, new analytical methods, and household surveys offered a spectrum of different visions for progress, which is broader than what specialized projects usually can do. The time frame of 10 years was shown to be conducive for strategic capacity building. The early connection of the research findings and data collected through multiple disciplines permitted a timely and permanent cross-checking of information with the project objectives and an optimization of the applicability of any of the proposed solutions. Linking the integrated research findings with practice further enhanced their usefulness for the end users. There were many lessons learned, and many are yet to be learned. It is imperative that the work started in this project be taken up by regional authorities and international donors, as failure to do so may one day lead to a breakdown in the agricultural sector and instability in the region.

References

Bekchanov M, Lamers JPA, Karimov A, Müller M (2012) Estimation of spatial and temporal variability of crop water productivity with incomplete data. In: Martius C, Rudenko I, Lamers JPA, Vlek PLG (Eds.) Cotton, water, salts and soums – economic and ecological restructuring in Khorezm, Uzbekistan. Springer: Dordrecht, pp. 329–344

Bobojonov I, Lamers JPA, Djanibekov N, Ibragimov N, Begdullaeva T, Ergashev A, Kienzler K, Eshchanov R, Rakhimov A, Ruzimov J, Martius C Crop diversification in support of sustainable agriculture in Khorezm. In: Martius C, Rudenko I, Lamers JPA, Vlek PLG (Eds.) Cotton, water, salts and soums – economic and ecological restructuring in Khorezm, Uzbekistan. Springer: Dordrecht, pp. 219–233

Djanibekov N, Van Assche K, Bobojonov I, Lamers JPA (2012) Farm Restructuring and Land Consolidation in Uzbekistan: New Farms with Old Barriers. Europe-Asia Studies, 64 (6): 1101–1126

Khamzina A, Lamers JPA, Vlek PLG (2012) Conversion of degraded cropland to tree plantations for ecosystem and livelihood benefits. In: Martius C, Rudenko I, Lamers JPA, Vlek PLG (Eds.), Cotton, Water, Salts and Soums – Economic and Ecological Restructuring in Khorezm, Uzbekistan. Springer: Dordrecht, pp. 235–248

Kats D M (1976) The influence of irrigation on the groundwater. Kolos Pub, Moscow. (in Russian). 271

Martius C, Rudenko I, Lamers JPA, Vlek PLG (2012) Cotton, water, salts and soums – Economic and ecological restructuring in Khorezm, Uzbekistan. Springer: Dordrecht, 419 pages

Mukhammadiev A (2010) Central Asia. UNESCO Science Report 2010. The Current Status of Science around the World, pp. 235–251. 520 pages

Musaev D, Yakhshilikov, Y Yusupov (2010) Food security in Uzbekistan. UNDP Mega Basim, Tashkent Uzbekistan, 64 pp.

Oberkircher L, Haubold A, Martius C, Buttschardt TK (2012) Water patterns in the landscape of Khorezm, Uzbekistan: A GIS approach to socio-physical research. In: Martius C, Rudenko I, Lamers JPA, Vlek PLG (Eds.) Cotton, water, salts and soums – economic and ecological restructuring in Khorezm, Uzbekistan. Springer, Dordrecht, pp. 285–307

O'Hara SL (1997) Irrigation and land degradation: implications for agriculture in Turkmenistan, central Asia, Journal of Arid Environments 37: 165–179

Orlovsky N, Glanz M, Orlovsky L (2000) Irrigation and Land degradation in the Aral Sea Basin. Pages 115–125 in Breckle S.W., Vesle M. and Wuecherer W. (Eds.) Sustainable Land Use in Deserts. Springer Verlag Heidelberg, Germany

Rakhmatullaev SA, Bazarov DR, Kazbekov JS (2003) Historical irrigation development in Uzbekistan from ancient to present: Past lessons and future perspectives for sustainable development. Pages 79–80 In Proceedings of the third International Conference of International Water History Association. Alexandria, Egypt

Scheer C, Wassmann R, Kienzler K, Ibraghimov N, Lamers JPA, Martius C (2008) Methane and nitrous oxide fluxes in annual and perennial land-use systems of the irrigated areas in the Aral Sea Basin. Global Change Biology 14, 1–15

Sommer R, Djanibekov N, Müller M, Salaev O (2012) Economic-ecological optimization model of land and resource use at farm-aggregated level. In: Martius C, Rudenko I, Lamers JPA, Vlek PLG (Eds.) Cotton, Water, Salts and Soums – Economic and Ecological Restructuring in Khorezm, Uzbekistan. Springer, Dordrecht, pp. 267–283

Spoor M, Visser O (2001) The State of Agrarian Reform in the Former Soviet Union. Europe-Asia Studies (Former Soviet Studies) 53: 885–901

Tolstov SP (1948) Following the tracks of ancient Khorezmian civilization. Academy of Sciences of the USSR, Moscow – Leningrad

Ul-Hassan M, Hornidge AK, van Veldhuizen L, Akramkhanov A, Rudenko I, Djanibekov N (2011) Follow the innovation: Participatory testing and adaptation of agricultural innovations in Uzbekistan. Universität Bonn, Zentrum für Entwicklungsforschung (ZEF)

UN (United Nations) (2008) *World Economic Situation and Prospects – 2012*. New York: United Nations publication

WFP (World Food Programme) (2008): Poverty and food insecurity in Uzbekistan, 57 pp.

World Bank 2010. Gross national income per capita 2010, Atlas method and PPP. Online database at www.worldbank.org

UN (2012) World Economic Situation and Prospects 2012, New York: United Nations

Wehrheim P, Schoeller-Schletter A, Martius C (Eds.) (2008) Continuity and change: land and water use reforms in rural Uzbekistan Socio-economic and legal analyses from the region Khorezm. Halle/Saale, Germany: Leibniz-Institut für Agrarentwicklung in Mittel- und Osteuropa (IAMO)

Worldbank (2010) http://en.wikipedia.org/wiki/List_of_countries_by_GDP_%28nominal %29_per_capita

WWF (2002) Living Planet Report 2002. 36 pp. http://www.wwf.org.uk/filelibrary/pdf/livingplanet2002.pdf

Section 2: Production Systems

Nodir Djanibekov, Ihtiyor Bobojonov, Kristof Van Assche,
Inna Rudenko, Kudrat Nurmetov, John P. A. Lamers

2.1 Farm restructuring in Uzbekistan through fragmentation to consolidation

Abstract

The state-induced farm consolidation in 2008 for boosting agricultural production in Uzbekistan was examined with a focus on the presently experienced opportunities for rural development. Farm consolidation as a stand-alone measure under the current constraints, e.g., infrastructure and policy regulations, is an insufficient incentive for increasing farm efficiency. In fact, the process can be referred to as "farm consolidation" rather than as a comprehensive land consolidation process as observed elsewhere. It is argued that the consolidation process is likely to improve and advance rural development only when a number of supplementing policies are introduced to relax existing production constraints, such as reducing the extent of the state procurement system, ensuring land ownership, and increasing access to auxiliary farm services.

Keywords: farm consolidation, farm efficiency constraints, Khorezm region, Central Asia

2.1.1 Introduction

The transition from a centrally planned towards a market economy, which Uzbekistan pursued after gaining independence from the former Soviet Union in 1991, has proceeded by what is assessed by many as a "gradual" reform path (Bloch 2002). In a nutshell, this path can be characterized as maintaining economic and social stability in the short run whilst taking advantage of operating market forces in the long run. Although the reformation targeted many sectors of the Uzbek economy, in particular the agricultural sector was restructured (Djanibekov 2008), with the objective to maintain the provision of income, food, feed, fiber and fuel for most of the rural households as well as to support a range of industries beyond the agricultural sector. The various stepwise land reforms divided, e. g., the large, previously inefficient agricultural production units into a large number of smaller, privately operated farms (Khan 2007). In effect, however, the farm restructuring did not meet the expectations of increasing land and water productivity, but rather illustrated the unsuitability of the existing irrigation and drainage infrastructure for the new form of farms (Djanibekov et al. 2012c). As a result, in 2008, a reverse reform, i. e., farm optimization resulting in farm consolidation, was initiated. This study analyzes the most recent (2008 – 2009) process of farm restructuring in Uzbekistan, extending the analyses of the latest farm restructuring process by Djanibekov et al. (2012a) with particular emphasis on the consolidation process.

2.1.2 Process of farm restructuring in Uzbekistan

2.1.2.1 First three stages of farm restructuring

The process of farm restructuring in Uzbekistan has been the subject of previous studies (e. g. Lerman 2009; Djanibekov 2008; Veldwisch, Spoor 2008; Djanibekov et al. 2012a). Table 2.1.1 summarizes the four stages of the farm and land reform process since 1991 namely: (i) transformation of state farms into collective units, and later (ii) into agricultural shareholding cooperatives, (iii) partial and continued disaggregation of large farms into smaller individual (private) farms, and (iv) consolidation of small farms into larger ones.

The farm restructuring process initiated and guided by the state was gradual. At the onset, the state-owned large-scale farms (*sovkhozes*) were transformed into collective farms (*kolkhozes*). Following this, *kolkhozes* were transferred into agricultural producer cooperatives (*shirkats*)[1] distributing and sharing property rights over agricultural income and output (except for cotton and part of the wheat harvests) with its members. Concurrently, the Uzbek legislation for the first time defined private farming as an individual ownership of agricultural income, output, and inputs with the exception of water and land. Furthermore, the legislation defined three types of private farms based on their production specialization: (i) cotton and wheat farms (the largest and dominant farm type) that also produce rice and vegetables on a small share of their farmland, (ii) horticultural and gardening farms (specialized in fruits, grapes and vegetables production), and (iii) livestock-rearing and poultry farms. The latter two farm types are not part of the state procurement system (Djanibekov et al. 2012a).

---

1 Law "on Agricultural Co-operatives (*Shirkats*)" (1998).

In all stages of the farm restructuring process, cropland remained under state ownership, meaning that no legal private land ownership has been introduced. To run a farm enterprise, individuals lease land from the state at zero rent with long-term usufruct rights. This implies that farmers cannot use their leased land, for instance, as collateral for accessing credit.

Table 2.1.1: Important characteristics of the farm restructuring stages in Uzbekistan since 1991

	First stage 1991–1997	Second stage 1998–2002	Third stage 2003–2007	Fourth stage 2008–2009	2009-present
	Transformation	Partial disbandment of large-scale farms	Complete disbandment of large-scale farms	Farm consolidation	
Main transformation process	Transformation of *sovkhozes* into *kolkhozes*	Transformation of *kolkhozes* into *shirkats*. Partly land leased from *shirkats* to private farms	Complete transformation of *shirkats* into private farms	Amalgamation of small farms into medium-sized farms	Amalgamation of medium-sized farms
Dominant farm types	*Kolkhozes*, *sovkhozes*	*Shirkats*, private farms	*Shirkats*, private farms	Private farms	Private farms
Land ownership	State ownership	State ownership and land lease			

Source: Updated from Djanibekov (2008)

The lease agreements also constrained market-driven changes in farm sizes. In the first stages of farm restructuring, private farms were established through the lease of the unproductive land of *shirkats*[2]. As the reform progressed, the pace of fragmentation of *shirkats* into smaller farms was accelerated. By the end of 2007, Khorezm experienced a dramatic shift towards the above-mentioned three types of private farms, which were allotted 87 % of all arable land (Djanibekov 2008). Concurrently, the agricultural equipment of the *shirkats* was transferred to machinery and tractor parks (MTPs)[3], whilst water distribution and canal management was (partially) transferred to water users associations (WUAs)[4]. The state continued to coordinate the distribution of other agricultural inputs,

---

2 Law "on Private Farm" (1998).
3 Decree of the Cabinet of Ministers of Uzbekistan "on Measures for Strengthening Operating Efficiency of Machinery and Tractor Parks" (1997).
4 Decrees of the Cabinet of Ministers of Uzbekistan "on Measures for Reorganization of Agricultural Co-operatives into Private Farms" (2002). From the end of 2009 onwards they were officially called Water Consumers Association.

and maintained the procurement system for cotton and wheat (Pomfret 2008). This meant that with the progress of farm restructuring, state-mandated production targets were fully transferred to private farms, and cotton and wheat production in 2007, for instance, was allotted to roughly 70 % of the farm land.

2.1.2.2 Fourth stage: "Consolidation" (end of 2008 until today)

The first steps of farm restructuring thus resulted in a vast number of small farms. Also, the idea of a farm as one production unit in the sense that parcels are located next to each other to form a single territory of a farm unit was not promoted. Instead, a private farm could consist of several scattered fields often far from each other. The parcels were 2–3 ha (from 5 to 20 ha in the case of cotton and wheat farms) of various soil qualities and shapes Djanibekov et al. 2012c).

To counterbalance the adverse effects of a large number of small-scale production units that were to operate within an infrastructural setup designed for a small number of large production units, WUAs were established for increasing irrigation performance. Yet the irrigation water supply to, for example, tail-end users in the irrigation system and administrative districts further away from the water source, was often delayed (Abdullaev, et al. 2008). The reconstruction of the existing irrigation system to suit the numerous small farms would have required large investments.

At the same time, the availability of a farm-serving infrastructure fell behind the demands of the large number of newly established farms (cf. Djanibekov et al. 2012c; Niyazmetov et al. 2012). As a consequence, and similar to the situation in central and eastern European countries (Pašakarnis, Maliene 2010), it became evident from the third stage of reforms onwards, that, due to the initial set-up of the infrastructure (roads, canal and drainage system) for the large-scale farms, the stability of agricultural production was endangered by the vast number of small farms with insecure access to key resources. In 2008, to cope with the problem of the unsuitable infrastructure, the state reversed the previous reforms. With the declared aim of the "optimization" of farm sizes, farm consolidation was rapidly implemented by merging small farms into large units (Fig. 2.1.1). At the end of 2008, the first phase of the consolidation process began[5] followed by a second phase[6] in 2009. The aim was to increase the economic

---

[5] Decree of the President of Uzbekistan "on Special Committee for Elaboration of Recommendations for Optimization of Fields of Private Farms" (2008).
[6] Decree of the President of Uzbekistan "on Measures for Further Optimization of Private Farms Fields" (2009).

efficiency of private farms by increasing farm size and in the process to reduce the problem of the widely scattered private farm fields. The driving force behind the consolidation of these fields into larger production units was to increase the profitability of agricultural production. Hence, lease contracts for small farms were revoked, and the lands were re-allotted and became part of larger production units, resulting in a greater concentration of production by fewer, but larger farms. Concurrently, the minimum size of cotton and wheat farms, as well as that of gardening and horticulture farms[7], was re-defined in the Uzbek legislation. The minimum size of cotton and wheat farms increased from 10 to 30 ha, and of horticultural and gardening farms from 1 to 5 ha. During the first phase of farm consolidation, the number of private farms in the Khorezm region reduced from 19,000 to around 10,000 by the end of 2008 and to 5,760 at the onset of 2010. The average farm area increased from 13 ha in 2007 to 24 ha in 2008 and more than 40 ha in 2010 (Fig. 2.1.1, left Y-axis).

Hence between 2005, i.e., the middle of the reforms when *shirkats* were to be abolished, and 2010, which represents the second phase of the consolidation process, the farm groups and land distribution among farm groups had changed significantly. Two farm sizes then dominated (Fig. 2.1.2). About 50 % of all farms in Khorezm in 2010 were less than 10 ha in size and leased about 5 % of all farmland, mainly specializing in gardening and horticulture. The second group (37 % of all farms occupying 87 % of all farmland in 2010) comprised farms larger than 50 ha. These were mainly cotton- and wheat-producing farms. The private farms with a size of 10–50 ha, prior to consolidation the main group, lately accounted for only 13 % of all farms and occupied the remaining 7 % of all farmland.

The consolidation process in Uzbekistan differed considerably from that in Western European (e.g., in The Netherlands known as "ruilverkaveling", in France as "remembrement", and in Germany as "Flurbereinigung") and in central and eastern European countries (van Dijk 2007). The consolidation process in Uzbekistan was implemented in the first place by rapidly optimizing farm size without addressing other infrastructural changes typical for the land consolidation processes elsewhere. Since the average size of farm fields before consolidation was about 2–3 ha, the consolidation of these fields into larger ones was not necessary. Hence, this process can be referred to as "farm consolidation" and not so much as the comprehensive land consolidation process as observed elsewhere.

---

7 Law "on Introduction of Changes and Additions to Legislative Acts of the Republic of Uzbekistan in Connection to Deepening of Economic Reforms in Agriculture and Water Sector"(2009).

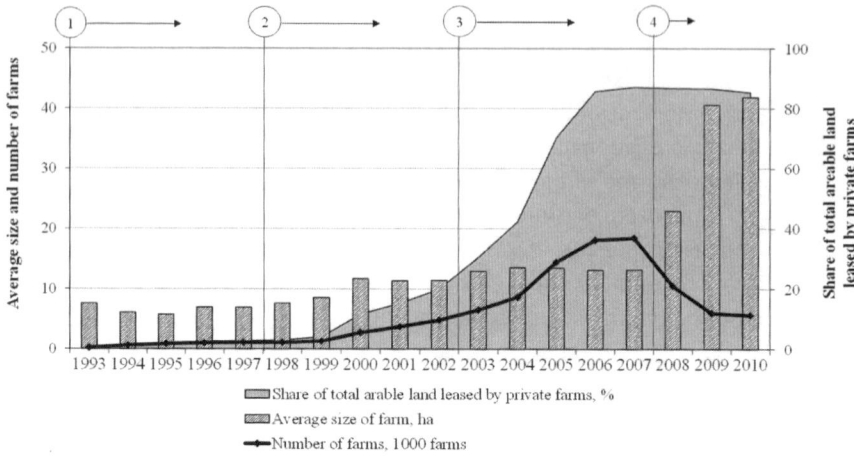

Figure 2.1.1: Evolution of size and number of private farms in Khorezm

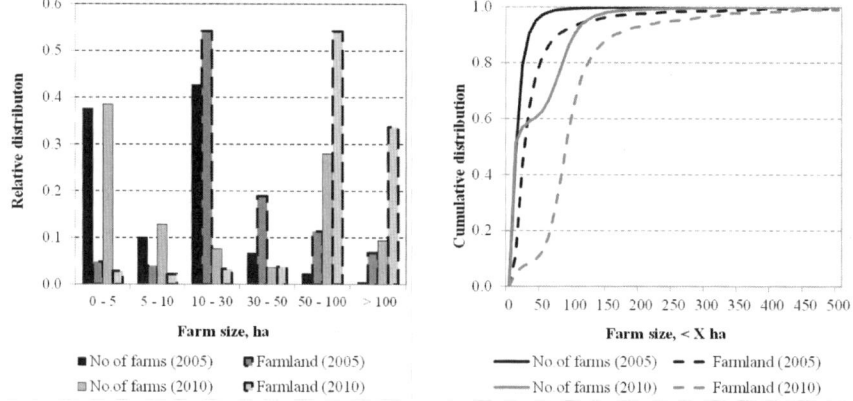

Figure 2.1.2: Relative (left) and cumulative (right) distribution of farms by size and number before and after farm consolidation in Khorezm

2.1.3 Farm sizes and productivity

The Uzbek legislation intended to provide equal access to land by rural households to prevent an increase in the number of rural, landless poor and to contribute to an increase in food and cotton production. As a result, since 2010 the agricultural production system in Uzbekistan has mainly consisted of a combination of large cotton and wheat farms. These have advantages regarding access to markets, infrastructure, and technology. The rural households, on the other hand, have the advantages in own-family labor and intensive production of vegetables, fruits and animal products (Djanibekov 2008). This combination of large farms and rural households can provide considerable benefits in terms of rural employment, income and food security and of the adoption of new agricultural technologies and maintenance of desired levels of cotton production.

Despite their important role in food security and poverty alleviation, rural households lack the ability to cope with an increasing variability of commodity prices and increasing input prices for which they do not have sufficient capital (Hazell et al. 2010). As a result, they may fall into what is called the "poor but efficient" trap (Fan,Chan-Kang 2010). In Uzbekistan, rural households heavily depend on earnings from employment in private farms in addition to the income from the non-agricultural sector (Veldwisch, Spoor 2008). In this respect, the economic performance of the private farms is essential in providing not only rural employment, but also in securing the rural sources of income and the food situation in rural households.

It recurrently has been argued that an increase in farm size may improve agricultural productivity (Deininger et al. 2004; Lerman, Cimpoies 2006; Lerman, Shagaida 2007; Eastwood et al. 2010; Fan, Chan-Kang 2010; Hazell et al. 2010). Although a generalization, this seems to have favored the recent farm optimization process in Uzbekistan. Yet there still is a recurring debate on the roles of small vs. large farms in fostering agricultural growth and economic development on which the empirical literature has failed to reach a consensus. The worldwide promotion of small, family-operated farms comes from their responsiveness to new markets and technologies and better development outcomes in terms of overall economic growth, employment generation, and poverty reduction. This places their productivity at the center of the development agenda (Fan, Chan-Kang 2010). Based on the experience worldwide it is recognized that small farms use their variable resources more efficiently and produce higher output per hectare compared to their larger counterparts (Hazell et al. 2010).

However, agriculture in Uzbekistan has a long history of collective farming based on intensive input use, employment of trained farm managers and engineers, and of operating within a specially designed infrastructure of irrigation canals and roads. Following several decades of collective farms, agricultural production in Uzbekistan became intensive in the use of machinery and chemical fertilizers. Moreover, the application of expensive inputs and machinery also caused the positive relationship between farm size and productivity. As a result, during the transition period and in the first years of private farm establishment, the economic efficiency of farms in Uzbekistan depended largely on access to more expensive inputs, and new sources of capital and extension services, which previously were subsidized by and provided through the central budget.

A global assessment underlines that, when compared to large farms, smaller farms with similar production patterns have advantages regarding acquisition of labor and local knowledge (Hazell et al. 2010). However, they have difficulty in acquiring capital. For instance, the high transaction costs for formal credits mean an inverse relationship between the unit costs of borrowing and the loan size. This creates a bias in the lending of money to small-scale farmers. Furthermore, the seasonality of labor requirements in agriculture is another factor that determines the relation between farm size and productivity (Rao, Chotigeat 1981). Such issues also apply to agricultural production in Khorezm, which largely depends on seasonal short-time requirements for labor for cotton, wheat, rice and maize production, which together occupy the largest share of arable land in Uzbekistan. As a result, crop-producing farmers in Khorezm depend more on seasonal than on permanent labor. In this respect, large farms have better potential to deal with financing, infrastructure, and technology constraints than smaller farms of similar specialization.

Furthermore, larger farms are efficient when high capital outlays, which do not return a positive cash flow for several years, are required to improve the quality of arable land and maintain a necessary infrastructure such as irrigation and drainage canals. This is particularly important in areas with increasing land degradation and poor condition of irrigation and drainage systems such as in Khorezm. Another advantage of consolidating farms in Uzbekistan is that larger farms can better adopt new agricultural technology (Eastwood et al. 2010), and, as was observed during the field survey in the Khorezm region of Uzbekistan, can compensate for the shortcomings in the maintenance of public goods such as roads. Also, the further and anticipated development of markets in urban areas of Uzbekistan is likely to go hand in hand with a greater emphasis on the quality of agricultural products and the adoption of higher quality control standards. In this respect, the larger farms will again have an advantage over smaller farms (Eastwood et al. 2010).

With the exception of the state-induced establishment of MTPs and WUAs within the boundaries of former *shirkats*, the physical infrastructure, e.g., irrigation canals, has not been redesigned and has largely remained intact throughout the various farm restructuring steps since the early 1990s (Table 2.1.1). Furthermore, the consolidation process centered on farms specialized in cotton and wheat production, while other farm types, i.e., those specialized in gardening, horticulture and livestock rearing, remained unchanged but still influence water distribution within the present irrigation infrastructure due to their independent demand and scattered location. Hence, key problems such as that of scattered fields belonging to one farm have exacerbated the existing lack of a reliable, timely and adequate supply of irrigation water due to the malfunctioning water supply and distribution system. The combination of different types of farms can offset the anticipated advantages of the economies of scale, which is usually the intention of farm consolidation efforts. Hence, changing the farm size alone without changing the farm infrastructure deems insufficient to provide incentives for increasing the economic efficiency of farms.

Moreover, following farm fragmentation and reforms in irrigation water distribution, people lacked the time to learn, and had insufficient skills in dealing with problems caused by the existence of a vast number of farms with different production specialization, e. g., increasing water allocation and distribution conflicts, decreasing water use efficiency, and low fee payment rates by WUA members, etc. (Djanibekov et al. 2012b). The underperformance of state-built and centrally managed irrigation infrastructures is common for most part of Asia (Mukherji 2009) including Uzbekistan. The introduction of participatory irrigation management practices as a remedy against the failures of the former centrally managed irrigation systems did not show the expected results not only in Uzbekistan, but also in other places of Central Asia, India and Pakistan (Mukherji 2009). Due to the initial design, the large-scale irrigation schemes in Uzbekistan lack the capacity to provide reliable, flexible and equitable water to all water users. This also underlines that farm consolidation alone is not capable of increasing agricultural productivity in the long term unless the farm and the irrigation infrastructure are adapted.

Moreover, agricultural policies need still to be developed further in the follow-ups of the farm restructuring process (Bobojonov et al. 2008). The state coordination of cropping activities on a large share of the farmland could, for instance, be relaxed with the aim of improving the economic efficiency of the newly consolidated farms. At present, public expenditures focus on irrigation and crop extension programs. Additional public investments in the development of processing and storage would contribute to the development of the entire agricultural sector (Bobojonov 2008). The changes in land tenure rights, or an ease of the limitations in land lease (not necessarily introducing private land ownership), would be a convincing signal to the present farmers with respect to, for example, access credits needed for purchasing machinery and soil-improving and water-wise technologies. The present set of regulations lacks the provision of such incentives. Clarification and security of the land users' rights is an essential precondition for fully realizing the potential benefits when operating large farms (Deininger, Byerlee 2011). The introduction of a voluntary and market-based approach, e. g., subleasing rather than legally enforced procedures, could be more successful in optimizing farm sizes and thus improving their production efficiency as experienced in former socialistic republics elsewhere (Dixon et al. 2001).

In Uzbekistan, even without changing the infrastructure, the large cotton and wheat farms have a higher potential to become economically more efficient than the smaller farms. Yet if the various frame conditions such as improved land tenure security, modification of the cotton procurement policy, and extended value chains, are not adapted simultaneously, the present land-consolidation is unlikely to yield the expected benefits. There is, therefore, a need for more in-depth knowledge on the productivity, welfare, social and environmental impacts of farm consolidation in Uzbekistan, on the economic efficiency of large cotton and wheat farms compared to smaller ones, and on the impact of policies on the different farms types and sizes.

2.1.4 Summary and conclusions

In Uzbekistan, the state farm restructuring process meant the end of the original structure of large farms functioning within boundaries designed during the Soviet planning system. However, along all stages of the restructuring process, private farms still had to follow the production boundaries assigned by the state. On the other hand, the accompanying adjustments in infrastructure and policies turned out to be insufficient to improve production and productivity. In this situation, the state had to cope with the dilemma of either promoting private farming with an unchanged infrastructure and one too expensive to adapt, or to reverse the process and create larger farms again to improve agricultural production. Since land is under state ownership, farm consolidation was the preferred option, and farm sizes were re-modified through a process of withdrawing and merging farmlands into larger farms.

As has been shown in many places worldwide, an increase in farm size can positively contribute to agricultural productivity. Although the consolidation was thus a predictable step in Uzbekistan for improving conditions for farm production, as a stand-alone measure it seems insufficient for achieving the fundamental goals set by the national administration. The irrigation and water infrastructure initially installed to serve *kolkhozes* and *sovkhozes* still needs to be adjusted to better suit larger numbers of smaller farms. In further steps, the consolidation process must also address other aspects such as change in infrastructure (van Dijk 2007). Hence, although farm size is an important aspect in increasing production and productivity, it is not the only one. A package of measures and policies that would relax some of the present constraints of farm production needs to be integrated into the entire consolidation process. Changing the current settings of state policy of cotton production, for instance, would be an option for increasing the productivity and efficiency of farms and would also promote crop diversification. The consolidation process combined with these potential initiatives could result in a higher economic efficiency of the newly established large farms.

References

Abdullaev I, Nurmetova F, Abdullaeva F, Lamers JPA (2008) Socio- technical aspects of water management in Uzbekistan: emerging water governance issues at the grass root level. In: Rahaman M, Varis O (Eds.) Central Asian Water. Helsinki, Helsinki University of Technology, Water and Development Publications

Bloch P (2002) Agrarian reform in Uzbekistan and other Central Asian Countries. Working Paper 49, Land Tenure Center, University of Wisconsin-Madison

Bobojonov I (2008) Modeling Crop and Water Allocation under Uncertainty in Irrigated Agriculture: A Case Study on the Khorezm Region, Uzbekistan. Rheinische Friedrich-Wilhelms-Universität Bonn

Bobojonov I, Rudenko I, Lamers JPA (2008) Optimal crop allocation and consequent ecological benefits in large-scale (shirkat) farms in Uzbekistan's transition process In: Wehrheim P, Schoeller-Schletter A, Martius C (Eds.) Continuity and Change: Land and Water Use Reforms in Rural Uzbekistan. Socio-economic and Legal Analyses for the Region Khorezm. IAMO, Halle/Saale

Deininger K, Byerlee D (2011) The rise of large farms in Land Abundant countries: do they have a future? In: Development Research Group AaRDT (Ed.) Policy Research Working Paper

Deininger K, Sarris A, Savastano S (2004) Rural Land Markets in Transition: Evidence from Six Eastern European Countries. Quarterly Journal of International Agriculture 43: 361–390

Dixon J, Gulliver A, Gibbon D (2001) Farming systems and poverty: Improving farmers' livelihoods in a changing world. FAO, Rome, World Bank, Washington DC

Djanibekov N (2008) A Micro-Economic Analysis of Farm Restructuring in the Khorezm Region, Uzbekistan. Rheinische Friedrich-Wilhelms-Universität Bonn

Djanibekov N, Bobojonov I, Djanibekov U (2012b) Prospects of agricultural water service fees in the irrigated drylands, downstream of Amudarya. In: Martius C, Rudenko I, Lamers JPA, Vlek PLG (Eds.) Cotton, water, salts and soums – economic and ecological restructuring in Khorezm, Uzbekistan. Springer, Dordrecht, pp. 389–411

Djanibekov N, Bobojonov I, Lamers JPA (2012a) Farm reform in Uzbekistan. In: Martius C, Rudenko I, Lamers JPA, Vlek PLG (Eds.) Cotton, Water, Salts and Soums – Economic and Ecological Restructuring in Khorezm, Uzbekistan. Springer, Dordrecht, pp 95–112

Djanibekov N, Van Assche K, Bobojonov I, Lamers JPA (2012c): Farm restructuring and land consolidation in Uzbekistan: new farms with old barriers. Europe-Asia Studies 64: 1101–1126

Eastwood R, Lipton M, Newell A (2010) Farm size. In: Pingali PL, Evenson RE (Eds.) Handbook of Agricultural Economics. Volume 4. Elsevier: Amsterdam

Fan S, Chan-Kang C (2010) Is small beautiful? Farm size, productivity, and poverty in Asian agriculture. Agricultural Economics 32: 135–146

Hazell P, Poulton C, Wiggins S, Dorward A (2010) The future of small farms: trajectories and policy priorities. World Development 38: 1349–1361

Khan AR (2007) Land system, agriculture and poverty in Uzbekistan. In: Akram-Lodhi H, Borras S, Kay C (Eds.) Land, Poverty and Livelihoods in an Era of Globalization: Perspectives from Developing and Transition Countries. Routledge, London, New York

Lerman Z (2009) Agricultural development in Central Asia: a survey of Uzbekistan, 2007 – 2008. Eurasian Geography and Economics 49: 481 – 505

Lerman Z, Cimpoies D (2006) Land consolidation as a factor for rural development in Moldova. Europe-Asia Studies 58: 439 – 455

Lerman Z, Shagaida N (2007) Land policies and agricultural land markets in Russia. Land Use Policy 24: 14 – 23

Mukherji A, Facon T, Burke J, de Fraiture C, Faures J-M, Fuleki B, Giordano M, Molden D, Shah T (2009) Revitalizing Asia's irrigation: to sustainability meet tomorrow's food needs. In: Colombo, FAO, Rome

Niyazmetov D, Rudenko I, Lamers JPA (2012) Mapping and analyzing service provision for supporting agricultural production in Khorezm, Uzbekistan. In: Martius C, Rudenko I, Lamers JPA, Vlek PLG (Eds.) Cotton, water, salts and soums – economic and ecological restructuring in Khorezm, Uzbekistan. Springer, Dordrecht, pp. 113 – 126

Pašakarnis G, Maliene V (2010) Towards sustainable rural development in Central and Eastern Europe: applying land consolidation. Land Use Policy 27: 545 – 549

Pomfret R (2008) Tajikistan, Turkmenistan and Uzbekistan. In: Anderson K, Swinnen J (Eds.) Distortions to Agricultural Incentives in Europe's Transition Economies, Washington DC

Rao V, Chotigeat T (1981) The inverse relationship between size of land holdings and agricultural productivity. American Journal and Agricultural Economics 63: 571 – 574

van Dijk T (2007) Complications for traditional land consolidation in Central Europe. Geoforum 38: 505 – 511

Veldwisch GJA, Spoor M (2008) Contesting rural resources: merging 'forms' of agrarian production in Uzbekistan. Journal of Peasant Studies 35: 424 – 251

Alexandra Conliffe

2.2 Geography Matters: Understanding Smallholder Livelihoods in Rural Khorezm[1]

Abstract

This chapter demonstrates that irrigation-based agricultural livelihoods in rural Khorezm are geographically differentiated. Arguing that spatial location within a single agroclimatic zone such as Khorezm is an important and often overlooked variable influencing livelihood opportunities, the chapter nevertheless demonstrates the need for a broad interpretation of the concept "geography matters" if we are to understand fully households' abilities to respond to political and environmental change. An examination of three livelihood types, namely cropping, entrepreneurship, and migration, illustrates this concept.

The chapter focuses on smallholder—"dehqon"—households, that is to say the 95 % of households that did not become "fermer" households when the Uzbek government fully dismantled collective farms (and the successors to collectives).[2] Although the concept "geography matters" is equally applicable to all rural household and farm types, this chapter provides insight into the least-studied dehqon livelihoods.[3]

Keywords: livelihoods, political ecology, political economy, environmental change, Uzbekistan

---

1 For a more detailed analysis, see Conliffe (2009).
2 *Dehqon* households, in theory, are entitled to approximately 0.25 ha of land for household agricultural production within the irrigated areas of the Khorezm region. In addition to this land, *fermer* households lease land from the state and engage in agricultural production in part to meet state quotas and in part for household benefit or private sale. In 2006 – 7, *fermers* typically leased between 1.0 – 100.0 ha of land, depending on the type of agricultural production for which their land was designated. Thus, the population was stratified between those who had land predominantly for subsistence purposes and those with land for profit making.
3 A household refers to all the people who typically live in a house, including those who leave sometimes for work or other purposes. A household's livelihood sources include all sources that benefit the household, whether or not contributed by a household member. See Conliffe (2009) for a full explanation of this definition.

2.2.1 Introduction

The Soviet Union's collapse led to a significant rollback of state benefits and state-funded nonfarm employment.[4] In many of the newly emerging states, including Uzbekistan, the deterioration of these important nonfarm livelihood opportunities triggered a process of "reagrarianization," whereby rural households became increasingly dependent on agricultural production (Kandiyoti 2003). In Uzbekistan, land reforms in the early and mid-2000s led the vast majority of the population (i. e. *dehqons*) to lose their use rights (both formal and informal) to formerly collective land. Consequently, the importance of production on household plots increased for much of the rural population (Djanibekov, et al. 2012).

While a substantial body of literature on "winners" and "losers" of land reforms in post-Soviet states demonstrates that land is often a liability in the absence of various assets (e.g. labour, agricultural equipment, credit, and market access), it often overlooks that land can be productive and profitable only if the natural resources to make it so are available (see, for example Verdery 1998; Hann, Property Relations Group 2003 ; Allina-Pisano 2004; O'Brien, et al. 2004). It follows that the literature also largely neglects how spatial differentiation of natural resource access affects agricultural livelihoods. For example, in Uzbekistan, agriculture is almost entirely reliant on irrigation, and upstream/downstream location affects water access and is thus one of the factors that influences cropping options.

This oversight is particularly common for agricultural livelihoods from household plots because data from the plots have been difficult to acquire and are often poor. For example, unable to obtain data in Uzbekistan during the Soviet era, Khan and Ghai (1979) assumed that because household plots were all the same size, their potential contribution to household wellbeing was similar for all households. Following the Soviet Union's collapse, Coudouel et al. (1997) acknowledged that land quality and access to irrigation water differed across Uzbekistan, and that this likely affected what households could grow on their plots. They hypothesized, however, that these differences were insignificant compared to other factors, given that 90 % of income inequality across their three research *oblasts*[5] existed at the intra-*oblast* level; they did not question whether differences in income might result from differences in natural resource access at this scale. These oversights were perhaps less important under the Soviet regime, during which households could access a range of collective farm

---

4 Development economics literature is followed, which defines nonfarm activities as non-agricultural activities.
5 Administrative territorial divisions akin to provinces.

and nonfarm livelihood options. Yet in the post-independence context whereby the importance of household agricultural production has increased, spatial location could prove an important variable determining natural resource access and agricultural production potential. This context motivated an exploration of the impact of geographically differentiated production potential on household plots at the intra-*oblast* scale, and of the impacts of such differentiation on the overall suites of livelihood options in which households engage.

Following an explanation of natural resource access across Khorezm and a brief description of the methods and study sites, the chapter presents data demonstrating that *dehqon* livelihoods are geographically differentiated, but that a broad conception of "geography matters" is needed to fully understand this result's implications. The chapter draws on examples from three livelihood types to illustrate this concept: cropping, entrepreneurship, and migration.

2.2.2 Natural resource access in Khorezm

Farmers in Khorezm face challenging environmental conditions. Khorezm's soils typically have low natural fertility (Akramkhanov et al. 2012). Soil salinization is common in regions such as Khorezm that have topographically low-lying lands with poor drainage and semi-arid and arid climates where evapotranspiration rates are high (O'Hara 1997); by 2000, 80 % of Khorezm's irrigated lands were considered "highly affected" by salinization (Akramkhanov 2005). Salinity is made worse by an irrigation system that is only about 30 % efficient and a poor drainage system that is largely in disrepair (Tischbein et al. 2012). Additionally, water flow from the *oblast's* main water source, the Amudarya River, is variable and can be severely reduced in water scarce years: in 2000, the flow to Khorezm reached only 40 % of the long-term average, and in 2001 only 34 % (Müller 2006).

As part of the project "Economic and Ecological Restructuring of Land and Water Use in the Khorezm Region (Uzbekistan)," much data has been collected on soil and water conditions in Khorezm. As demonstrated in previous publications, the results exhibit regional trends, even within Khorezm's single agroclimatic zone. So too, however, the research demonstrates significant local heterogeneities (Martius et al. 2012).

Key environmental factors that affect agricultural productivity and vary spatially within Khorezm include water access, groundwater levels, and groundwater and soil salinity. Research summarized in Martius et al. (2012) suggests that water accessibility and prevalence of soils with better lateral water movement and less salinity are greatest in upstream regions and close to central irrigation canals, with conditions worsening downstream (i.e. towards the south, west, and northwest regions of the *oblast*; fig. 2.2.1). Empirical data further demonstrate that although downstream regions are allotted more water than upstream ones to compensate for expected conveyance losses, which are greater over longer distances, the likelihood of actually receiving sufficient irrigation water decreases from upstream to downstream (Müller 2006). The large conveyance losses also contribute to the shallower groundwater levels and higher levels of groundwater salinity and soil salinity found in the southern, western, and northwestern parts of Khorezm (Ibrakhimov et al. 2007).

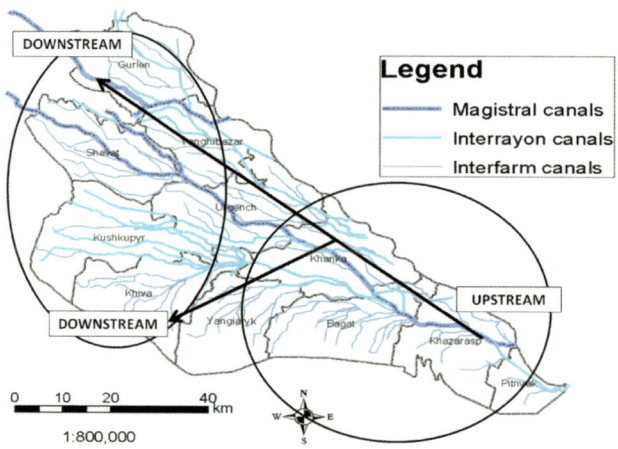

Figure 2.2.1: Schematic of upstream and downstream WUAs locations in Khorezm

In addition to these regional trends, significant heterogeneities exist at localized scales. The majority of Khorezm's soils are stratified, and groundwater movement and salinity levels depend on the textural layers (Tischbein et al. 2012). Consequently, soil hydraulic characteristics show strong vertical and horizontal heterogeneity. Distance from irrigation canals, topography, and stratigraphic characteristics of parent materials strongly influence groundwater tables, causing heterogeneity even within Water Users Associations (WUAs)[6] and at the field scale (Tischbein et al. 2012).

Crop management practices, including leaching, irrigation, and cropping

---

6 Each *oblast* in Uzbekistan is divided into administrative districts called *rayons*. *Rayons* are further divided into Water User Assocations (WUAs). In Khorezm, WUA boundaries often match those of former collective farms. The state allocates each *oblast*, *rayon*, and WUA a certain amount of water for agricultural production. The WUA, on one hand a territorial sub-district, is also an organization of stakeholders responsible for distributing water within the WUA for agricultural production, both for *fermers* and *dehqons*. In 2009, WUAs were renamed Water Consumers Associations. Because this research took place in 2006–7, this chapter uses the terminology of that time, namely WUA.

patterns (particularly mixed cropping patterns that include basin-irrigated rice, as in Khorezm; Kitamura et al. (2006)), also lead to significant heterogeneities. These management practices can affect crop productivity more than do soil and water conditions; working at the field level, Forkutsa (2006) found that even when salinity declines, productivity may decrease due to late irrigation, lack of weeding, and lack of pesticide and insecticide application. They concluded that poor management rather than salt stress is the biggest contributor to low cotton yields in Khorezm.

These results suggest that conditions and decisions at the WUA and field scale are at least as important as regional trends (see also Awan, et al. 2012). Yet regional trends nevertheless impose certain constraints on cropping potential. "Geography," then, is about more than spatial differentiation, and yet spatial differentiation requires consideration.

2.2.3 Research Design

2.2.3.1 Research Sites

On the basis of differentiated natural resource access and proximity to markets and centres of power, I selected two upstream and two downstream WUAs in which to conduct livelihoods research. To ensure anonymity, I label these WUAs WUA-U1 and WUA-U2 (the two upstream WUAs, whereby WUA-U1 is the most upstream) and WUA-D1 and WUA-D2 (the two downstream WUAs, whereby WUA-D2 is the most downstream).

As an upstream WUA, overall water availability in WUA-U1 is considered sufficient, although access is variable because the water comes from multiple sources. In general, the soil is considered among the most fertile in the *oblast*. The WUA is well-linked to the main *rayon* market and is a 25-minute drive from Khorezm's capital, Urgench, on main roads.

Located in the middle of Khorezm, WUA-U2's close proximity to one of Khorezm's three primary irrigation canals makes it an upstream WUA. Pumps provide all water access, which is thus rarely a problem for *dehqons*. Soil quality is typically good, particularly in the region where *dehqons* have household plots. WUA-U2 is located within a 15-minute drive from Urgench, and has the best access to markets and employment opportunities in the capital.

WUA-D1 is the most populous and densely-populated of the research WUAs. Water access is generally low but varies tremendously, as do soil salinity and fertility. WUA-D1 is a 30-minute drive from Urgench and Khiva, Khorezm's second-largest city, both on main roads.

WUA-D2 is one of Khorezm's most downstream WUAs. Water access varies tremendously. While soils in WUA-D2 are considered among the most marginal in the *oblast*, they are very heterogeneous: low soil fertility and wind erosion create problems near the desert, which borders this *rayon*, while waterlogging and high salinity cause problems in low-lying areas. WUA-D2 is poorly linked to the main *rayon* market and to Urgench, located a 45-minute drive away.

2.2.3.2 Research Methods

Livelihoods data were collected using a survey conducted June-July, 2007 with 68–73 randomly selected *dehqon* households in each research WUA. Complementing the survey were 8–13 follow-up, semi-structured interviews with purposively selected households from among those surveyed, and 302 informal interviews with 108 officials, *dehqons*, and *fermers* in 2006 and 2007.

The survey collected information on household demographics, land access and cropping patterns, and activities that contributed to household wellbeing. The survey also requested respondents to rank their three most important livelihood sources (i.e. those that generated the most "income"—cash or in-kind—for their households). Livelihood activities were divided between those derived from agricultural production, including livestock production, and from "nonfarm" options, namely social security benefits, state sector employment, private sector employment, entrepreneurship, and migration. Purposively selected semi-structured interviews enabled detailed, qualitative data collection on natural resource access and livelihood decision making. Details on methodologies, definitions of "income" and categorizing livelihood types are presented in Conliffe (2009).

2.2.4 Results and discussion

2.2.4.1 General Overview

Fig. 2.2.2 identifies key livelihood sources that generated household "income" or, in the case of agricultural activities, products that were additional to those required for household subsistence.[7] As in the Soviet era, the most common income sources remained those from the state, through employment and social security benefits, and from working for *fermers* (akin to working for collective farms under the Soviet regime). Other livelihood strategies included agricultural

---

[7] Data on illicit activities were not collected due to the sensitive nature and potential harm to respondents.

production on a variety of land types, seasonal migration, entrepreneurship and—new to the post-Soviet era—work in the private sector.

While fig. 2.2.2 presents the most common livelihood sources, Table 2.2.1 highlights sources that households cited as their most important income source.[8] The data corroborate findings in post-Soviet Russia, which underline that the state remains the population's main income source. Although its capacity has greatly diminished since independence, it appears that, as in post-Soviet Russia, the state nevertheless prevents a significant proportion of the population from slipping below the poverty line (Clarke 2002).

The data also corroborated those from livelihoods literature from other world regions, which find that farm wages are significantly lower than nonfarm incomes (Reardon 1997; Berdegue et al. 2001). Thus, although 43 % of *dehqon* households worked in some capacity for *fermers*, only 1 % said this was their top income source; only 13 % said it was one of their top three sources. This suggests a marked shift from the Soviet era, during which the rural population derived significant benefits from working on collective farms.

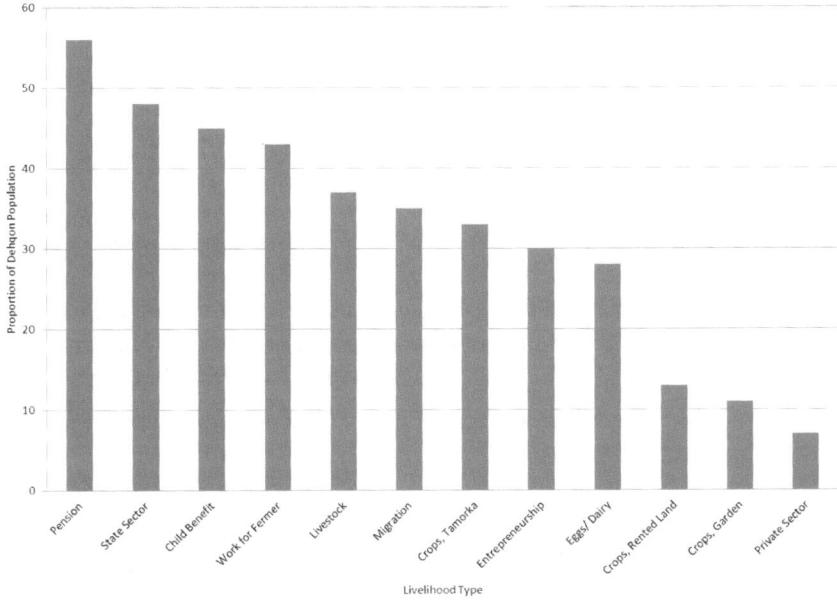

Figure 2.2.2: Most common livelihood sources for *dehqon* households in 2007 (n=280)

---

8 These data mask other income sources that households considered among their top three income sources. While these data are sufficient for the purposes of this chapter, Conliffe (2009) presents a more detailed analysis.

The figures highlight that, beyond state sources and agricultural production, seasonal migration and entrepreneurship are the most common income sources. Although these sources are not new, their nature has changed in the post-Soviet era, as discussed below. The low prevalence of the private sector supports the literature denoting the slow emergence of this sector in Uzbekistan.

Table 2.2.1: Most important income sources for *dehqon* households in 2007 (n=280)[9]

Livelihood Type	Proportion of *dehqon* households that ranked livelihood type as their most important income source (%)			
	WUA-U1	WUA-U2	WUA-D1	WUA-D2
Pension	3	25	8	29
State Employment	19	20	32	25
Seasonal Migration	2	20	29	19
Crops, *Tamorka*	15	4	13	4
Entrepreneurship	9	14	5	7
Crops, Karakalpakstan	46	0	0	0

Table 2.2.1 suggests significant spatial variation in the importance households afford to different livelihood strategies. While a detailed examination of how households build their livelihoods is beyond the scope of this chapter[10], using examples from three livelihood types, namely cropping, migration, and entrepreneurship, the remainder of the chapter illustrates how livelihoods vary across space and highlights some of the potential implications.

2.2.4.2 Cropping

Agricultural production on household plots is a dominant livelihood strategy for the majority of rural households in Khorezm. Households produce crops, livestock, and fowl for subsistence purposes and to generate cash and in-kind income (Djanibekov et al. 2012). Cropping livelihoods are spatially differentiated in two key ways: (1) cropping patterns and (2) access to plot types. In each instance, however, a broad range of factors affect how cropping activities contribute to household wellbeing.

Historically, households in Khorezm have owned gardens of approximately 0.12 ha located behind their homes to grow fruits, vegetables, potatoes, beans, and fodder[11] for predominantly subsistence purposes (Conliffe 2009). As former

---

9 The table includes all income sources that at least 10 % of households in at least one WUA considered as their most important income source.
10 See Conliffe (2009) for an in-depth analysis.
11 The term fodder is used to denote maize and other forage crops, such as sorghum, that have

collective land was leased to *fermers*, so *dehqon* households moved their livestock into their gardens. Beginning in 1991, the state started to take land from former collectives and divide it into additional household plots. In Khorezm, these plots, located in the irrigated areas, are called *tamorkas* and are 0.12 – 0.13 ha in size. They are often located within walking distance of a household's village, and groups of 10 – 20 *tamorkas* comprising 1 – 2 ha of land are scattered throughout the WUA amidst what are now *fermers*' fields.

While soil and water access in gardens vary across and within WUAs, because gardens are clustered within the village and water comes from the village water system, natural resource access is less variable than for *tamorkas*, which are geographically disbursed and which access water through the same irrigation canals as *fermers*' fields and thus compete with *fermers* for water. Consequently, cropping patterns are more differentiated on *tamorkas* than in gardens.

In principle, households have the opportunity to grow a variety of crops on their *tamorkas*. Originally *tamorkas* were intended to enable households to grow wheat to ensure food production for the household scale and to compensate households for the loss of payments from state and collective farms (Kandiyoti 2003; Trevisani 2007). Because households grow winter wheat, they have the opportunity to grow a second crop in the summer. In 2006 – 07, many households preferred to grow rice. As the most remunerative crop, growing rice helped save money by reducing the need to purchase it, while selling rice made the most profits. Rice, however, was also by far the most water-intensive crop (Table 2.2.2). Fruits and vegetables also generated reasonable profits; they were less water-intensive but required relatively fertile soils and low salinity. Fodder required less water and was relatively salt-tolerant.

Table 2.2.2: Crop water norms for common crops grown on *tamorkas*

Crop	Crop Water Norm (m^3/ha)
Rice (Paddy)	26,206
Vegetables and Potatoes	8,467
Maize	5,233
Orchards	5,133
Melons and Gourds	3,933
Wheat	3,600

Source: Adapted from Veldwisch (2008).

Based on the regional trends outlined earlier, we might expect that more households in upstream WUAs grew rice than in downstream WUAs, where a greater proportion of households might produce fodder. Indeed, while the majority of

---

low to medium irrigation requirements relative to other crops in Khorezm and that Khorezmian farmers consider to be relatively salt-tolerant.

households in all WUAs grew wheat, 67 % and 26 % grew rice in WUA-U1 and WUA-U2 compared to 6 % and 29 % in WUA-D1 and WUA-D2 (Table 2.2.3).

Table 2.2.3: Crops commonly grown on *tamorkas*, by WUA in 2007 (n=280)[12]

Crop Type	Proportion of *dehqon* households that grew each crop type (%)			
	WUA-U1	WUA-U2	WUA-D1	WUA-D2
Wheat	78	88	86	67
Rice	67	26	6	29
Fodder	28	71	86	64
Fruits/ Vegetables	10	29	32	8

The higher-than-expected proportion of households growing rice in WUA-D2 resulted from the large number of households that had *tamorkas* in low-lying, waterlogged fields suitable only for paddy rice production. Thus, only 8 % of these households double-cropped wheat and rice, whereas 21 % grew rice as a single crop and had to access wheat via alternative means (Table 2.2.4).

Table 2.2.4: Common crop mixes on *tamorkas*, by WUA in 2007 (n=280)[13]

Crop Type	Proportion of *dehqon* households that grew each crop type (%)			
	WUA-U1	WUA-U2	WUA-D1	WUA-D2
Wheat/ rice	50	12	2	8
Wheat/ fodder	12	29	59	46
Rice only	10	2	2	21

By contrast, only 10 % and 2 % of households in WUA-U1 and WUA-U2 single-cropped rice, while 50 % and 12 % double-cropped it with wheat. As expected, a higher proportion of households in WUA-D1 and WUA-D2 double-cropped wheat with fodder (59 % and 46 %) compared to WUA-U1 and WUA-D2 (12 % and 29 %). Fruit and vegetable production was most common in WUA-U2 and WUA-D1 (29 % and 32 %), where much of the soil was relatively fertile, compared to WUA-U1 (10 %), where many households preferred to grow rice and had the ability to do so and WUA-D2 (8 %), where soils were often too marginal for this purpose.

Although Tables 2.2.3 and 2.2.4 suggest that cropping patterns matched general regional variation in natural resource access, regional trends are never-

---

12 The sum of crops grown is greater than 100 % in each WUA because the majority of households grow a mix of crops on their *tamorkas* and/or double-crop (for example, winter wheat followed by rice, vegetables, or fodder).
13 The crop mixes presented are illustrative only; households engage in a wider range of mixed and double cropping than presented here. Consequently, the sum of crop mixes in each WUA is less than 100 %.

theless insufficient to fully explain household cropping patterns. First, the regional trends mask local heterogeneities in soil quality and water access within each WUA. Furthermore, households accounted for a broader range of factors when making cropping decisions. For example, households in WUA-D1 had a long tradition of selling vegetables to Central Asian traders who came to the WUA to purchase them from the collective. Following the Soviet Union's collapse, and the distribution of *tamorkas*, households started to coordinate vegetable production on these new plots so that they could continue to sell their produce in bulk and maintain this longstanding niche market. Available labour also mattered; female-led households (or households in which all the men were seasonal migrants) could not always grow rice even if they had good water access because it was culturally unacceptable for them to irrigate at night when water was available. Propensity to take risk was also a factor; in 2007 state officials warned households not to grow rice due to a likely water shortage. Although there was no shortage, many households chose to grow fodder instead.

The explanatory power of spatial location is also limited by the possibility to build agricultural livelihoods in multiple locations. For example, 55 % of households in WUA-U1 rented a median of 5 ha of land to grow rice in the autonomous republic of Karakalpakstan, north of Khorezm. 72 % of these households considered the rice profits as their most important income source[14]. In WUA-D1, 21 % of households rented additional land from *fermers* (compared to 6 %, 8 %, and 1 % in WUA-U1, WUA-U2, and WUA-D2). Many of these households were renting land in other WUAs or even *rayons*, in part to overcome the land shortage resulting from WUA-D1's high population density, but also to grow additional vegetables to sell to foreign traders; market access encouraged them to acquire additional land. In each of these examples, however, only households with sufficient financial capital and labour could take advantage of the multi-locational opportunities.

2.2.4.3 Entrepreneurship

Having determined that regional trends related to natural resource access partly contribute to differentiated cropping patterns across WUAs, the question follows: is there a relationship between regional natural resource access trends and associated cropping patterns on the one hand and engagement in non-agricultural livelihood strategies on the other?

Engagement in entrepreneurship did not vary tremendously across WUAs.

---

14 This equates to 46 % of all *dehqon* households in WUA-U1, as denoted in Table 2.2.1 as "Crops, Karakalpakstan."

Between 24 % and 38 % of households in each research WUA derived some of their livelihood from entrepreneurial activities. Entrepreneurship was likely under-reported, either because some activities generated in-kind benefit that households did not think to include in their list of livelihood strategies (for example, selling scrap pieces of metal or bottles to collectors) or because some income sources were illicit (for example, cross-border sales of goods with Turkmenistan or small-scale sale of car fuel).

Many entrepreneurial activities were common across WUAs, including traditional activities such as sewing, hairdressing, and construction. Some households in all WUAs were making reasonable profits from providing services that collective farms or the state had previously provided either formally or informally. Thus, almost one quarter of the types of entrepreneurial activities cited supported agricultural activities such as cultivation (e.g. providing pesticides and fertilizers or ploughing *fermer* or household land with tractors purchased from dismantled collectives), processing (e.g. milling wheat and rice from *tamorkas* and *fermer* land), and business management (e.g. accounting for *fermers*, who are now responsible for the financial management of their farms).

Regional trends emerged for those entrepreneurial activities that were most lucrative. Thus, a larger proportion of households in WUA-U1 generated significant income by providing agricultural support services to *dehqon* and *fermer* households. Importantly, however, while a market for such services may in part have resulted from the generally superior natural resource access there, so too it supported the large proportion of households growing significant amounts of rice in a different location entirely, namely Karakalpakstan. The largest proportion of households stating that entrepreneurship was their most important income source (14 %) was in WUA-U2. Here, the stated activities related mainly to provision of services to the capital city.

Despite its downstream location, a small proportion of households in WUA-D1 were making significant profits by transporting agricultural produce to markets throughout Uzbekistan in cargo trucks that households had purchased from the dismantled collectives. More remote from major cities, establishing such market links appeared harder for households in WUA-D2. Households in WUA-D2 thus seemed doubly hard-hit: their downstream location meant that agricultural productivity was limited by marginal natural resource access, while their remote location limited their ability to use multi-locationality to tap into more lucrative markets outside their WUA.

2.2.4.4 Migration

In 2007, seasonal migration abroad was increasing.[15] While temporary migration was common within the Soviet Union, the formation of borders between newly independent states after the regime's collapse limited such movement. By the mid-1990s, however, a new trend was emerging: men were migrating seasonally, often to Russia and Kazakhstan, mainly to provide unskilled labour to the rapidly growing construction industry (Sadvskaya 2007). The jobs were often informal, and lacked security regarding pay, benefits, and protection in the instance of accidents on the job.

At least one person in 35 % of *dehqon* households engaged in migration, which, in 2007, was the highest level that had been reported in Khorezm.[16] 87 % of households that obtained remittances said this was from construction for a median migration period of 5.5 months.[17]

To further examine spatial differentiation of livelihoods, engagement in seasonal migration was analysed according to upstream/downstream location. Migration was certainly a response to the lack of access to agricultural land: regression analysis demonstrated that whether a household was a *dehqon* or *fermer* household was the dominant variable determining whether it engaged in seasonal migration: at least one male in 35 % of *dehqon* households engaged in seasonal migration compared to only 12 % in *fermer* households.

Engagement in seasonal migration also displayed a spatial dimension: at least one male in 48 % and 52 % of *dehqon* households in WUA-D1 and WUA-D2 migrated compared to only 4 % and 33 % in WUA-U1 and WUA-U2. This corroborates research by Veldwisch (2008) in Khorezm in 2006, which found migration rates of 8 % and 16 % in two upstream WUAs and 21 % and 30 % in downstream WUAs for *dehqon* and *fermer* households combined. Further analysis, however, revealed a more complex picture. For instance, availability of male labourers between the ages of 17–30, a common age bracket for seasonal migrants, was more significant than spatial location in determining *dehqon* household engagement in seasonal migration.[18] Additionally, where spatial lo-

---

15 Migration within Uzbekistan was 3 %. This low figure likely resulted from the Uzbek government's control over internal movement and the significant distance between Khorezm and the Uzbek capital, Tashkent (the most common destination for internal migrants), which made illicit migration more difficult.
16 Based on a similar survey conducted in 2006 in four WUAs in Khorezm, Veldwisch (2008) found migration levels of only 19 %.
17 Results are from 93 households (i.e. 76 % of surveyed households) that were asked the duration of the migration period.
18 Households in WUA-U1 that grew rice in Karakalpakstan were excluded from the analysis because their inclusion masked the impact of other variables on households' engagement in seasonal migration.

cation influenced households' decision to engage in seasonal migration, the linkage to upstream/downstream location was non-simplistic. First, migration rates were low in WUA-U1 not because of the superior agricultural production potential as an upstream WUA, but rather because of households' extensive engagement in rice production in Karakalpakstan. Second, a major impact of differentiated natural resource access is the ability to grow rice. Yet, as a variable in the regression analysis, whether or not a household grew rice on its *tamorka* had no significant impact on sending labourers abroad. This is likely because *tamorkas* are too small in size to generate profits at a scale that would negate the need to generate additional income via other means. Given that a significant difference between upstream and downstream WUAs is the ability to grow rice on *tamorkas*, spatially differentiated natural resource access cannot fully explain spatial trends related to migration.

In contrast, the ability to grow rice on land other than *tamorkas* (e.g. on rented *fermer* land or abandoned land, often at further distances from the villages) was a significant factor in determining involvement in migration. Yet while the ability to grow rice on alternative land types depends on natural resource access, many factors in addition to spatial location affect a household's ability to cultivate additional land (see analysis above on cropping). Thus, spatial differentiation must be considered as only one of many variables that affect whether a household can (1) access additional land and (2) grow rice on it.

Third, access to other forms of nonfarm employment, in particular state sector employment, proved significant in determining households' engagement in migration. This suggests that the benefits that some households derive from state sector employment may be sufficient to enable them to build their livelihoods within Khorezm.

Entrepreneurship was not a significant variable determining a household's engagement in migration as long as "entrepreneurship" was not disaggregated into classes indicating different scales of entrepreneurial activities. Cursory disaggregation suggested that households that engaged in entrepreneurial activities that generated significant income were less likely to engage in seasonal migration. For example, of the nine households in WUA-D1 that transported agricultural produce in cargo trucks, none engaged in seasonal migration. While it is true that the peak season for buying and selling agricultural produce conflicted with the construction season abroad, that these households chose truck driving over seasonal migration appears noteworthy.

2.2.5 Conclusions

Three important implications emerge from the analysis: first, while livelihood strategies in Khorezm varied from upstream to downstream, the concept "geography matters" must be interpreted with care. Households in upstream WUAs were more able to double-crop the staple crop wheat with the region's most lucrative and water-intensive crop, rice, on their *tamorkas*. This general trend, however, masked significant differences in natural resource access within each WUA, while other factors, including availability of male labour, propensity to take risks, and access to niche markets, also influenced household cropping decisions. Households in downstream WUAs were more likely to engage in seasonal migration, yet to suggest that they did so predominantly to overcome poor natural resource access is simplistic. For example, households in WUA-U1 engaged little in migration not because of superior natural resource access in their own WUA, but because they adopted a strategy of multi-locationality and grew rice in a different location altogether. More likely, households in downstream WUAs engage in seasonal migration because they are doubly hard-hit, both by inferior agricultural potential due to poor natural resource access, and by their remoteness, which makes access to markets and multi-locational livelihood opportunities more difficult.

Second, early trends suggest that farm privatization has led to stratification in wellbeing, even among *dehqon* households. For example, wealthier households were able to access additional plots of land or invest in lucrative entrepreneurial activities, such as cargo truck driving. Again, this appeared more difficult in remote downstream locations, where access to lucrative opportunities was limited.

Finally, spatial differentiation and stratification could have implications for households' abilities to respond to shocks that adversely affect certain livelihood strategies. For example, we might expect that the financial crisis that greatly reduced seasonal migration opportunities in Russia and Kazakhstan starting in 2008 affected downstream regions most. Droughts might have the greatest impact on those households reliant on rice production—including households in WUA-U1—that invested significant capital into rice production in Karakalpakstan. Time series data could better inform these hypotheses and increase our understanding of household resilience. This in turn could help inform what resilience-building strategies are required and how to target them.

References

Akramkhanov A (2005) The spatial distribution of soil salinity: Detection and prediction. PhD dissertation, Bonn University

Akramkhanov A, Kuziev R, Sommer R, Martius C, Forkutsa O, Massucati L (2012) Soils and soil ecology in Khorezm. In: Martius C, Rudenko I, Lamers JPA, Vlek PLG (Eds.) Cotton, water, salts and soums – economic and ecological restructuring in Khorezm, Uzbekistan. Springer, Dordrecht, pp 37–58

Allina-Pisano J (2004) Land reform and the social origins of private farmers in Russia and Ukraine. Journal of Peasant Studies 31 (3): 489–514

Awan UK, Tischbein B, Kamalov P, Martius C, Hafeez M (2012) Modeling irrigation scheduling under shallow groundwater conditions as a tool for an integrated management of surface and groundwater resources. In: Martius C, Rudenko I, Lamers JPA, Vlek PLG (Eds.) Cotton, water, salts and soums – economic and ecological restructuring in Khorezm, Uzbekistan. Springer, Dordrecht, pp 309–327

Berdegue JA, Ramirez E, Reardon T, Escobar G (2001) Rural nonfarm employment and incomes in Chile. World Development 29 (3): 411–425

Clarke S (2002) Sources of subsistence and the survival strategies of urban Russian households. In: Rainnie A, Smith A, Swain A (Eds.) Work, Employment and Transition: Restructuring Livelihoods in Post-Communism. Routledge, London

Conliffe A (2009) The Combined Impacts of Political and Environmental Change on Rural Livelihoods in the Aral Sea Region of Uzbekistan. PhD dissertation, University of Oxford

Coudouel A, Marnie S, Micklewright J, Shcherbakova G (1997) Regional differences in living standards in Uzbekistan. In: Falkingham J, Klugman J, Marnie S and Micklewright J (Eds.) Household Welfare in Central Asia. Macmillan Press Ltd, London

Djanibekov N, Bobojonov I, Lamers JPA (2012) Farm reform in Uzbekistan. In: Martius C, Rudenko I, Lamers JPA, Vlek PLG (Eds.) Cotton, water, salts and *Soums* – economic and ecological restructuring in Khorezm, Uzbekistan. Springer, Dordrecht, pp 95–112

Forkutsa I (2006) Modeling Water and Salt Dynamics Under Irrigated Cotton with Shallow Groundwater in the Khorezm Region of Uzbekistan. University of Bonn

Hann CM, Property Relations Group (2003) The Postsocialist Agrarian Question: Property Relations and the Rural Condition. LIT Verlag, Münster

Ibrakhimov M, Khamzina A, Forkutsa I, Paluasheva G, Lamers JPA, Tischbein B, Vlek PLG, Martius C (2007) Groundwater table and salinity: Spatial and temporal distribution and influence on soil salinization in Khorezm region (Uzbekistan, Aral Sea Basin). Irrigation and Drainage Systems 21 (3–4): 219–236

Kandiyoti D (2003) The cry for land: agrarian reform, gender and land rights in Uzbekistan. Journal of Agrarian Change 3 (1–2): 225–256

Khan AR, Ghai D (1979) Collective Agriculture and Rural Development in Soviet Central Asia. Macmillan, London

Kitamura Y, Yano T, Honna T, Yamamoto S, Inosako K (2006) Causes of farmland salinization and remedial measures in the Aral Sea basin – research on water management to prevent secondary salinization in rice-based cropping system in arid land. Agricultural Water Management 85 (1–2): 1–14

Martius C, Rudenko I, Lamers JPA, Vlek PLG (2012) Cotton, Water, Salts and Soums – Economic and Ecological Restructuring in Khorezm, Uzbekistan. Springer, Dordrecht, 419 p

Müller M (2006) A general equilibrium approach to modeling water and land use reforms in Uzbekistan. PhD dissertation, Rheinischen Friedrich-Wilhelms-Universität Bonn

O'Brien DJ, Wegren SK, Patsiorkovski VV (2004) Contemporary rural responses to reform from above. The Russian Review 63: 256–276

O'Hara S (1997) Irrigation and land degradation: Implications for agriculture in Turkmenistan, Central Asia. Journal of Arid Environments 37: 165–179

Reardon T (1997) Using evidence of household income diversification to inform study of the rural nonfarm labor market in Africa. World Development 25 (5): 735–747

Sadvskaya E (2007) International labour migration, remittances and development in Central Asia: towards regionalization or globalization? In: Proceedings of Migration and Development, 13–15 September, Moscow, Lomonosov State University

Tischbein B, Awan UK, Abdullaev I, Bobojonov I, Conrad C, Forkutsa I, Ibrakhimov M, Poluasheva G (2012) Water management in Khorezm: current situation and options for improvement (hydrological perspective). In: Martius C, Rudenko I, Lamers JPA and Vlek PLG (Eds.) Cotton, water, salts and *Soums* – economic and ecological restructuring in Khorezm, Uzbekistan. Springer, Dordrecht, pp 69–92

Trevisani T (2007) After the kolkhoz: rural elites in competition. Central Asian Survey 26 (1): 85–104

Veldwisch GJ (2008) Cotton, rice & water: transformation of agrarian relations, irrigation technology and water distribution in Khorezm, Uzbekistan. PhD dissertation, Bonn University

Verdery K (1998) Property and power in Transylvannia's decollectivization. In: Hann CM (Ed.) Property Relations: Renewing the Anthropological Tradition. Cambridge University Press, Cambridge

Krishna P. Devkota, Ahmad M. Manschadi, John P. A. Lamers,
Erkin Ruzibaev, Mina K. Devkota, Oybek Egamberdiev,
Raj K. Gupta, Paul L. G. Vlek

2.3 Exploring innovations to sustain rice production in Central Asia: A case study from the Khorezm region of Uzbekistan

Abstract

The Khorezm region is a major rice growing region in Uzbekistan, Central Asia. The demand for irrigation water in the region keeps increasing due to among others the excessive and inefficient water use. Increasing demands stem also from the urgent need of farmers to increase rice production, which is one of the most profitable crops. Yet, present cropping practices have contributed to the ongoing land degradation and desertification that is also threatening the sustainability of rice production in the irrigated lowlands of Khorezm and beyond in Central Asia. Various innovations in rice production applied worldwide, e.g. water saving and conservation agriculture (CA) practices, optimizing seeding time, using appropriate varieties and transplanting can help counterbalance the present threats. A series of different experiments during two years focusing on the analysis of options for sustaining rice production in the Khorezm region showed that frequent wet and dry irrigation (WAD) led to a 70 % water-saving potential but reduced rice yield by 42 %. Such rice yield reduction can be minimized by applying the conventional flood irrigation up to the stage of panicle initiation and then changing to intermittent irrigation. Furthermore, grain yields of eight rice varieties subjected to direct seeding and transplanting did not differ. The most promising seeding dates were up to the end of June for short- and from May 25 to June 10 for medium- and long-duration rice varieties. Various innovations tested as stand-alone measures or in combination showed that dry direct seeding of medium- (Allanga-3) or short- (Nukus-2) duration rice within the period from May 25 to June 10 on permanent beds or under zero tillage combined with WAD irrigation could substantially save irrigation water and sustain rice production without compromising yields.

Keywords: conservation agriculture, wet and dry irrigation, rice transplanting, rice varieties, seeding date

2.3.1 Background

In the five Central Asian countries, rice is the most important food crop after wheat. It is an important part of the national diet in Uzbekistan, Kazakhstan and Tajikistan, where it forms the basis for the national dish plov. The per capita consumption of rice varies between 20 and 25 kg (Ismail 2006). The total rice area in Central Asia was about 341,000 ha after independence in 1991, but it decreased progressively to 241,254 ha in 2010. In that year, 39 % of the total rice area in Central Asia was in Kazakhstan followed by 27 % in Turkmenistan, 25 % in Uzbekistan, 6 % in Taijkistan, and 3 % in Kyrgyzstan (FAOSTAT 2012). Rice is mostly grown in the irrigated areas bordering the Amudarya and Syrdarya rivers. Currently, most of these countries meet their rice needs through imports.

Water management is one of the most important factors determining the sustainability of rice cultivation in Central Asia (Gupta et al. 2009), as rice production under arid climatic conditions often requires more than 3,000 mm of irrigation water (FAO 1997; UNESCO 2000). Wet-direct seeding (broadcasting of pre-germinated rice seed into the standing water) of mainly Japonica rice cultivars, after harrowing (2 – 3 times), chiseling and leveling, is the most common farmer practice in the Khorezm region, and is representative for the rice producing zones in Central Asia. The present cultivation practices include the maintenance of a continuous collar of 20 – 30 cm standing water from the time of field preparation to crop maturity.

In Central Asia, rice is a remunerative crop that fetches a several times higher price than wheat and cotton, and even 2 – 3 times higher than the world market price, making it attractive to farmers (Ismail 2005; Djanibekov 2008). However, this important source of income is endangered by a diminishing supply of irrigation water (Christmann et al. 2009). The decreasing water availability is compelling rice farmers to develop and adopt innovations that demand less irrigation water, increase water productivity and sustain soil fertility. Many of these innovations, such as alternate wetting and drying (Dong et al. 2004) and dry-direct seeding in zero-tillage flat land and permanent beds (Sayre and Hobbs 2004), have been successfully implemented elsewhere including in regions with similar agro-climatic conditions as in the study region Khorezm. The objective of this study was to explore the potential of various innovations to reduce water application and in turn sustain irrigated rice production in the Khorezm region of Uzbekistan.

2.3.2 Materials and methods

2.3.2.1 Khorezm region

Khorezm lies in the north-western part of Uzbekistan (41°32'12" N, 60°40'44" E) in the lower reaches of the River Amudarya in the transition zone of the Karakum and Kyzylkum deserts. The climate is arid with hot dry summers and very cold winters. The average precipitation is less than 100 mm year^{-1}. Potential evapotranspiration greatly exceeds precipitation. The mean annual temperature is 13.6° C. The soils are dominated by sandy, sandy- loam and loamy-sandy textures, low in organic matter, and prone to salinization (Akramkhanov et al. 2012).

2.3.2.2 Experimental site

A series of experiments, stepwise and parallel, were conducted during 2008 – 2009 to explore the different innovations for reducing water application in irrigated rice production. The soil at the experimental site is an irrigated alluvial meadow (Russian classification) with a sandy-loam to loamy soil, low in soil organic matter (0.3 – 0.6 %), saline (EC1:1 ranging from 2 – 16 dS m-1) and with a shallow groundwater table (0.5 to 2 m) and thus representative for the dominating soil characteristics in the region. Rice-growing farmers exclusively use Japonica cultivars (Ismail 2006). Rice is cultivated in the region on up to 20 % of the available cropland (Djanibekov 2008). Short-duration varieties are used in double-cropping systems, i.e., growing two cereal crops (wheat and rice) in the same growing season of roughly eight months under the prevailing climatic conditions. Here, rice is seeded following the harvest of winter wheat in June. Long-duration rice varieties are planted in April/May and harvested in September. More details are presented in Devkota (2011).

2.3.2.2.1 Experiment I

This experiment was conducted in order to quantify rice yield under different irrigation methods combined with conservation agriculture (CA) technologies (tillage methods and residue levels). Six frequent intermittent wet and dry (WAD) rice treatments involving dry-direct seeded rice (dry-DSR) on raised beds (BP), and with zero tillage (ZT) were combined with three levels of residue retention, i.e., farmers' method of residue harvest (RH), 50 % residue retention (R50) and 100 % residue retention (R100). These innovative practices were compared with wet-direct seeded rice (wet-DSR) grown under conventional tillage, continuous flood irrigation (CT-FI), i.e., paddy cultivation, and conventional tillage, intermittent irrigation (CT-II). The experiment was conducted in a randomized complete block design in four replications. The WAD rice and CT-II treatments were irrigated using intermittent flood irrigation when the soil water potential in 15–20 cm soil depth dropped below -10–20 kPa, i.e., exhibiting 5–10 % decrease in volumetric moisture content compared to that at field capacity. The CT-FI treatment was irrigated continuously to keep a 5–20 cm standing water collar. Rice was seeded on newly established beds in 2008, which was considered as the effective start of the CA experiment. Thereafter, all BP and ZT plots were kept permanent and untilled. In 2009, two additional treatments, i.e., zero-tillage-continuous flood irrigation up to the time of panicle initiation and intermittent irrigation thereafter (ZT-FI-II), and zero-tillage-continuous flood irrigation throughout the crop growing period (ZT-FI) were also evaluated. For additional details see Devkota (2011).

2.3.2.2.2 Experiment II

The method of preparing rice seedlings in a nursery and transplanting these after 25–35 days in the field (TPR) allows avoiding water applications on larger areas during the seedling raising period. This could thus save a significant amount of irrigation water compared to the conventional wet-DSR method. However, the observed yield difference in Japonica cultivars under wet-DSR and transplanted conditions is not yet understood. In 2009, this experiment was conducted with four popular local rice varieties (Shoternboy-1, Nukus-2, Allanga-3 and Avangard) to determine the best rice planting technique. In both methods, rice was planted without limitations of water and nitrogen.

2.3.2.2.3 Experiment III

Adapted, high yielding and optimum maturing rice varieties could increase crop yields. However, the performance of rice varieties could differ under water-saving and conventional irrigation methods. Therefore, eight locally adopted rice varieties were evaluated under a late sown (seeded on 20 June) dry-DSR treatment using WAD irrigation in 2008 and under a timely sown (seeded on 1 June) wet-DSR treatment using CT-FI in 2009. The seed rate was 130 kg ha^{-1} whilst fertilizer applications were 250:120:80 kg NPK ha^{-1}.

2.3.2.2.4 Experiment IV

In arid and semi-arid regions, the time frame for better growth and development of rice is limited under the dominating double-cropping practices. Hence, an appropriate seeding date is the key for obtaining high rice yields (Christmann et al. 2009). The seeding time of rice is, therefore, very crucial, as the flowering period may coincide with peak maximum temperatures during early seeding, whilst during late seeding rice crops may be affected by low-temperature stress (Devkota 2011). Commonly cultivated local short-maturity duration (Shoternboy-1, Nukus-2), medium maturity (Setora, Allanga-3, Novbahor and Shoternboy-2) and long-maturity (Mustakillik, Avangard) rice varieties were evaluated for their phenological development following five sowing dates starting from May 5 to July 6, 2009, under wet-DSR conditions.

2.3.3 Results

2.3.3.1 Experiment I: Water application and yield of rice under conservation agriculture technologies combined with water-saving irrigation

The total amount of irrigation water necessary for keeping the defined soil moisture under CT-FI was 6,690 and 5,905 mm in 2008 and 2009, respectively for Nukus-2 (100-days rice variety dominant in the region). In contrast, the amount of water applied in the WAD treatments was 28 % (1,900 mm) and 33 % (1,928 mm) of the total amount applied under CT-FI in 2008 and 2009, respectively. Also, BP showed a higher potential to reduce water applications than ZT, and received 19 % less irrigation water in both 2008 and 2009. Water application under CT-II was higher than for BP but not significantly different from ZT. Likewise, water application was not affected by residue retention.

Compared to CT-FI, rice yield was lower in the WAD treatments by 30 % in 2008 and 56 % in 2009 (fig. 2.3.1 A). Irrespective of the residue level, rice yield was not significantly different between BP and ZT in both years. No differences in yield were observed between RH and residue levels in 2008, whereas yields were significantly reduced under residue retention in 2009. Rice grain yield in the RH treatment was 28 % lower in 2008 and 40 % lower in 2009 compared to CT-FI. Yield reduction in the WAD treatments increased with increasing residue level. Averaged over the tillage methods, R50 had 2 % lower yields in 2008 but 21 % in 2009 than RH. The R100 treatments had 10 % lower yields in 2008 and 59 % lower in 2009 compared to R50; CT-II had 39 % lower yields than CT-FI, while yield differences were insignificant between BP-RH, ZT-RH and CT-II in 2009 (fig. 2.3.1 A).

No significant difference in grain yields was observed between ZT-II and CT-II or between ZT-FI and CT-FI (fig. 2.3.1B). In contrast, grain yield under flood irrigation (ZT-FI and CT-FI) was significantly higher than under WAD (ZT-II and CT-II), while it was in-between in flood irrigation up to panicle initiation and subsequent intermittent irrigation (ZT-FI-II).

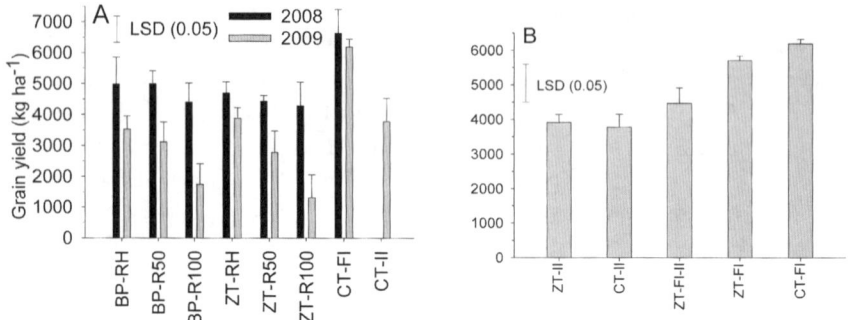

Figure 2.3.1: Grain yield (kg ha^{-1}) of rice as affected by irrigation, tillage and residue level in 2008 and 2009 (A) and irrigation and tillage methods in 2009 (B). BP-RH=bed planting residue harvested, BP-R50=bed planting 50 % residue, BP-R100=bed planting 100 % residue, ZT-RH=zero tillage residue harvested, ZT-R50=zero tillage 50 % residue, ZT-R100=zero tillage 100 % residue, CT-FI=conventional tillage continuous flood irrigation, CT-II=conventional tillage intermittent irrigation, ZT-II=zero tillage intermittent irrigation, ZT-FI-II=zero tillage continuous flood irrigation up to panicle initiation and intermittent irrigation thereafter, and ZT-FI=zero tillage continuous flood irrigation. LSD (0.05) is difference between treatments at P=0.05. Bars indicate standard error of the mean.

2.3.3.2 Experiment II: Direct seeded vs. transplanted rice

Grain yields of the four rice varieties under wet-DSR and TPR conditions were not significantly different except for Avangard that showed an 18 % higher yield under the TPR than under the wet-DSR treatment (fig. 2.3.2 A). Under both planting methods, Allanga-3 (medium-duration variety) followed by Nukus-2 (short-duration variety) produced higher grain yields than the other two varieties except for the transplanted Avangard. Under transplanted conditions, physiological maturity of all varieties was delayed by 10 – 12 days compared to wet-DSR (fig. 2.3.2B).

2.3.3.3 Experiment III: Performance of local rice varieties

Yields of all eight rice varieties screened were significantly lower (p<0.001) in 2008 under dry-DSR with WAD irrigation than in 2009 under wet-DSR with CT-FI (Table 2.3.1). Also, yield components were significantly different among the tested varieties in both years.

Nukus-2 followed by Allanga-3 and Shoternboy-1 produced higher grain yields in 2008, whereas in 2009 the order was Allanga-3 followed by Novbahor and Nukus-2 under wet-DSR conditions. Under dry-DSR conditions in 2008, the long-duration varieties Mustakillik and Avangard produced significantly lower

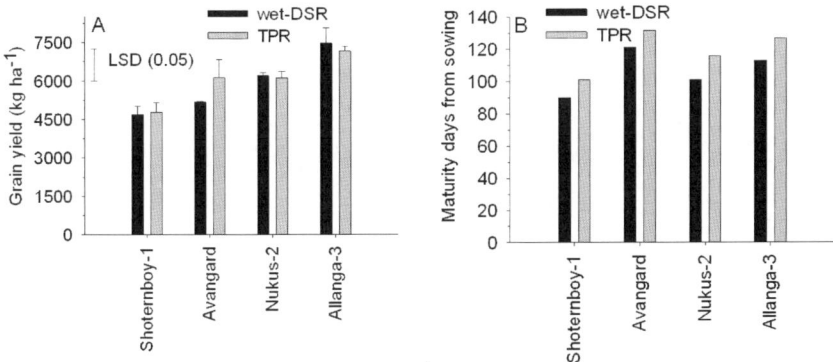

Figure 2.3.2: Grain yield (kg ha^{-1}) (A) and days from emergence to maturity (B) of four local rice varieties as affected by the rice planting method. DSR=direct seeded rice, TPR=transplanted rice. LSD (0.05) is difference between varieties and crop establishment methods at $P=0.05$.

yields mainly due to a reduced number of spikelets per panicle (Table 2.3.1). In terms of maturity, in both years Shoternboy-1 and Nukus-2 were the shortest duration varieties, and Mustakillik and Avangard the longest. Allanga-3 had the highest number of filled grains per panicle. Across the years, the highest harvest index was estimated for Shoternboy-1 followed by Nukus-2, and the lowest for Mustakillik, Avangard, and Novbahor. Setora had the highest 1000-grain weight and Shoternboy-1 the lowest. Allanga-3 followed by Avangard and Mustakillik were the tallest rice varieties, and Shoternboy-1 and Nukus-2 the shortest ones.

2.3.3.4 Experiment IV: Phenological development of rice varieties seeded at different dates

When seeded on the May 5, rice germination and emergence were slowed down by low soil temperature, and emergence occurred only 13 days after sowing (Table 2.3.2). In contrast, flowering and grain-filling periods were not affected in the short-duration varieties, while in the medium-duration varieties (Allanga-3, Setora and Novbahor), these stages were affected by hot and dry weather conditions. This led to the formation of a higher percentage of rudimentary, sterile, and unfilled spikelets. Similarly, in the long-duration varieties, panicle initiation stage was affected by the hot and dry weather, which reduced the number of spikelets per panicle.

When seeded on the May 18 and on June 1, all rice varieties emerged 3 days earlier compared to those seeded on May 5. Phenological development stages were not affected in any variety. However, Allanga-3 and Novbahor showed a higher percentage of sterile spikelets compared to the other six varieties.

Table 2.3.1: Yield and yield components of eight Uzbek rice varieties grown under dry direct-seeded rice with frequent wet and dry irrigation in 2008 and conventional wet direct-seeded rice with continuous flood irrigation in the Khorezm region, Uzbekistan, in 2009.

Variety	Grain yield (kg ha^{-1})		AGB (kg ha^{-1})		Maturity (DAS)		Effective panicles (m^{-2})		Grains panicle^{-1}		1000-grain weight (g)		Harvest index (%)		Plant height (cm)	
	2008	2009	2008	2009	2008	2009	2008	2009	2008	2009	2008	2009	2008	2009	2008	2009
Allanga-3	4008b†	7482a	9947ab	16720a	115	113	181c	238b	113a	133	26.4b	27.6	41b	45c	108a	142a
Avangard	2606cd	5176cd	9000bc	16439a	125	121	338a	237b	40e	95	25.5bc	27.7	29c	32d	99c	133b
Mustakillik	2310d	5913bcd	8416c	16628a	124	122	335a	258ab	37e	90	24.4cd	26.6	27c	35d	97c	128c
Novbahor	2925cd	6689ab	10670a	14644a	110	106	250b	249b	66c	115	23.3d	25.8	28c	46bc	109a	127c
Nukus-2	4940a	6301abc	9954ab	12386bc	105	101	316a	307a	81b	109	25.3bc	24.1	50a	51ab	72d	81d
Setora	3171c	4392de	8289c	10864c	123	108	223b	248b	64cd	77	29.3a	29.8	38b	41c	104b	124c
Shoternboy-1	4075b	4695de	8107c	8966c	93	90	310a	290a	75bc	81	23.0d	25.2	50a	52a	70d	80e
Shoternboy-2	-	4685de	-	14013ab	113	114	-	211b	-	119	-	23.2	-	34d	-	116d
Mean	3434	5709	9198	14189			279	255	68	102	25.3	26.6	37.6	42	94	116
P (0.05)	***	***	***	**			***	*	***		***		***	***	***	***
LSD	621	1231	1497	3309			38	53	9.7		1.3		4.1	5.9	3.2	4.4
CV (%)	18	12.4	16.2	13.5			12.2	11.9	14.1		4.8		10.8	7.9	5.4	7.1

AGB=aboveground biomass; DAS=days after sowing; Significance levels: *$P \leq 0.05$; **$P \leq 0.01$; ***$p \leq 0.001$; † values for rice varieties indicated by same letter are not significantly different.

When seeded on June 13, emergence was delayed by one day compared to the June 1 seeding. A few germinated seedlings even died due to the high temperatures of the standing water (>30 °C). Flowering was delayed in the long-duration varieties, e.g., Mustakillik showed only 50 % flowering whereas such effect was not observed in the other varieties.

When seeded on July 6, emergence in all varieties was delayed by 3 days compared to the June 1 seeding due to the hot and dry weather conditions. In all varieties, flowering started only after September 25. Hence, the flowering stage was prolonged, which negatively affected grain filling. Except for Shoternboy-1, where a few filled grains were observed, the other seven varieties had no grain-filling phase.

Table 2.3.2: Effect of date of seeding on phenological development (days after sowing) of eight Uzbek rice varieties in the Khorezm region, Uzbekistan, in 2009.

Sowing date	Growth stage*	Short duration (<100 days)		Medium duration (100–115 days)			Long duration (>115 days)		
		Shotern-boy-1	Nukus-2	Setora	Allan-ga-3	Novba-hor	Shotern-boy-2	Avan-gard	Musta-killik
May 5	E	13	13	13	13	13	13	13	13
	PI	44	51	51	56	68	61	63	72
	F	73	83	83	88	102	93	97	109
	PM	97	111	111	118	130	123	127	139
May 18	E	10	10	10	10	10	10	10	10
	PI	41	45	48	54	54	56	62	63
	F	69	77	84	86	85	86	97	97
	PM	95	109	109	117	113	119	127	127
June 1	E	5	5	5	5	5	5	5	5
	PI	39	42	46	51	49	50	51	59
	F	64	73	76	81	78	83	92	95
	PM	90	104	106	116	111	115	121	126
June 13	E	6	6	6	6	6	6	6	6
	PI	42	45	48	57	50	49	60	63
	F	74	82	82	92	80	89	93	93
	PM	99	107	110	121	111	118	123	-
July 6	E	8	8	-	8	8	-	8	-
	PI	41	49	-	50	46	-	-	-
	F	76	84	-	85	80	-	-	-
	PM	-	-	-	-	-	-	-	-

*E=emergence, PI=panicle initiation, F=flowering, PM=physiological maturity

2.3.4 Discussion

2.3.4.1 Conservation agriculture technologies combined with water saving irrigation

The direct seeded rice using WAD (frequent wet and dry irrigation) needed about 70 % less water but at the price of an average 42 % reduction in yield of the dominant rice variety in the study region (fig. 2.3.1 A). This is in line with the previously observed trends in other regions (Singh et al. 2011; Belder et al. 2004). Furthermore, the findings show the potential of reducing more than 70 % of the current water applications in the Khorezm region through shifting rice cultivation from the current flood-irrigation practices to a water-saving irrigation method such as WAD (Devkota 2011). Our findings on grain production under conservation and conventional practices suggest that conventional tillage, extensive field preparation, and continuous flood irrigation may not be needed to maintain the present level of average rice production, i. e., 3.6 t ha^{-1} in Uzbekistan and 3.1 t ha^{-1} in Central Asia (FAOSTAT 2012). The absence of yield differences between CT-II (conventional tillage-intermittent irrigation) and ZT-II (zero tillage-intermittent irrigation) and CT-FI (conventional tillage-continuous flood irrigation) and ZT-FI (zero tillage-continuous flood irrigation) treatments (fig. 2.3.1B) is a further argument that the present intensive soil tillage practices in rice cultivation can be avoided without compromising yields but with reduced water demand and production costs. The most favorable comprise for irrigation and cultivating rice in the study region would be continuous flood irrigation up to panicle initiation and intermittent irrigation thereafter (fig. 2.3.1B). This has the potential of saving 30–40 % water without yield reduction.

Reduced rice yields with residue retention under BP and ZT indicate that the type of crop residue retention examined (standing residues) does not benefit rice cultivation in the region. The practice of BP, with a higher water-saving potential and similar yields to ZT, could be the better rice-planting strategy. For detailed discussion see Devkota (2011).

The application of a higher amount of water in conventionally tilled, flood-irrigated rice in an experimental field in Khorezm than in other regions (UNESCO 2000) could be due to the presence of a sand layer below 45 cm soil depth (70 % sand, 28 % silt and 2 % clay), a shallow groundwater table (1.6 m in 2008 and 0.9 m in 2009), deep plowing for laser leveling before the start of the experiment, and no history of rice cultivation in the field in the past 20 years (Devkota 2011). Stand-alone or a combination of these factors may explain the higher water loss through seepage and percolation which demands higher water application.

2.3.4.1.1 Direct seeded vs. transplanted rice

Under arid conditions, aside from the soil tillage practices and irrigation methods, the seeding method alone can already save considerable amounts of irrigation water. For instance, transplanted rice needed 500 – 1,000 mm less irrigation water than wet-DSR, without yield reduction. Thus, as opposed to the commonly practiced continuous-flood irrigation from seeding to harvest, in wet-DSR water can be saved by growing seedlings in a small separate nursery for 25 – 35 days.

Due to the long and cold winters, the seeding of rice under the double-cropping practice is time limited. Hence, timely planting of both rice and wheat is crucial to achieving the yield potential of both crops. Under such conditions, transplanted rice can fit better into the double-crop rice-wheat systems, as it provides a one-month time advantage. Transplanting the short-duration rice varieties, such as Shoternboy-1 or Nukus-2 (Table 2.3.2), further allows a timely planting of wheat before the winter season. On the other hand, transplanting is highly labor and cost intensive and could lead to deterioration of the soil structure (Timsina and Conner 2001). Therefore, the alternatives to the conventional method of transplanting, such as mechanical un-puddled transplanting, should be studied.

2.3.4.1.2 Suitable rice varieties

Not only the seeding method but also the choice of the most appropriate rice varieties bears the potential of water saving when they optimally fit the rice-wheat system. Rice production under water-saving irrigation modes generally requires early seedling vigor, drought and submergence tolerance, thick and deep-penetrating roots, and short-duration, input-responsive, and high-yielding varieties (Farooq 2009). Among the tested varieties, Nukus-2 followed by Shoternboy-1 and Allanga-3 can be sown under dry-DSR conditions using intermittent wet and dry irrigation, whereas using this combination of cultivation practices for the other varieties is not advisable. These three short-duration varieties showed high yield stability under both DSR and TPR conditions, and thus have the highest potential to suit the rice-wheat systems in the study region.

2.3.4.1.3 Planting time

The appropriate choice of the planting time also turned out to be a promising stand-alone measure to save water. Timely planting is particularly crucial under arid conditions since both early and late planting of rice (especially of the long-duration varieties) showed higher spikelet sterility due to the adverse effect of high temperatures during flowering (Devkota 2011). Compared to the long- and medium-duration varieties, short-duration varieties suffered less from the high temperatures during flowering when planted both early and late. Rice germination was also affected under early planting conditions, i.e., before the second week of May due to the low and after the third week of June due to the high soil temperatures (Table 2.3.2). Thus, in the study region only a relatively small window of opportunity exists for all rice varieties seeded between May 20 and June 15 when there is only little danger of low or high temperatures. Short-duration rice varieties can give comparatively high yields under early as well as under late seeding conditions compared to the long-duration rice varieties. The best seeding dates could be up to the end of June for short-, and up to May 25 and June 10 for medium- and long-duration rice varieties, respectively. For detailed discussion see Devkota (2011).

2.3.5 Conclusions

Under the arid conditions of the Khorezm region, water-saving irrigation methods combined with CA practices showed a high water-saving potential in rice production compared to the present practices. Since this option turned to be at the expense of yields, as a stand-alone measure it seems unacceptable for rice producers given the high market price for rice and the presently low prices for irrigation water (Devkota 2011). Yield reduction under water-saving irrigation could, however, be compensated for when flooding until the grain-setting phase.

Straight-forward, stand-alone cultivation methods to reduce water demand without compromising yields were achieved through adapted seeding/planting methods, the right choice of the varieties and the optimal seeding date. Grain yield between transplanted and wet-direct seeded rice did not differ. Thus, future studies should focus on mechanical un-puddled transplanting, as this would counterbalance the present labor- and cost-intensive transplanting practices. The selection and the use of appropriate rice varieties and seeding dates made a great difference. Medium- (Allanga-3) or short-duration rice varieties (Nukus-2 or Shoternboy-1), for example, could be a better choice than the long-duration variety at all seeding dates. The appropriate seeding date turned out to be up to the end of June for short-duration, and May 25 to June 10 for medium- and long-duration rice varieties in such climate conditions.

References

Akramkhanov A, Kuziev R, Sommer R, Martius C, Forkutsa O, Massucati L (2012) Soils and Soil Ecology in Khorezm. In: Martius C, Rudenko I, Lamers JPA, Vlek PLG (Eds.) Cotton, water, salts and Soums – economic and ecological restructuring in Khorezm, Uzbekistan pp 37–58

Belder P, Bouman BAM, Cabangon R, Guoan L, Quilang EJP, Li Y, Spiertz JHJ, Tuong TP (2004) Effect of water-saving irrigation on rice yield and water use in typical lowland conditions in Asia. Agricultural Water Management 65:193–210

Christmann S, Martius C, Bedoshvili D, Bobojonov I, Carli C, Devkota K, Ibragimov Z, Khalikulov Z, Kienzler K, Manthrithilake H, Mavlyanova R, Mirzabaev A, Nishanov N, Sharma RC, Tashpulatova B, Toderich K, Turdieva M (2009) Food security and climate change in Central Asia and the Caucasus, Tashkent, Uzbekistan 2009. CGIAR-PFU, ICARDA Tashkent, P.O.Box 4564, Tashkent, 100 000, Uzbekistan

Devkota K (2011) Resource Utilization and Sustainability of Conservation-based Rice-Wheat Cropping Systems in Central Asia. Ph D Thesis, Bonn University, Centre for Development Research (ZEF)

Djanibekov N (2008) A Micro-Economic Analysis of Farm Restructuring in the Khorezm Region, Uzbekistan. Ph D Thesis, Bonn University, Centre for Development Research (ZEF)

Dong B, Molden D, Loeve R, Li YH, Chen CD, Wang JZ (2004) Farm level practices and water productivity in Zhanghe Irrigation System. Paddy and Water Environment 2: 217–226

FAOSTAT (2012) FAO Statistics Division. http://faostat.fao.org/site/567/DesktopDefault.aspx?PageID=567#ancor

Farooq M, Kobayashi N, Wahid A, Ito O, Basra SMA (2009) Strategies for producing more rice with less water. Adv Agron 101:351–388

Gupta R, Kienzler K, Martius C, Mirzabaev A, Oweis T, de Pauw E, Qadir M, Shideed K, Sommer R, Thomas R, Sayre K, Carli C, Saparov A, Bekenov M, Sanginov S, Nepesov M, Ikramov R (2009) Research Prospectus: A Vision for sustainable land management research in Central Asia. ICARDA Central Asia and Caucasus Program. Sustainable Agriculture in Central Asia and the Caucasus Series No.1. CGIAR-PFU, Tashkent, Uzbekistan, p 84

Ismail AM (2005) Activities of IRRI in Central Asia and the Caucasus (CAC) region in 2005. Annual Report submitted to IRRI, Metro Manila Philippines

Ismail AM (2006) Activities of IRRI in Central Asia and the Caucasus (CAC) region in 2006. Annual Report submitted to IRRI, Metro Manila Philippines

Sayre KD, Hobbs PR (2004) The raised-bed system of cultivation for irrigated production conditions. In: Lal R, Hobbs PR, Uphoff N, Hansen DO (Eds.) Sustainable Agriculture and the Rice-Wheat System. Chapter 20, Ohio State University. Columbus, Ohio, USA, p 337–355

Singh Y, Singh VP, Singh G, Yadav DS, Sinha RKP, Johnson DE, Mortimer AM (2011) The implications of land preparation, crop establishment method and weed management on rice yield variation in the rice–wheat system in the Indo-Gangetic plains. Field Crops Research 121 (1):64–74

Timsina J, Connor DJ (2001) Productivity and management of rice-wheat cropping systems: issues and challenges. Field Crops Research 69:93–132

UNESCO, United Nations Educational, Scientific and Cultural Organization (2000) Water related vision for the Aral Sea basin for the year 2025, United Nations Educational, Fontenoy, France

Inna Rudenko, John P. A. Lamers, Utkur Djanibekov,
Sanjar Davletov

2.4 Virtual water along the Uzbek cotton value chain

Abstract

In the last decades, water in Uzbekistan in the Aral Sea Basin (ASB) has become scarce due to numerous internal and external factors, such as the growing demand for water resources by the upstream countries, expansion of the irrigated areas to feed the growing population, and the poor condition of irrigation and drainage networks, which causes high water losses. Most ASB countries have an agricultural profile and are challenged to consolidate their efforts in identifying and introducing suitable management practices to ease the advancing deterioration of natural resources. In Uzbekistan, agriculture diverts up to 95 % of the country's available water for irrigating agricultural crops, specifically cotton, commonly causing elevated groundwater tables and secondary soil salinization. However, cotton is one of the major sources of income for food and energy for most of the rural households and forms the basis of the entire cotton value chain, including cotton processing. The combined findings of the value chain analysis (VCA) of the Uzbek cotton sector and the water footprint analysis (WFA) can provide valuable information for decision making, as they include both the economic and environmental aspects of crop production, in contrast to the individual findings of these analyses. The integrated approach reveals two practicable options: (i) reducing agricultural water use by upgrading irrigation and drainage networks and introducing water saving technologies on the field level, and (ii) promoting the shift of water use from the high water-consuming agricultural sector to an industrial sector with lower water consumption with the focus on yarn production within the cotton chain. Increasing water use efficiency, processing products with a higher value added, and raising water users' awareness on water shortage and "real" water value are recommended for achieving food and water security in the region.

Keywords: Central Asia, Uzbekistan, water footprint, value chain analysis, cotton sector

2.4.1 Introduction

Cotton (*Gossypium hirsutum* L.) production in Uzbekistan, located in the ecologically threatened Aral Sea Basin (ASB) utilizes about 41 % of all irrigation water and about the same share of all irrigated land (Rudenko et al. 2009). To intensify cotton production, a vast irrigation and drainage network has been constructed since the 1960s to divert water from the two rivers that used to feed the Aral Sea, the Amudarya with an annual flow of around 75 km^3, and the Syrdarya with 34 km^3 of annual water flow (Tischbein et al. 2012). Since cotton continues to be a centerpiece of Uzbekistan's agriculture and its national and regional economy, a considerable effort over the long term is required to reduce the country's dependence on this crop while combating ecological degradation.

Not only Uzbekistan but also all countries in the ASB are urged to proactively confront the risks and vulnerabilities. They are thus challenged to identify feasible options for a more efficient water use not only on various management levels but also within various sectors of the economy. Taking into account the present economic and ecological situation, it is compulsory to assess options that not only give high return on investments but also prioritize the environment and, in particular, increase the resilience of the rural landscape and livelihoods. A comprehensive assessment therefore demands a methodological approach that considers financial and ecological aspects of regional development while focusing on interrelated agricultural and subsequent industrial activities rather than on single sectors.

The value chain analysis (VCA) describes the full range of activities in a commodity chain that are required to bring a product or service from the design through the different phases of production to the delivery to final consumers and the disposal after use (McCormick and Schmitz 2001). While the VCA of cotton includes the products, describes the underlying production cycles, and estimates the financial gains of each sector in the chain, it does not consider the water use during each stage (Rudenko et al. 2009). The water use and the virtual water content in various products can be estimated through the water footprint analysis (WFA) (Hoekstra and Hung 2002; Chapagain and Hoekstra 2004), which thus complements the VCA by considering the key environmental indicator. In this study, the cotton VCA and WFA were combined to: (i) quantify the water footprint of the entire cotton value chain (CVC) in the Khorezm region in Uzbekistan, (ii) estimate the virtual water content of various cotton products, and (iii) calculate the water footprint value index.

2.4.2 Materials and methods

2.4.2.1 Study site

The case study region Khorezm is a 680,000 ha large administrative district located in the lower reaches of the Amudarya River in northwest Uzbekistan (41°41' N latitude, 39°40' E longitude and altitude 113 m). In the sharply continental, arid climate of the region characterized by low precipitation (annually 100 mm) and high evaporation rates (up to 1,400 – 1,600 mm), agriculture is made possible through irrigation. Annually, 4.5 – 5 km^3 of water are diverted from the sole source, the Amudarya river (Tischbein et al. 2012). The probability of receiving sufficient water for irrigation has been decreasing in the last decades (Müller 2008). In addition, it is expected that as a result of climate change, the availability of water in the Syrdarya and Amudarya rivers may decrease by as much as 30 % and 40 %, respectively (Perelet 2007). The recurring water scarcity is caused not only by external factors, such as the impact of climate change and the growing demand for water resources in the upstream countries (Djanibekov et al. 2012), but also by internal factors, including the expansion of irrigated areas to support the growing population in the region, and the poor condition of the irrigation and drainage networks, which causes high water losses (Tischbein et al. 2012).

2.4.2.2 Calculation of value added and virtual water content

Value added methodologies measure the increase in wealth for a nation as a whole, and include remuneration for labor, interest charges, taxes and the net margin (profit) of the producers. From a financial point of view, the value added represents the worth that has been added to a product or a service at each stage of production. In simple words, value added is the difference between the value of the product and the value of the purchased inputs (McCormick and Schmitz 2001). The value added along the Uzbek cotton chain was calculated as the difference between sales price of a product and its primary cost, i.e., its total production costs (Rudenko et al. 2008).

The virtual water content of cotton products was calculated according to Hoekstra and Hung (2002) and Chapagain and Hoekstra (2004). In total, the virtual water content (m^3 ton^{-1}) and financial indicators of eleven types of cotton products (raw cotton, fiber, yarn, fabrics, T-shirt, absorbent cotton, cottonseed, cottonseed oil, cottonseed meal and cake, soap) were calculated. The virtual water of the raw cotton was calculated based on its irrigation water requirements in Khorezm. The virtual water content of all processed cotton products was calculated based on product and value fractions. The value chain method helped to trace the flow of cotton products and to estimate water use at each stage of production and processing.

From the field to semi-finished or finished products, cotton passes through a number of production stages, which in the study were subdivided into agricultural and industrial stages. Agricultural water use (AWU) under the irrigated practices in Uzbekistan includes: (i) water for leaching salts from the crop root zone, (ii) water used for irrigation during the entire crop-growing period, (iii) water conveyance losses in the main and on-field canals, and (iv) the water virtually needed to dilute, for instance, pollutants such as pesticides and fertilizers percolated to the groundwater, the so-called "grey water component"[1].

$$AWU = Leaching + Irrigation + Losses + Grey \qquad (1)$$

Total virtual water (TVW) was calculated as the sum of AWU and the water amount used at each industrial stage (IWU):

$$TWV = AWU + IWU \qquad (2)$$

Only that fraction of agricultural or industrial water was assumed for each subsequent processed product, the corresponding fraction of which was used to produce this subsequent product. For example, the total virtual water of cotton fiber was calculated as the sum of 33 % of agricultural water use plus industrial water use at ginning.

$$TWVfiber = (Leaching + Irrigation + Losses + Grey) * 33\% + IWUginning \qquad (3)$$

Finally, the water footprint value index was calculated as the ratio of a value added to a certain cotton product to its virtual water content, as an indication of the monetary return on each m^3 of virtual water spent for producing various cotton products.

2.4.2.3 Data collection

The data collection methods, borrowed from formal and informal survey methods, allowed generating a data set on product fractions and flows. Semi-structured interviews using questionnaires were conducted with the main actors

---

1 The calculation of the grey water content is described in detail in Chapagain et al. (2005).

of the chain. Technical coefficients, parameters, and water use requirements at various processing stages and the processing organizations along the cotton value chain in Khorezm were previously estimated by Rudenko et al. (2009).

2.4.3 Results

2.4.3.1 Cotton Value Chain: From raw cotton to textile products

The CVC consists of cotton farmers, ginneries (cotton refining plants), textile companies, and oil extracting plants. The flow of products (fig. 2.4.1) starts with raw cotton being transferred to the ginneries. Ginneries produce cotton fiber that, in Uzbekistan, is up to 33 % of the raw cotton. Cotton fiber is mostly exported, while the remaining part is forwarded by the ginneries to the domestic textile companies for further processing.

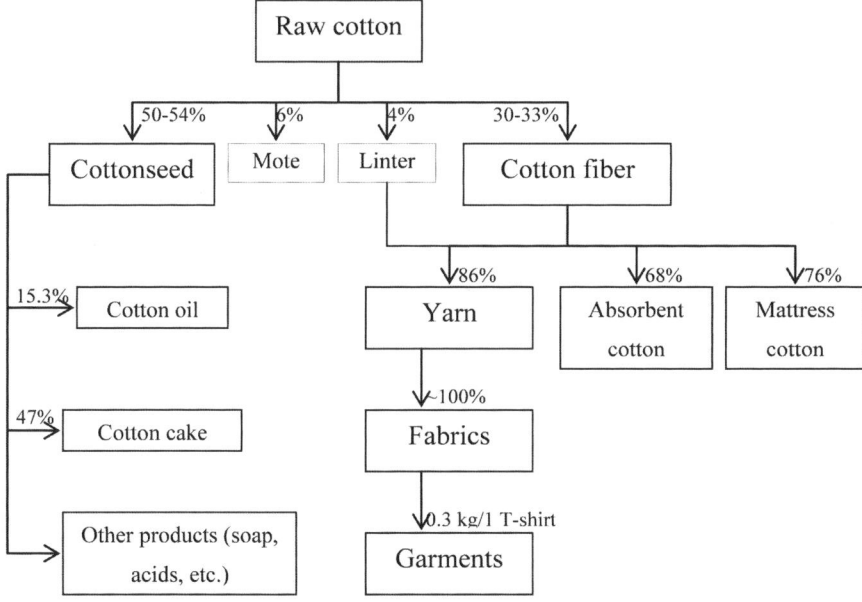

Figure 2.4.1: Cotton product flows and output fractions

Cottonseed, constituting up to 54 % of the Uzbek raw cotton, partly flows back to farmers as seeding material for the next agricultural season but to a larger extent it is used as an input for oil extracting plants. Cotton oil and cottonseed meal and cake from these oil-extracting plants are then sold locally or exported. Finally, textile products, such as yarn (29 % of raw cotton), fabrics

(28.5 %) and ready-made garments (28.5 %)[2] from textile producers are either consumed in the country or exported. However, since a larger part of the produced cotton fiber is exported, domestic cotton processing factories (spinning, weaving and textile) are not functioning to their full capacities, especially in the remote provinces of Uzbekistan such as Khorezm, where the capacities utilized are below 50 % (Rudenko 2008; Rudenko et al. 2012).

2.4.3.2 Cotton Value Chain: Value added and virtual water of cotton products

The newly industrialized countries have built up a diversified export and industrial structure, since this strategy is more sustainable due to the access to large, diversified markets, the economy of scale, and the competitive effects (Stamm 2004). The underlying argument is that the more processing stages raw materials pass through, the higher are the prices that can be obtained, and that the higher value is added with every additional production cycle. In line with this argument, value added to the raw cotton produced was about USD 50 per ton, whereas it doubled for cotton fiber and continued to increase for all the subsequent products such as yarn, fabrics and garments (Table 2.4.1). The highest value added (0.66 USD unit^{-1} or 2,000 USD ton^{-1}) was estimated for T-shirts produced domestically and then exported.

The agricultural stage consumes the most water due to the irrigation practices and the deficiencies of the irrigation and drainage networks. According to BUIS[3] (2006), the water demand of cotton production in Khorezm ranges from 6,000 to 8,000 m^3 ha^{-1} including the water for leaching, and yields are on average 2.6 tons. In contrast, in this study, total AWU (considering also water seepage loss in the canals) for cotton production in 2006 was estimated at 17,729 m^3 ha^{-1} or 6,819 m^3 of water ton^{-1} of produced raw cotton (Table 2.4.1).

Table 2.4.1: Value added and virtual water of cotton products in Khorezm, Uzbekistan

	Value added	AWU	IWU (cumulative)	TVW	Water footprint value index
	USD ton^{-1}	m^3 ton^{-1}	m^3 ton^{-1}	m^3 ton^{-1}	USD per m^3
Raw cotton	50	6,819	0	6,819	0.007
Fiber	112	6,819	1	6,820	0.016
Yarn	284	7,759	(1)+0.7	7,761	0.037
Fabrics	313	7,895	(1+0.7)+789	8,686	0.036
T-shirt (1 T-shirt – 0.3 kg)	2,000	2,074	(1+0.7+789)+0	2,865	0.698

---

2 Ready made garments are analyzed in this study with the example if a simple T-shirt, which is produced from 300 grams of cotton fabrics (0.3 kg T-shirt^{-1})

3 The Basin Department of Irrigation Systems (Russian abbreviation)

Once the volume of water used at each subsequent production stage, i.e., the IWU, was added to the virtual water of the preceding cotton product, the findings showed that the processing of raw cotton into cotton fiber consumed about 1 liter of water kg^{-1}, rendering the virtual water content of fiber equal to ca. 6,820 m^3 of water ton^{-1} (Table 2.4.1). The most water-intensive processing stage along the cotton chain (which refers to the textile industry) was weaving or producing cotton fabrics. The IWU in the weaving process amounted to 789 litres of water kg^{-1} of fabrics, since the bleaching, washing and dyeing consumed a large amount of water and produced much waste water.

The water footprint value index, illustrating the monetary return on each m^3 of virtual water spent for producing various cotton products, showed, as expected, the highest monetary return at the level of finished textile products (T-shirt, Table 2.4.1). Interestingly, the index for yarn was somewhat higher than that of cotton fabrics. This means that from a financial consideration, the virtual water is more efficiently spent in yarn production than in the following fabrics production.

2.4.4 Discussion

Experience shows that nations specializing in exports of primary commodities (e.g., cotton) are vulnerable to fluctuations in the world markets (Stamm 2004). Uzbekistan, for example, lost about USD 1.5 billion due to the low global cotton prices in 1998–2001. In the light of such experience, a shift from the primary commodity exports to the export of value-added commodities and removal of trade barriers with the aim of facilitating trade became an important part of the reform package in Uzbekistan, which among others targeted at reviving light (textile) industry. In order to achieve the benefits of the value chain development and of producing and exporting goods (agriculture based) with higher value added, it is important to create a favorable environment for increasing exports. For this, the Uzbek export regime has to be properly defined as to include financial, fiscal and other instruments in compliance with international settings, rules, and standards (Abdurazakov 2006). In addition to the creation of favorable export settings, it is also necessary to support the industrial upgrading of local producers and also the subsequent product upgrading, which could lead to higher competitiveness and world recognition of the Uzbek (cotton) products. Some lessons could be learnt from the textile and clothing sectors in the European Union, which had responded to a highly competitive and demanding world market by factors other than price, i.e., the quality of production and "fashion content", the capacity to develop the highly demanded brands, the ability to deliver the products in a fast and reliable way, and finally the sustainability and safety of industrial systems with respect to the environment and the employed workers (Commission of the European Communities 2003).

Our findings confirm that the highest water use along the CVC occurred at field level for leaching and irrigation, as was postulated earlier (Aldaya et al. 2010). Due to the deficiencies in the irrigation network resulting in high conveyance losses along the entire irrigation network, additional water is supplied to compensate for these expected losses (Bekchanov et al. 2012). Hence, decreasing the conveyance losses and simultaneously increasing water use efficiency in the entire CVC would be needed for sustainable use of this natural resource.

Agricultural stage of the CVC. Practices at the field level can be improved not only by technical solutions or resorting to water-saving irrigation techniques, but also by economic-oriented (e. g., water pricing) and management-oriented solutions (Bekchanov et al. 2010). The potential benefits of water-wise innovations are estimated to be huge. However, the impact strongly varies with the technology, and more water-efficient technologies usually are more capital-intensive (Bekchanov et al. 2010; Bekchanov et al. 2012). For instance, the introduction of irrigation water-saving technologies in the Khorezm region could reduce water demands by 1.5 – 3.0 km^3 annually (Bekchanov et al. 2010). With an estimated 70 % water saving potential, drip irrigation is most efficient but needs relatively high financial investments. Considerable amounts of irrigation water (between 0.4 – 0.9 % annually) in the field could be saved with improved irrigation methods such as double-sided furrow irrigation, alternate dry furrows, and shorter furrows, which are simple and low-cost solutions (Bekchanov et al. 2010). Between 25 – 30 % of water can be saved by employing a laser-guided land leveler, an expensive equipment that might be introduced through extension service providers or farmers' cooperatives (Egamberdiev et al. 2008). Management-oriented solutions for reducing AWU include introducing alternative, less water-demanding crops such as maize or aerobic rice (Devkota 2011). However, these would reduce not only irrigation water demand, but also farmers' income due to lower yields and the associated profits.

A reduction of AWU can also be tackled beyond the field level. Since irrigated agriculture is with 85 – 95 % by far the largest water-consuming sector in Uzbekistan, a reduction in water demand can be expected when investing in improvements to the irrigation and drainage networks. At present, less than 30 % of the canals are lined, and only 12 % considered waterproof, resulting in high seepage losses and rising groundwater tables (Tischbein et al. 2012). Furthermore, the average irrigation system efficiency is with about 30 – 40 % low (Tischbein et al. 2012). Rehabilitating and renovating the irrigation and drainage systems, e. g., by concrete lining of channels, could reduce irrigation water losses, but would require high investments in human resources and materials (Micklin 2002).

Industrial stage of the CVC. Our findings suggest that increasing overall water use efficiency along the CVC can be achieved by diversifying the economy and processing of agricultural output by less water-consuming, domestic industrial sectors. For instance, total water use can be reduced by encouraging the manufacture of yarn or ready-made garments. Such a shift from the present strategy of exporting mainly cotton fiber to that of exporting manufactured products could maintain similar cotton export revenues while reducing the land area currently used for cotton production by up to almost 70 %. Additionally, this alternative strategy bears the potential to free about 0.5 km^3 of irrigation water annually and to reduce the present state subsidies by about 14,000,000 USD (Rudenko et al. 2009). The value chain study of the cotton sector in Khorezm reveals that cotton products (yarn, fabrics and garments) produced by the local manufacturers (equipped with modern German, Swiss and Turkish machinery) were highly demanded beyond Uzbekistan, particularly in the neighboring CIS countries but also in Turkey, to where virtually 100 % of the Khorezm yarn was exported (Rudenko 2008). According to interviews[4] with producers, the largest share of ready-made cotton garments is also exported from Khorezm to Russia. These locally produced cotton products are rarely available at local markets, and local consumers buy either expensive Turkish textiles or cheap Chinese textiles of low quality. Thus, the demand for Uzbek textile products exists both inside the country and abroad, but local cotton processors need strong support (including from the state) in order to operate at full capacity.

However, what needs to be stressed is that 'clean' production has to be intensified in the region, especially in the light of the presently low capacities for wastewater treatment. Despite the low IWU in the total cotton chain, which is below 10 % of the total water use, water pollution is an important issue, as its magnitude depends on the processing stage and its intensity. To avoid environmental contamination by industrial pollutants in the cotton processing, the scope for treating and re-use of wastewater in the industrial sector has to be explored. For example, combining the VCA and WFA via the water footprint value index indicated the feasibility of exploring the cotton value chain up to the point of the production of cotton yarn. The latter does not require much IWU, hence does not produce much wastewater but brings higher export earnings than cotton fiber and higher financial gain per unit of water used. This calls for the development of the CVC up to the stage of yarn production, which has the second highest water footprint value index. Development of the weaving industry, although economically beneficial, would require much more water and

---

4 Interviews conducted in the framework of a doctoral research by Rudenko (2008)

would lead to extensive pollution. Therefore, this is not recommended in the case of remote provinces like Khorezm unless additional efforts and investments are made for upgrading their cleaning facilities, i. e., treatment of wastewater.

2.4.5 Conclusions

The combined findings of the economic-based value chain analysis and water footprint analysis provide a more comprehensive view to support decision making on land and water use as compared to the recommendations based on the two approaches separately. The results indicate two paths for reducing water use and coping with water scarcity. One is to reduce agricultural water use through upgrading irrigation and drainage networks and to introduce innovations that have a high potential for increasing water use efficiency in the field. The other is to shift water use from the high water-demanding agricultural sector to a more economic water consuming industrial sector. This shift should, however, be made with care. Increasing water use efficiency at the agricultural level, moving towards the production of processed products with higher value added, and raising water users' awareness on water shortage/real water value should help Uzbekistan to adapt to the increasing water scarcity without negative effects on the economy.

References

Abdurazakov A (2006) Eksportnaya Prioritetnost. Uzbek Journal 'Economic Review' 3 (78): 62–68
Aldaya MM, Muñoz G, Hoekstra AY (2010) Water footprint of cotton, wheat and rice production in Central Asia. Value of Water Research Report Series, vol. 41. UNESCO-IHE. Delft, The Netherlands
Bekchanov M, Lamers JPA, Karimov A, Müller M (2012) Estimation of spatial and temporal variability of crop water productivity with incomplete data. In: Martius C, Rudenko I, Lamers JPA, Vlek PLG (Eds.) Cotton, water, salts and soums – economic and ecological restructuring in Khorezm, Uzbekistan. Springer, Dordrecht pp 329–344
Bekchanov M, Lamers JPA, Martius C (2010) Pros and cons of adopting water-wise approaches in the lower reaches of the Amu Darya: A socio-economic view. Water 2: 200–216
Chapagain AK, Hoekstra AY (2004) Water footprints of nations. Value of Water Research Report Series, vol. 16. UNESCO-IHE, Delft, The Netherlands
Chapagain AK, Hoekstra AY, Savenije HHG, Gautam R (2005) The water footprint of cotton consumption. Value of Water Research Report Series No. 18, UNESCO-IHE, The Netherlands. http://www.waterfootprint.org/Reports/Report18.pdf

Commission of the European Communities (2003) Economic and competitiveness analysis of the European textile and clothing sector in support of the Communication The future of the textiles and clothing sector in the enlarged Europe. Commission Staff Working Paper, SEC 1345, Brussels

Devkota K (2011) Resource Utilization and Sustainability of Conservation-based Rice-Wheat Cropping Systems in Central Asia. PhD Dissertation, Bonn University

Djanibekov N, Bobojonov I, Djanibekov U (2012) Prospects of agricultural water service fees in the irrigated drylands, downstream of Amudarya. In: Martius C, Rudenko I, Lamers JPA, Vlek PLG (Eds.) Cotton, water, salts and *Soums* – economic and ecological restructuring in Khorezm, Uzbekistan. Springer, Dordrecht pp 389–411

Egamberdiev O, Tischbein B, Lamers JPA, Martius C, Franz J (2008) Laser land levelling: More about water than about soil. Science brief from the ZEF-UNESCO project on Sustainable Management of Land and Water Resources in Khorezm, Uzbekistan. The Center for Development Research (ZEF), Germany, ZUR no.1

Hoekstra AY, Hung PQ (2002) Virtual water trade: a quantification of virtual water flows between nations in relation to international crop trade. Value of Water Research Report Series, vol. 11. UNESCO-IHE, Delft, the Netherlands

McCormick D, Schmitz H (2001) Manual for Value Chain Research on Homeworkers in the Garment Industry. http://www.ids.ac.uk/ids/global/pdfs/wiegomanualendnov01.pdf

Micklin P (2002) Water in the Aral Sea Basin of Central Asia: Cause of conflict or cooperation? Euroasian Geography and Economics 43 (7): 505–528

Müller M (2008) Where has all the water gone? In: Wehrheim P, Shoeller-Schletter A, Martius C (Eds.) Continuity and change: Land and water use reforms in rural Uzbekistan. IAMO series 43: 89–104

Perelet R (2007) Central Asia: background paper on climate change. UNDP, pp 17

Rudenko I (2008) Value chains for rural and regional development: The case of cotton, wheat, fruit, and vegetable value chains in the lower reaches of the Amu Darya River, Uzbekistan. PhD dissertation, Hannover University

Rudenko I, Djanibekov U, Lamers JPA (2009) Cotton water footprint. Is rational water use possible in Khorezm? Vestnik the Journal of Uzbek Academy of Science, Nukus

Rudenko I, Grote U, Lamers JPA (2008) Using a value chain approach for economic and environmental impact assessment of cotton production in Uzbekistan. In: Qi J, Evered KT (Eds.) Environmental problems of Central Asia and their economic, social, and security impacts. NATO Science for Peace and Security Series – C: Environmental Stability. Springer, Dordrecht pp 361–380

Rudenko I, Nurmetov K, Lamers JPA (2012) State order and policy strategies in the cotton and wheat value chains. In: Martius C, Rudenko I, Lamers JPA, Vlek PLG (Eds.) Cotton, water, salts and soums – economic and ecological restructuring in Khorezm, Uzbekistan. Springer, Dordrecht pp 371–387

Stamm A (2004) Value Chains for Development Policy. Challenges for Trade Policy and the Promotion of Economic Development. GTZ Concept study. Eschborn

Tischbein B, Awan UK, Abdullaev I, Bobojonov I, Conrad C, Forkutsa I, Ibrakhimov M, Poluasheva G (2012) Water management in Khorezm: current situation and options for improvement (hydrological perspective). In: Martius C, Rudenko I, Lamers JPA, Vlek PLG (Eds.) Cotton, water, salts and *Soums* – economic and ecological restructuring in Khorezm, Uzbekistan. Springer, Dordrecht pp 69–92

Section 3: Natural Resource Management

Bernhard Tischbein, Usman Khalid Awan, Fazlullah Akhtar, Pulatbay Kamalov, Ahmad M. Manschadi

3.1 Improving irrigation efficiency in the lower reaches of the Amu Darya River

Abstract

Improvements in irrigation efficiency and irrigation scheduling are needed to overcome the widespread occurrence of water and salt stress at farm level, adverse impacts on soil and groundwater resources, and low water productivity in the Khorezm region. Such measures should aim at reducing the vulnerability of the region to water supply shortages caused by a combination of natural (e.g., meteorological droughts) and human (e.g., increased upstream water withdrawal) factors, which are imminent given the projected adverse impacts of climate change on river water flow and a sharpening competition for water. The assessment of irrigation performance in a representative Water Consumers Association (WCA) in Khorezm revealed various deficits in water management including the delivery performance ratio during the season exceeding the target value of 1 (1.39 in the vegetation season) and showing a wide range (0.88 – 1.79 on a monthly basis). Despite high water inputs, actual evapotranspiration (ETa) was below the potential level throughout the season, in particular during the peak irrigation month July. The low relative evapotranspiration (ratio between actual and potential evapotranspiration) of 0.63 indicated severe water stress. The depleted fraction of on average 0.4 over the vegetation period dropped in the peak irrigation periods to 0.23 (July) and 0.27 (August), illustrating the inefficient use of irrigation water. The technical overall irrigation efficiency in the entire WCA was estimated at 30 – 35 % only.

Besides improving the application process, the key interventions for achieving an appropriate and efficient irrigation should include strengthening the coordination of irrigation activities within the network and at the field-network-interface (e.g., by strengthening WCAs and the participation of water consumers), abolishing hydraulic bottlenecks in the irrigation system (e.g., by intensified maintenance and targeted re-design), and flexible irrigation scheduling that takes into account the time-dependent and site-specific demand and supply. A tool was therefore developed by linking models of irrigation scheduling (CROP-WAT and AquaCrop), groundwater dynamics (FEFLOW) and water fluxes between groundwater table and soil surface (HYDRUS). Under limited water resources, AquaCrop made it possible to derive optimal deficit irrigation schedules while minimizing the impact of unavoidable under-supply of irrigation water. These technical approaches should be combined with institutional and socio-economic factors.

Keywords: Irrigation performance assessment, irrigation scheduling, shallow groundwater, deficit irrigation, water saving

3.1.1 Current situation of water use in Khorezm

The huge withdrawals of water from the Amudarya River for irrigation and the occurrence of water and salt stress at field level are both typical for the water management in Khorezm. The two major problems are the low irrigation efficiency and the inappropriate water supply at farm level. Improving the water supply at farm level and hence raising irrigation efficiency is urgently needed to (i) reduce the currently enormous waste of water, (ii) contribute to overcoming the underutilization of agricultural yield potentials, and (iii) lower adverse impacts on the soil and groundwater resources (Manschadi et al. 2010). Furthermore, effective and appropriate irrigation management is a promising strategy to mitigate Khorezm's vulnerability to limited water availability, which is alarming due to the upstream water utilization in the Amudarya basin (Conrad et al. 2012). Water management in the upstream regions exacerbates the reduction in the water supply caused by hydrological factors in the river's lower reaches such as in the Khorezm region, particularly in drought years. This becomes obvious when comparing the ratio between (a) the withdrawals (SIC-ICWC 2010) in the vegetation season of each year in the period from 1999 to 2009 and (b) the average withdrawal in the vegetation season over this period for the irrigation schemes located upstream (irrigated areas in Tajikistan), midstream (areas fed by the Amu Bukhara and the Karakum canal) and Khorezm as a downstream location (fig. 3.1.1). In drought years (e.g., 2000, 2001, 2008), the ratio for Khorezm dropped below that in the midstream regions, whereas water supply in the upstream parts of the basin was not impacted by the droughts in these years.

It is expected that climate change and increasing water use in upstream regions of the basin will change the quantity and temporal behavior of the water resources available in Khorezm for the worse (Martius et al. 2008). This un-

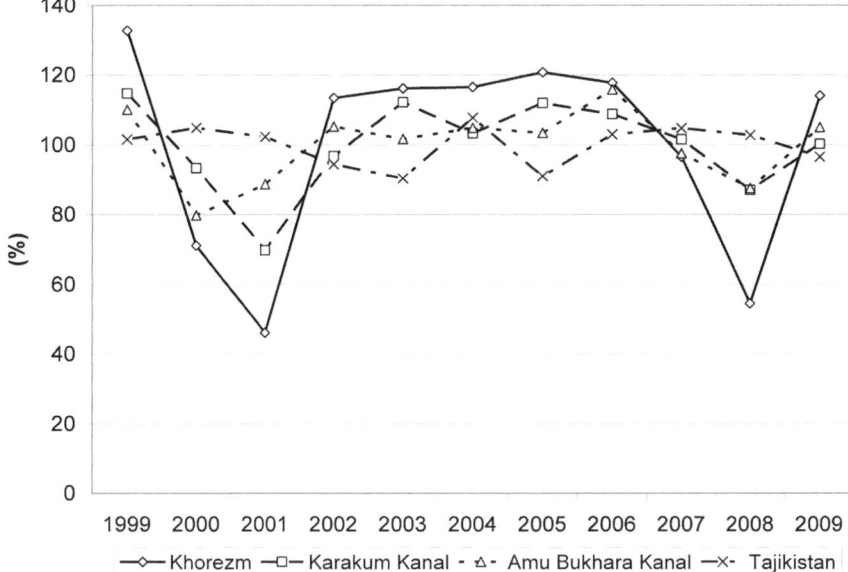

Figure 3.1.1: Ratio of annual withdrawal in the vegetation season of the years 1999 until 2009 and the average during the same period for Khorezm, the areas fed by the Amu Buhkara and the Karakum canal and Tajikistan (data source: SIC ICWC 2010)

derlines the need for improving irrigation efficiency and adequacy of water management in Khorezm, which presently are notoriously low (Tischbein et al. 2012).

Based on the estimated evapotranspiration derived from remote sensing and inflow-outflow monitoring in the irrigation and drainage network, Conrad (2006) established a Khorezm-wide water balance for the vegetation period 2005. The gross irrigation depth amounted to 2,240 mm. Analyzing the irrigation strategy at the level of a hydrologically defined sub-unit fed by the Serchalli canal (606 ha irrigated area) allowed estimating the gross irrigation depth for the vegetation period in the range of 2,110 mm for 2005 and around 1,940 mm for 2004 (Tischbein et al. 2007). About 700 mm water was applied additionally as gross input for pre-seasonal leaching in this sub-unit.

In addition, a comprehensive analysis of the water balancing and irrigation performance assessment was carried out in the Water Consumers Association (WCA) Shomakhulum covering an irrigated area of 1,885 ha. According to Awan (2010), the gross irrigation depth amounted to 1,589 mm in the vegetation period 2007. This lower irrigation depth can be attributed to a relatively small share of rice cultivation (ca 1 %) in comparison to around 10 % and 20 % monitored in previous studies (Tischbein et al. 2007; Conrad 2006) and to a reduced total water supply to the Khorezm region in 2007 compared to 2004 and 2005 (SIC-ICWC 2010).

An inadequate water supply at field level is evident when the actual evapotranspiration is lower than the potential values, which was observed at various scales. Forkutsa et al. (2009a) observed a reduction in actual transpiration of cotton at the beginning and in the middle of its vegetation period to less than 60 % of the potential values. This was caused mainly by an ill-timing of irrigation events, which did not match the time-dependent crop water requirements. Studies referring to mid-level scales showed a reduction in actual evapotranspiration in the range of 82 % in a WCA (Awan 2010) and to 90 % in the sub-unit (Tischbein et al. 2007) even in average and water-rich years. According to Conrad (2006), actual evapotranspiration in tail-end locations was reduced to 75 % of the potential values in 2005 despite the huge gross irrigation depths in the range of 2,240 mm.

The technical overall irrigation efficiency[1] estimated for a sub-unit was 27.5 % in the monitoring years 2004 and 2005 (Tischbein et al. 2007). In the WCA Shomakhulum, this efficiency was slightly higher with 33 % in 2007, which appears to be associated with the lower total water availability in Khorezm during 2007 compared to 2004 and 2005, since water productivity tends to increase with reduced irrigation water supply (Bekchanov et al. 2010). Parallel to the low technical irrigation efficiencies, high drainage ratios and low depleted fractions were monitored. In the vegetation period 2005, a drainage ratio[2] of 58 % was monitored in an area covering 82 % of Khorezm (Conrad 2006). A ratio of 55 % was measured in the WCA Shomakhulum in the vegetation period 2007 (Awan 2010). When including the leaching period and regarding the entire year in the sub-unit Serchalli, a drainage ratio of 65 % between drainage outflow and irrigation inflow was estimated (Table 3.1.1)

---

[1] This is defined as the ratio between irrigation water stored in the root zone for crop use in relation to the amount of water supplied to the sub-unit (Bos and Nugteren 1974).
[2] This is defined as the drainage outflow in relation to the sum of irrigation inflow and effective rainfall.

Table 3.1.1: Indicators of irrigation performance at different scales in Khorezm

Location	Period	Magnitude	Method
Gross Irrigation depth (mm)			
Shomakhulum	Vegetation period 2007	1,589	monitored
Sub-unit Serchalli	Vegation period 2004	1,940	monitored
Sub-unit Serchalli	Leaching period 2004	700	monitored
Sub-unit Serchalli	Vegetation period 2005	2,110	monitored
Sub-unit Serchalli	Leaching period 2005	700	monitored
Khorezm	Vegetation period 2005	2,240	monitored
Overall irrigation efficiency (%)			
Shomakhulum	Vegetation period 2007	33	monitored
Sub-unit Serchalli	Average of the vegetation periods 2004 and 2005	27.5	monitored
Drainage ratio (%)			
Khorezm	Vegetation period 2005	58	monitored
Shomakhulum	Vegetation period 2007	55	monitored
Sub-unit Serchalli	Nov 2003 – Oct 2004	62	monitored
Sub-unit Serchalli	Nov 2004 – Oct 2005	67	monitored
Delivery performance ratio (/)			
Shomakhulum	Vegetation period 2007	1.39 (range: 0.88 – 1.79)	monitored
Relative evapotranspiration (/)			
Shomakhulum	Vegetation period 2007	0.82 (minimum in July with 0.63)	modeled
Sub-unit Serchalli	Average of the vegetation periods 2004 and 2005	0.9 – 0.95	estimated from water balance
Depleted fraction (/)			
Shomakhulum	Vegetation period 2007	0.40 (range: 0.23 – 0.92)	monitored, ETa modeled

Sources: Awan 2010; Conrad 2006; Tischbein et al. 2007.

3.1.2 Analyzing the irrigation performance in the WCA Shomakhulum

Awan (2010) approached the performance of water management from a technical perspective with selected indicators of irrigation performance, i.e., delivery performance ratio, relative evapotranspiration, depleted fraction, drainage ratio, and irrigation efficiencies (field application ratio, conveyance ratio and overall consumed ratio) (Bos et al. 2005; ICID 1978). These indicators were used to assess irrigation performance in a representative WCA and to identify the problems of the current water management and their causes in more detail. The case-study WCA was characterized by a cotton-dominated cropping pattern, loamy to sandy-loamy soils, mid-location within the irrigation system, and shallow groundwater, which are typical for Khorezm (Ibrakhimov et al. 2011).

3.1.2.1 Delivery performance ratio (DPR)

The relation between actual and intended amount of water directed to a scheme or part of a scheme is the most important indicator to assess the operational performance of water distribution (Bos et al. 1991; Clemmens and Bos 1990; Molden and Gates 1990). Actual delivery was determined at the inflow points to the case-study WCA by transferring measured water levels into discharges with the help of rating curves, which were established from simultaneous measurements of the discharge (using a current meter) and water level (Awan 2010). WCA officials provided the information on the intended amount of irrigation water to be delivered to the WCA. Comparing the actual inflow during the vegetation period (referenced to the unit of irrigated area) of 1,589 mm to the intended inflow (1,147 mm) led to an average DPR of 1.39. Especially high DPR values in the peak irrigation season (July: 1.46; August: 1.79) revealed that much more water was directed to the WCA than intended, and actually needed. In September, the DPR reached its minimum (0.88). The limited infrastructure for discharge control and insufficient knowledge on the actual losses in the network were the main reasons for the high excess of the planned schedule in the peak irrigation season.

3.1.2.2 Relative evapotranspiration (RET)

Relative evapotranspiration, i.e., the ratio of the actual evapotranspiration to the potential level, indicates the adequacy of water supply at field level (Perry 1996).

Actual evapotranspiration was determined by the Surface Energy Balance Algorithm for Land (SEBAL) (Bastiaanssen et al. 1998) and potential evapotranspiration according to Allen et al. (1996). With the exception of June 2007, the RET was below the potential level in the entire vegetation period of 2007 reaching a low value of 0.63 in the peak irrigation month (July) and averaging 0.82 for the whole vegetation period. The oversupply at the scheme border, indicated by the high DPR in July, and the low RET in the same month underline the low efficiencies at both network and field level.

3.1.2.3 Depleted fraction (DF)

According to the water accounting approach, the DF is the ratio between actual evapotranspiration aggregated for the considered scheme to the sum of gross water input at the border of the scheme and effective precipitation (Molden and Sakthivadivel 1999). The DF was low in the vegetation period, averaging 0.4. This value is numerically in line with the overall irrigation efficiency in the range of 30 % (see section 3.1.2.5) when taking into account that around 25 % of the actual evapotranspiration is met by the capillary rise from shallow groundwater (Forkutsa et al. 2009a, b). The poor DF in July (0.23) and August (0.27) indicates huge inefficiencies in irrigation water use, especially in the period of highest requirements.

3.1.2.4 Drainage ratio (DR)

The DR is the relation of the water volume drained out from the irrigation and drainage scheme and the input consisting of the sum of irrigation water directed to the scheme and effective rainfall. The drained volume was derived by aggregating discharges from the collectors taking away the water from the boundaries of the scheme. These discharges were determined from the water-level monitoring in combination with the rating curves. An average drainage ratio of 55 % for the whole season indicated high losses at irrigation network and at field level. Whereas high values in April are due to delayed drainage discharge caused by leaching, a value of 55 % in July strongly underlines inefficiency of water management. The DR and DF monitored in the WCA sum up to 0.95 during the vegetation period. The slight deviation of the value from 1 can be explained by the water storage changes (soil moisture, groundwater).

3.1.2.5 Irrigation efficiencies

Applying water in the field and conveying and distributing irrigation water in the network are relevant processes of irrigation water management. The technical performance can be assessed by the field application ratio (FAR), the conveyance ratio (CR) and their product, which represents the overall irrigation efficiency (OCR). The FAR relates the water stored in the root zone for use by the crops to the water directed to the field. The CR is the ratio between the water volume available at field level and the volume fed into the irrigation system or the sub-system under consideration.

The FAR was determined at two typical sites (loamy and sandy-loamy soils, both furrow irrigated) in the WCA by monitoring the spatial distribution of the soil moisture in the fields before and after irrigation, and by measuring the inflow during respective irrigation events. Measurements carried out for 5 irrigation events revealed a poor FAR averaging not more than 0.43 with slightly higher values at the loamy site (0.45 compared to 0.4 at the sandy-loamy fields).

Interviews with farmers indicated that besides limited information on the site-specific requirements and technical deficiencies (lack of infrastructure for discharge control, poor leveling), over-irrigation was caused by the farmers' attempt to compensate for an unreliable water supply in the network (e. g., Awan 2010; Forkutsa et al. 2009a, b). Deficits in the coordination between farm and network levels are a further reason for water losses. As the capacity of the pumps is fixed, the discharge made available at field level often exceeds the capacity of an individual farmer to handle the discharge, and therefore a part of the discharge is directed immediately to the drainage system.

Ponding tests, carried out at the level of inter-farm, farm and field canals revealed losses (by seepage, percolation and evaporation) of 2, 4 and 18 %, respectively, of the design discharge (related to representative canal length in the WCA). Furthermore, the analyses of the ponding findings showed a dependency of the losses on specific site conditions. As the canals at the higher hierarchy level (inter-farm canals) had been designed as dug-in canals (low bottom and water level height in relation to surrounding topography), the losses monitored are under the influence of the time-dependent groundwater level. This can even lead to temporary exfiltration from the aquifer to the canals. In the lower hierarchy, seepage losses differ with soil conditions. Such short-term and spatial variability is hard to cover by the static efficiency factors. Taking the representative lengths of the canals in the WCA into account, the CR amounts to 76 %, which indicates moderate physical losses through seepage, percolation, and/or evaporation.

Multiplying the CR values by the FAR (0.43) resulted in an overall efficiency of not more than 33 %. Combined with the estimated operational losses of 10 % of the conveyance losses, this suggests a realistic network efficiency in the range of 65 to 70 % and a resulting technical overall efficiency of around 30 %.

3.1.3 Causes of current problems

The estimated poor irrigation efficiency (OCR of 30–33 %), low depleted fraction (average DF = 0.4), and high drainage ratio (DR=55 %), indicate inefficient water management at irrigation network level and especially at field level. Furthermore, the relative evapotranspiration is below 1 (RET= on average 0.82) although it varied across the irrigation schemes, revealing an inadequate supply of irrigation water to the field and inequities in supply. In the following, we identify the reasons for these deficits, because tackling the reasons is used as a starting point to conceive options for improvement (see: section 'Approaches and tools for improving the irrigation efficiency')

3.1.3.1 Inappropriate information on site-specific and time-dependent irrigation requirements at field level

The application of rigid norms used by the water managers and farmers leads to irrigation timing and amounts that do not match site-specific and time-dependent crop water requirements (Forkutsa 2009a). The established norms may be sufficient to determine the requirements of huge uniform production units as were in place during the Soviet Union time. But an appropriate irrigation scheduling needed in a variable environment (dynamic water availability, groundwater levels) and in increasingly diversified schemes must rely on more flexible approaches than those based on the rigid norms. Improved and more appropriate approaches that consider the interrelation between surface and groundwater are especially relevant in the case of the shallow groundwater tables prevailing in the Khorezm region during the vegetation season. These approaches can thus benefit by the use of flexible modeling tools.

3.1.3.2 Insufficient measures to optimize the application process

The low application efficiencies monitored are a result of inadequate adjustment of the application discharge to the conditions influencing uniformity of the water application in the field, be it soil conditions, field size and geometry, irrigation method, net irrigation depth or even irregular micro-topography due to insufficient field leveling. The limited adjustment of the application discharge is caused primarily by insufficient information of the farmers about the optimal or at least appropriate discharge, and by the lack of infrastructure for proper handling of the application discharge. The resulting unreliable supply of irrigation water motivates farmers to over-irrigate whenever water is available.

3.1.3.3 Shortcomings in the irrigation network operations on the field and irrigation network level

The high DPRs estimated for the systems, deviating clearly from the target value 1, and the reduced RET indicate severe operational deficiencies in the network. Further analyses showed that these are caused mainly by limited information availability on spatially variable and time-dependent efficiencies of the network, and insufficient coordination of irrigation activities between field and network level. This explains the high operational losses monitored (Awan 2010) and deducted (Veldwisch 2008). A major reason for the operational losses at the interface between field and network level is that the water pumps used for taking water from the higher-level channels to lower-level channels cannot be regulated. Therefore, the discharge through the pumps often exceeds the capacity of farmers to handle this discharge, which as a consequence is then partly directed to the drains. The inadequacy of the water availability at field level refers therefore to both amount and timing of the irrigation water supply.

3.1.3.4 Infrastructure and maintenance deficits

The current layout and status of the irrigation and drainage infrastructure also hinder appropriate operation. The present situation is characterized by a large number of water users with diversified requirements, a situation that differs substantially from the conditions relevant at the design stage that was typified by large production units with high uniformity (Djanibekov et al. 2012). Furthermore, the widespread occurrence of deteriorated hydraulic structures and lowered discharge capacity of the canals are consequences of a low maintenance intensity as postulated earlier (Awan 2010, Tischbein et al. 2012).

3.1.3.5 Institutional and economic context

Due to negative factors such as limited functioning of the WCA, insufficient participation of water users in the processes governing water allocation (cf. Saravanan et al. this book), missing transparency of decision making on water distribution, strict regulations (e.g., ensuring state order), limiting space for innovations driven by the water users and insufficient support regarding the introduction of water-saving strategies (Bekchanov et al. 2010), farmers are not willing to increase efficient use of water. Furthermore, economic incentives for water saving are missing (Djanibekov et al. 2010).

Promising approaches to solve or at least reduce these widespread challenges need to tackle the above-summarized reasons. For this purpose, various tools and approaches were developed.

3.1.4 Approaches and tools for improving irrigation efficiency

3.1.4.1 Linking irrigation scheduling and groundwater models

As an alternative to the presently used static norms, a water management tool for a case study WCA was developed (Awan 2010). This tool links the irrigation scheduling model CROPWAT (Clark et al. 1998) and the groundwater model FEFLOW (Diersch 2002). As CROPWAT does not consider the groundwater contribution, the HYDRUS model (Simunek et al. 2005) was used to determine the capillary rise from the groundwater, which then was introduced in the water balancing procedure of CROPWAT. The model configuration allows (i) establishing a field water balance that considers all relevant components, (ii) deriving irrigation schedules from the water balance thus avoiding water stress, (iii) minimizing impacts of unavoidable water stress on yield, and (iv) quantifying the spatio-temporal behavior of the groundwater. Therefore, the tool can be used to work out site-specific irrigation schedules that take into consideration a variable environment and integrate surface and groundwater use. Geographical information systems (GIS) and remote sensing (RS) techniques were applied to support the data provision. The concept of Hydrological Response Units (HRU) was used for up-scaling of the information (Awan 2010).

As an advantageous feature, HYDRUS allows simulating water fluxes and salt dynamics. It is thus possible to develop strategies that include the most appropriate leaching period and irrigation events during the vegetation season. Forkutsa et al. (2009b) simulated strategies with HYDRUS for leaching-water savings that could amount to as much as 25 % without lowering the effectiveness of leaching. The authors also suggested using the water saved during leaching in the vegetation period to reduce water stress (cf. Akramkhanov et al. this book). The implementation of such strategies could also be supported by the construction of decentralized reservoirs (cf. Bhaduri and Djanibekov this book).

3.1.4.1.1 Refining the irrigation scheduling model with respect to deficit irrigation

The CROPWAT model allows estimating the crop response to water stress in terms of a relative reduction in yield based on a linear response factor ky (Doorenbos and Kassam 1979). To raise the reliability of irrigation scheduling in case of deficit irrigation strategies, the more advanced AquaCrop model enables an appropriate compromise between the rather simple CROPTWAT approach and sophisticated but input-demanding crop models (APSIM, CROPSYST).

Based on local conditions, a water-driven simulation model, AquaCrop (Steduto et al. 2009), was used to plan deficit irrigation scheduling of cotton in Khorezm. The yield response factor (Ky) in AquaCrop was adopted from the FAO Irrigation and Drainage Paper 33 (Doorenbos and Kassam 1979) as a ratio of yield decline and evapotranspiration deficit. AquaCrop advances from the Ky approach by (i) dividing crop evapotranspiration (ET) into soil evaporation (E) and crop transpiration (T) to avoid the effect of the non-productive consumptive use of water (E), (ii) obtaining biomass (B) from the product of water productivity (WP) and cumulated crop transpiration (the core of the AquaCrop growth engine), and (iii) expressing the final yield (Y) as the product of B and Harvest Index (HI). Since the conventional version of the AquaCrop model ignores the groundwater contribution, the HYDRUS-1D model (Simunek et al. 2005) was used to determine the daily capillary rise contribution, which was introduced as pseudo-precipitation in the AquaCrop model (Akhtar 2011).

3.1.4.2 Deriving measures to improve the application process

Interventions to address the current causes of low efficiency of the water application process at field level include laser-guided land leveling (Ergamberdiev et al. 2008) or double-side irrigation at sites with zero-slope (Paluasheva 2005), optimization of application discharge based on surface irrigation models as well as on field test experiences in the region (Horst et al. 2005), refinement of the conventional furrow irrigation by surge-flow strategies, and the introduction of modern irrigation methods (Awan 2010).

3.1.4.3 Improving operation of the network

Only when site-specific knowledge is available on the current level of the seepage losses depending on the canal hierarchy, soil conditions and groundwater dynamics as provided by ponding tests can the requirements at field level be matched more closely. Information exchange on water requests (quantity and time) among the water users and between water users and water managers of the WCA is the first step to improve the coordination of their irrigation activities. As mentioned above, a major part of the present operational losses is caused by a fixed discharge through pumps, where the surplus can be handled by individual farmers only when discharging this to the drains. Such losses can be reduced by adapting the number of fields to be irrigated at the same time to the discharge capacities of the pumps while taking into account the optimal application discharge. For that purpose, irrigation dates of different fields need to be streamlined while accepting that some fields will be irrigated only slightly before or after the optimal date. To assess the impact of adjusting the irrigation dates, the linked models (see section 3.1.4.1) can be used and adjustments can be made such that water stress is avoided or minimized.

3.1.4.4 Decentralized reservoirs

Inappropriate water supply at field level is not only a question of the water quantity, but also of the timing of the supplies from the irrigation network. Besides improving the performance of the network operation, the introduction of decentralized reservoirs is a promising strategy for tackling the problem of temporarily unreliable water supplies (cf. Bhaduri and Djanibekov this book). Filling the reservoirs during periods of oversupply (as indicated above by DPR values higher than 1) makes the farmers more independent from the water supply by the network. The use of additional water-saving techniques would allow increasing the application efficiency, therefore fulfilling net water requirements with lower gross water inputs and, as a consequence, adjusting the quantity of the water supply at farm level. Yet the vulnerability to a temporal, unreliable water supply will not be solved by such measures and, from the farmers' perspective, could hinder investments in water-saving techniques (Djanibekov et al. 2012). Therefore, combining the introduction of water-saving measures with decentralized reservoirs would create a win-win situation by tackling both shortcomings that affect the water productivity, i.e., unreliable water supplies and irrigation timing. This should go hand in hand with raising the willingness of farmers to introduce and invest in water-saving technologies.

3.1.5 Achievements and case studies

The developed irrigation-scheduling groundwater model was used to derive irrigation schedules for the case study WCA Shomakhulum. This hypothetical irrigation schedule was compared to the empirical schedules. The model enabled the development of an optimal irrigation schedule taking into account the relevant daily capillary rise, which depends on site-specific conditions such as soil, crop, and groundwater level. When taking this dependency into consideration, the irrigation schedule could be substantially improved, because the capillary rise was highly variable in the WCA, ranging from 28 % of the actual evapotranspiration (cotton, shallow groundwater, silt loam) to 0 % (vegetables, deeper groundwater, silt-clay loam).

Assessing the empirical irrigation scheduling against the linked-model scheduling revealed that the former led to a 7 % yield loss. This loss was caused by water stress as a result of inappropriate irrigation timing and amounts. Furthermore, under the empirical schedule practices, 9 % of the water was wasted. Following the irrigation schedule derived by the linked model enabled both water savings and avoidance of yield reduction. The schedule presently practiced by the monitored farmers indicates a reduction in cotton yield in the range of 42 %.

A further advantage of the model is that deficit irrigation strategies could be assessed and optimized, which is particularly important under a limited water supply. For the scenario of a water supply reduction by 25 %, the yield losses varied from 10 to 18 %, while a water reduction of 50 % could cause yield losses of 22 to 30 %. A change in cropping pattern by introducing vegetables in the place of cotton can result in a 9 % water saving. About 15 to 20 % of water can be saved by leaving marginal lands out of the cotton production.

Akhtar (2011) applied the AquaCrop model with data collected at two typical cotton fields in Khorezm to analyze deficit irrigation strategies in two ways. In the first case, yield response was simulated against a proportional water deficit of 20, 40 and 50 %, which was introduced in each irrigation event of the optimal schedule. In the second approach, an artificial stress was introduced into a specific crop growth stage to help identifying periods when crop yield is sensitive to water stress, which in turn allows minimizing the impact of water stress.

The results of the AquaCrop simulations show that a 20 % deficit in irrigation application did not reduce cotton yield. The biomass loss of 10 % was compensated for by an increased harvest index. A 40 % reduction in irrigation water during each irrigation event (proportionally deficit) resulted in a yield loss of 14 – 29 %. Yield losses against a 50 % proportional deficit were in the range of 30 – 45 %. Simulating the water stress at the early crop development stage under a water supply reduced by 8 – 9 % resulted in yield losses of 17 – 18 %. The late vegetative stage was found to be a feasible time window for water saving and raising water productivity, because a 12 – 13 % reduction in irrigation supply during this period increased cotton yields by as much as 8 %. Water stress at this stage slows down the vegetative growth and induces flowering, thus resulting in higher yields (Kienzler 2010). However, in the currently uncertain temporal availability of water provided by the irrigation network, a targeted practice of deficit irrigation at a specific growth stage is hard to realize and risky compared, for instance, to a proportional reduction of the irrigation amount at each event.

AquaCrop is also able to simulate achievable yields. For instance, by adopting an optimal irrigation schedule, cotton yield can be raised up to 4.8 tons ha^{-1} (optimum). This simulated yield is in line with the results of Kienzler (2010), who measured a 4.2 tons ha^{-1} cotton yield under optimal conditions in the Khorezm region. Therefore, to meet the maximum achievable cotton yield, AquaCrop can be suggested for simulating different irrigation scenarios for a better informed decision making by the water managers at both scheme and field level.

3.1.6 Conclusions

The deficits of water management in Khorezm are mainly caused by (i) currently applied approaches to determine irrigation schedules taking the spatio-temporal variability of the requirements insufficiently into account, (ii) limited capacity to optimize the process of water application and distribution at the field, (iii) insufficient coordination of irrigation activities in the irrigation network and at the interface farm-network, (iv) shortcomings of the irrigation network (hydraulic bottlenecks, low maintenance intensity), and (v) weak institutions and missing economic incentives for increasing water use efficiency.

Besides introducing measures for improving the water application at field level (e. g., laser-guided leveling, equipment for discharge dosage), much can be expected from tools that enable flexible irrigation scheduling. Such a tool was developed by linking models of irrigation scheduling (CROPWAT and Aqua-Crop), groundwater dynamics (FEFLOW) and water fluxes between groundwater table and soil surface (HYDRUS). The tool allows deriving flexible irrigation schedules closely matching the spatio-temporal requirements. In case of an unavoidable under-supply, the tool enables minimizing the impact of water stress on crop yields. Prerequisites for implementation of flexible schedules consist of (i) re-structuring the irrigation network by adapting hydraulic structures to the new situation characterized by a high number of diversifying water users, abolishing hydraulic bottlenecks, and creating and using storage capacities, (ii) strengthening the WCAs and the participation of water users, and (iii) creating incentive systems supporting efficient and effective use of water for irrigation and salt management.

References

Akhtar F (2011) Combining AquaCrop and HYDRUS models to improve current irrigation scheduling under shallow groundwater conditions in Khorezm, Uzbekistan. Master Thesis, Faculty of Agriculture, University Bonn

Allen RG, Pruitt WO, Businger JA, Fritschen LJ, Jensen ME, Quinn FH (1996) ASCE Manuals and Reports on Engineering Practice, No. 28. ch. 4. Hydrology Handbook, Second edition. Heggen, R.J., (Ed) 125–252

Awan UK (2010) Coupling hydrological and irrigation schedule models for the management of surface and groundwater resources in Khorezm, Uzbekistan. Dissertation, University Bonn

Bastiaanssen WGM, Menenti M, Feddes RA, Holtslag AAM (1998) A remote sensing surface energy balance algorithm for land (SEBAL)-1. Formulation. J Hydrol 212–213: 198–212.

Bekchanov M, Lamers JPA, Martius C (2010) Pros and cons of adopting water-wise approaches in the lower reaches of the Amu Darya: A Socio-Economic View. Water, 2(2), 200–216

Bekchanov M, Karimov A, Lamers JPA (2010) Impact of water availability on land and water productivity: A temporal and spatial analysis of the case study region Khorezm, Water 2 (3), 668–684

Bos MG, Nugteren J (1974) On Irrigation Efficiencies. Publication no. 19, International Institute for Land Reclamation and Improvement (ILRI), Wageningen, The Netherlands

Bos MG, Burton MA, Molden DJ (2005) Irrigation and Drainage Performance Assessment: Practical Guidelines. CABI Publishing, Trowbridge, US

Bos MG, Wolters W, Drovandi A, Morabito JA (1991) The Viejo Retamo secondary canal –performance evaluation case study: Mendoza, Argentina. Irrig Drain Syst (5): 77–88

Clarke D, Smith M, El-Askari K (1998) CropWat for Windows: User Guide, University of Southampton.

Clemmens AJ, Bos MG (1990) Statistical methods for irrigation system water delivery performance evaluation. Irrig Drain Syst (4): 345–365

Conrad C (2006) Fernerkundungsbasierte Modellierung und hydrologische Messungen zur Analyse und Bewertung der landwirtschaftlichen Wassernutzung in der Region Khorezm (Usbekistan). Dissertation, University of Wuerzburg

Conrad C, Schorcht G, Tischbein B, Davletov S, Sultonov M, Lamers JPA (2012) Agrometeorological trends of recent climate development in Khorezm and implications for crop production. In: Martius C, Rudenko I, Lamers JPA, Vlek PLG (Eds.) Cotton, water, salts and soums – economic and ecological restructuring in Khorezm, Uzbekistan. Springer: Dordrecht pp 25–36

Djanibekov N, Rudenko I, Lamers JPA, Bobojonov I (2010) Pros and Cons of Cotton Production in Uzbekistan. Case Study #7–9, In: Per Pinstrup-Andersen and Fuzhi Cheng (eds.), Food Policy for Developing Countries: Case Studies. 13 pp

Djanibekov N, Bobojonov I, Lamers JPA (2012) Farm reform in Uzbekistan. In: Martius C, Rudenko I, Lamers JPA, Vlek PLG (Eds.) Cotton, water, salts and soums – economic and ecological restructuring in Khorezm, Uzbekistan. Springer: Dordrecht pp 95–112

Diersch H-JG (2002) Reference manual for FEFLOW – finite element subsurface flow and transport simulation system. User's Manual/Reference Manual/White Paper Release 5.0. Wasy Ltd., Berlin

Doorenbos J, Kassam A H (1979) Yield response to water. FAO Irrigation and Drainage Paper 33, Food and Agriculture Organization of the United Nations, Rome

Ergamberdiev O, Tischbein B, Franz J, Lamers JPA, Martius C (2008) Laser land leveling: more about water than about soil. Zentrum für Entwicklungsforschung (ZEF), Bonn, Germany

Forkutsa I, Sommer R, Shirokova Y, Lamers JPA, Kienzler K, Tischbein B, Martius C, Vlek PLG (2009a) Modeling irrigated cotton with shallow groundwater in the Aral Sea Basin of Uzbekistan: I. Water dynamics. Irrigation Sci 27(4): 331–346

Forkutsa I, Sommer R, Shirokova Y, Lamers JPA, Kienzler K, Tischbein B, Martius C, Vlek PLG (2009b) Modeling irrigated cotton with shallow groundwater in the Aral Sea Basin of Uzbekistan: II. Soil salinity dynamics. Irrigation Sci 27(4): 319–330

Horst MG, Shamutalov SS, Pereira LS, Gonçalves JM (2005) Field assessment of the water saving potential with furrow irrigation in Fergana, Aral Sea Basin. Agr Water Manage 77(1–3): 210–231

Ibrakhimov M, Martius C, Lamers JPA, Tischbein B (2011) The dynamics of groundwater table and salinity over 17 years in Khorezm. Agricultural Water Management 101, 52–61

ICID (1978) Standards for the calculation of irrigation efficiencies. International Commission on Irrigation and Drainage Bulletin 1(27): 91–101

Kienzler KM (2010) Improving the nitrogen use efficiency and crop quality in the Khorezm region, Uzbekistan. Ecology and Development Series No. 72, ZEF/University of Bonn

Manschadi AM, Oberkircher L, Tischbein B, Conrad C, Hornidge A-K, Bhaduri A, Schorcht G, Lamers JPA, Vlek PLG "White Gold" and Aral Sea disaster – Towards more efficient use of water resources in the Khorezm region, Uzbekistan. Lohmann information, Vol. 45 (1), April 2010, pp 34–47

Martius C, Froebrich J, Nuppenau E-A (2008) Water Resource Management for Improving Environmental Security and Rural Livelihoods in the Irrigated Amu Darya Lowlands. In: Brauch HG, Grin J, Mesjasz C, Dunay P, Chadha Behera N, Chourou B, Oswald Spring U, Liotta PH, Kameri-Mbote P (Eds.): Globalisation and Environmental Challenges: Reconceptualising Security in the 21st Century. Hexagon Series on Human and Environmental Security and Peace, Volume 3, Berlin, Springer-Verlag

Molden DJ, Gates TK (1990) Performance measures for evaluation of irrigation water delivery systems. J Irrig Drain E-ASCE: 116: 804–823

Molden DJ, Sakthivadivel R (1999) Water accounting to assess Use and Productivity of Water. Water Resources Development, 15(1/2): 55–71

Paluasheva G (2005) Dynamics of soil saline regime depending on irrigation technology in conditions of Khorezm oasis. In: Proceedings of the International Scientific Conference: 'Scientific Support as a Factor for Sustainable Development of Water Management': Taraz, Khazakhstan, 20–22 October 2005 (in Russian)

Perry CJ (1996) Quantification and Measurement of a Minimum Set of Indicators of the Performance of Irrigation Systems. International Irrigation Management Institute, Colombo, Sri Lanka

SIC-ICWC (2010) On-line data of the BWO Amudarya. Scientific Information Center – Interstate Commission for Water Coordination of Central Asia, (http://www.cawater-info.net; assessed October 2010)

Simunek J, Van Genuchten MT, Sejna M (2005) The HYDRUS-1D software package for simulating the one-dimensional movement of water, heat, and multiple solutes in variably-saturated media. Version 3.0, HYDRUS Software

Steduto P, Hsiao TC, Raes D, Fereres E (2009) AquaCrop-The FAO Crop Model to Simulate Yield Response to Water: I. Concepts and Underlying Principles. Agronomy Journal, Volume 101, Issue 3, pp 426–437

Tischbein B, Rücker G, Martius C (2007) Research Area N (Natural Resource Management Strategies) of the Final Report on Phases 1 and 2 (2001–2007) of the ZEF-UNESCO Project: Economic and Ecological Restructuring of Land- and Water Use in the Region Khorezm (Uzbekistan): A Pilot Project in Development Research

Tischbein B, Awan UK, Abdullaev I, Bobojonov I, Conrad C, Jabborov H, Forkutsa I, Ibrakhimov M, Poluasheva G (2012) Water management in Khorezm: current situation and options for improvement (hydrological perspective). In: Martius C, Rudenko I, Lamers JPA, Vlek PLG (Eds.) Cotton, water, salts and soums – economic and ecological restructuring in Khorezm, Uzbekistan. Springer: Dordrecht pp 69–92

Veldwisch GJA (2008) Cotton, Rice & Water. The Transformation of Agrarian Relations, Irrigation Technology and Water Distribution in Khorezm, Uzbekistan. Dissertation, University of Bonn

Mirzakhayot Ibrakhimov, Bernhard Tischbein, Usman Khalid Awan

3.2 Knowledge on groundwater – A prerequisite for water management in Khorezm

Abstract

Groundwater in Khorezm, Central Asia, is shallow fluctuating between 2.2 m below surface at the end of the non-irrigation period (February) and 1.15 m during the peak irrigation season (August). Shallow groundwater contributes moisture to meet crop water requirements (24 – 32 % of actual evapotranspiration), but enhances the process of salt accumulation also which in turn lowers the effectiveness of leaching. Analyzing 16 years of data stemming from around 2,000 groundwater observation wells in Khorezm revealed (I) a typical intra-annual pattern of the groundwater table, (II) a clear dependency of the groundwater table on the water directed to Khorezm (0.4 m difference of average groundwater level during the vegetation period comparing water ample and drought years) and (III) a trend towards a slower dropping of groundwater levels in autumn driven by the expansion of the winter wheat area. The average salt content of groundwater was moderate with around 1.75 g l^{-1}. The knowledge on the spatial variability of groundwater salinity (between 1.3 and 15 g l^{-1}) can be utilized to identify localizations promising for conjunctive use options of surface and groundwater.

Monitoring the groundwater table at a typical site in Khorezm with a high spatial and temporal resolution revealed that the groundwater table follows with a delay of several days the leaching and irrigation events. Furthermore, the impact of rice cultivation on the groundwater table in the vicinity of permanently flooded fields was quantified. Analyzing groundwater slopes evidenced a limited impact of drainage ditches (due to insufficient depth, blocking and outlet problems) and a de facto-drainage function of main irrigation canals.

Based on simulations with the groundwater model FEFLOW the impact of improved irrigation efficiencies (and reduced losses recharging the groundwater) on the groundwater dynamics was analyzed in a Water Consumers Association

(WCA). *Raising the overall irrigation efficiency from currently 33 % in the WCA to 56 % would lower the groundwater level by 44 cm in average over the vegetation period. As a consequence, capillary rise would be reduced and current irrigation schedules would need to be adapted to compensate for reduced groundwater contribution. Providing spatial and temporal information on the groundwater on a regular basis to land users and water managers has a high potential to contribute to overall irrigation water management.*

Keywords: Groundwater monitoring, modeling, surface-groundwater interaction, Central Asia, Aral Sea Basin

3.2.1 Introduction

Shallow groundwater recharged by water stemming from the high irrigation losses at field and network level and leaching is a typical feature of Khorezm, in Central Asia (Tischbein et al. 2012). Capillary rise contributes to meet crop water requirements, but enhances soil salinization (Forkutsa et al. 2009). The current irrigation and leaching strategies in Khorezm rely on shallow groundwater. Therefore, understanding irrigation water management and introducing eventual modifications needs to be based on information regarding the groundwater situation and its interrelationships with irrigation, leaching and drainage strategies. This study therefore aims at (i) summarizing the spatio-temporal behavior of groundwater and analyzes the driving factors at Khorezm-wide scale, (ii) illustrating the impact of water management on the spatio-temporal behavior of the groundwater at small-scale and (iii) modeling the groundwater level and assessing the impact of various scenarios aiming at improving irrigation efficiency on the groundwater table at the level of a Water Consumers Association (WCA).

3.2.2 Spatio-temporal behavior of groundwater in Khorezm and factors influencing the groundwater table

3.2.2.1 Spatio-temporal behavior of groundwater table and salinity

Our study analyzed long-term data on the groundwater (GW) level and salinity collected by the Khorezmian Hydrogeological Melioration Expedition (GME) of the Ministry of Agriculture and Water Resources Management (MAWR) through a network of around 2,000 observation wells distributed over Khorezm. The analysis of the 16-year data set (1990–2006) showed a similar intra-annual pattern of both GW table and its salinity levels as previously concluded (Ibrakhimov et al. 2011). The deepest GW levels were 1.8 – 2.6 m (±0.55 – 0.64) below the soil surface in January/February meaning outside the irrigation season. The pre-seasonal leaching events made the GW rise till the region-average of 1.1 – 1.4 m (±0.48 – 0.66) being thus only a few decimeters below the surface especially shortly after leaching events. Except for a minor drop till 1.2 – 1.5 m (±0.46 – 0.56) due to the low irrigation intensities in April and May, the GW levels remained at 0.9 – 1.4 m (±0.43 – 0.63) during the remainder of the irrigation season.

The GW salinity was measured thrice a year: in April (2.35 dS m^{-1}), July (2.29 dS m^{-1}) and October (2.10 dS m^{-1}). The findings would categorize these as moderately-saline (Rhoades et al. 1992). The seasonal fluctuations in GW salinity were insignificant despite different irrigation intensities.

Shallow GW tables are contributing up to 24 to 32 % of the crop water requirements in the region (Forkutsa et al. 2009; Awan 2010). According to Awan (2010) reporting the results of several surveys with farmers, discussions with the irrigation officials, and to Forkutsa et al. (2009), farmers and irrigation planners actively tend to prevent the decline in GW tables. Shallow GW is considered as a safety net against the present unreliable delivery of irrigation water to individual farms and fields. This perception often results in an active blocking of drains by farmers to even rise the GW table as to mitigate the spatial and temporal water shortage in anticipation of obtaining more secure and even higher yield (Awan 2010). Although such blocking is postulated by some as an effective risk-avoiding strategy (e.g., Bobojonov 2008), such practices increase concurrently the risk of secondary soil salinization. Water and salt balance estimates by Ibrakhimov et al. (2007) at sites with comparatively high GW salinity (> 3.5 g l^{-1}) revealed that salt input to the rooting zones driven by capillary rise of the GW exceeded the salt input in the same zone via irrigation water applications.

3.2.2.2 Khorezm-wide analyses of factors influencing the groundwater table

The GME dataset was used to detect the magnitude of factors knowing to influence the spatio-temporal behavior of the GW table such as topography, soil texture and distance to irrigation canals and to drainage ditches/collectors. These were considered as spatial predictors whilst seasonal irrigation water input including the winter wheat irrigation events (October) was used as an explanatory variable to understand the inter- and intra-annual GW behavior in Khorezm.

The region's topography, estimated from topographic images (based on NASA Shuttle Radar Topographic Mission (SRTM)), was classified into plain areas, depressions and elevations. The flat areas dominated the regional relief (87 %). The analysis revealed no significant differences of GW levels in all three types of the areas despite differing in seasonal or annual irrigation intensities (see also Ibrakhimov et al. 2011).

During periods of intensive irrigation, the GW was shallower in areas with heavy and moderate soil texture in comparison to locations characterized by light texture. Outside the irrigation periods, no texture-specific differences could be detected (Ibrakhimov et al. 2011). Heavy soil texture with attributed low hydraulic conductivity does not let the GW table subside during March (massive leaching) and July (peak irrigation), whereas the period outside of the irrigation season (September to February) is long enough for GW levels to drop irrespective of soil texture. This is consistent with the assessment that the problem of shallow GW is of a temporary nature: it can be regulated by interventions tackling the input side (e. g. lowering irrigation losses), addressing the output side (e. g. improving drainage) or by a combination of the two. An influence of large or small irrigation canals or drains on the GW table at differing distances from observation wells could not be detected (Ibrakhimov et al. 2011) which may very well be caused by the small GW slopes in the region. This influence was however revealed in GW profiles at locations with higher slopes (see: 3.2.2.3).

The seasonal water input to Khorezm strongly influences the GW table. In water-ample years such as 1992, 1998, and 2005 where the water input to the entire region during the vegetation period alone amounted to 3.8 km^3 (SIC-ICWC 2010), GW table in the period from May to October averaged 1.1 m below the surface. In the drought years 2000 and 2001 (2.1 and 1.4 km^3 input during the vegetation period according to SIC-ICWC), GW was with 1.5 m almost 30 % deeper.

The extension of winter wheat cultivation since 1990 delayed the dropping of GW table in autumn indicated by shallower GW in October since 1990.

3.2.2.3 GW monitoring: the influence of water management on the spatio-temporal behavior of GW at small scale

A network of 24 GW observation wells covering an area of around 220 ha research site (CRS) enabled detailed analyses of the GW situation in this area. Figure 3.2.1 (upper part, left) depicts the location of the wells and the field layout. Water management activities over the entire research area dominated the spatio-temporal behavior of the GW, which is mirrored especially in pre-seasonal leaching and rice cultivation.

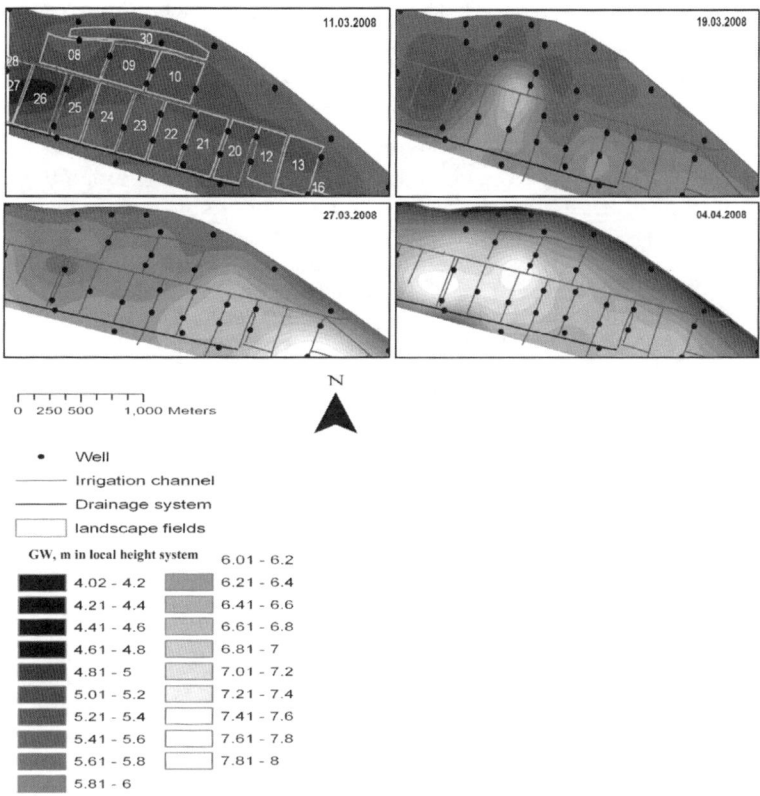

Figure 3.2.1: Groundwater table in experimental site as influenced by the sequence of leaching events

Due to a limited water supply and weather-imposed short windows of opportunities only two leaching events could be possible (March and April 2008). Figure 3.2.1 (part "11.03.2008") indicates the situation before leaching by GW tables referenced in a local height system. In the first round, leaching started at the fields in the eastern part (fields 12, 20, 21) on March 12 and was carried out at the western fields (fields 23, 24, 25, 26) in the period March 15 – 19. The second event started in the east (March 20 – 27) as well. Leaching in the western part was completed between March 25 and April 4. Seepage losses from the canals and especially the heavy leaching amounting to 250 mm per event raised the GW table. The spatio-temporal behavior of the GW tables follows the sequence of the leaching events. Figure 3.2.1 depicts on March 19 high GW tables in the western part due to leaching performed in that period, whereas the tables in the western part already dropped slightly after the first events. On March 27 GW tables in the eastern part rose again with the second round of leaching and tables in the western part lowered (Figure 3.2.1; part 27.03.2008). With leaching proceeding to the western part (April 4) the GW tables rose whereas in the eastern part a GW dropping became obvious (Figure 3.2.1; part 04.04.2008).

Figure 3.2.2 depicts the spatial distribution of GW table in the peak irrigation season (July/August) 2008. The extreme high GW tables seen at field 13 and in the southern part of field 22 were caused by permanently flooded paddy rice cultivated at these spots.

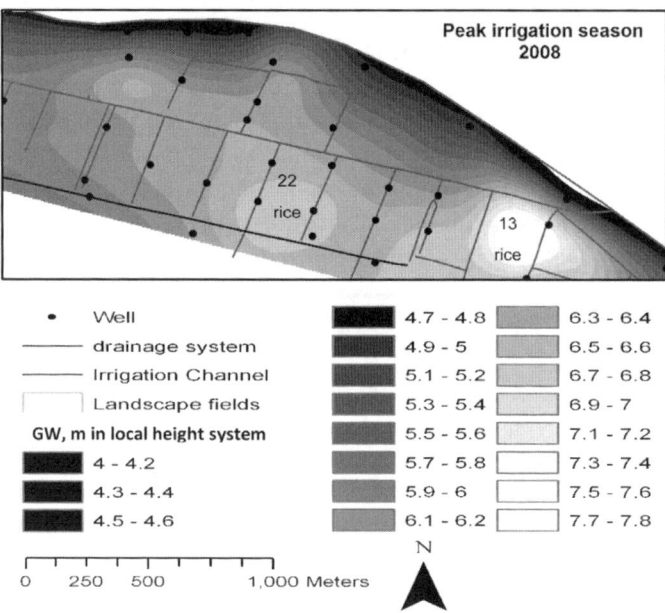

Figure 3.2.2: Groundwater table influenced by rice cultivation

Figure 3.2.2 reveals a clear slope of GW towards the Shavat canal, a magistral canal at the northern border of the experimental area. Due to low water table (dug-in canal), the Shavat canal is acting as a drain during most of the vegetation period.

3.2.3 Modeling of the GW behavior in the WCA Shomakhulum

Awan (2010) applied the GW finite-element model FEFLOW-3D (Version 5.1) to the conditions in the Shomakhulum WCA. FEFLOW is a Finite Element Subsurface Flow & Transport Simulation System for modeling subsurface water resources and has a fully graphics-based and interactive user interface (Diersch and Kolditz 2002). In combination with the irrigation scheduling model

CROPWAT (Clarke et al. 1998), FEFLOW was used to simulate the change of GW tables under the influence of different irrigation regimes. In scenario building of four regimes, the irrigation efficiency concept was used. It was considered that either by increasing the application or conveyance efficiencies or both of them from the current situation, overall efficiency would raise from initially 33 % to 36 %, and from 51 % and 56 %, respectively. As the losses recharge the GW, improved efficiencies contribute to lower the GW table.

3.2.3.1 Parameterizing of FEFLOW

The FEFLOW parameters for estimating the GW flow include the horizontal and vertical (spatial) distribution of permeable and impervious layers, hydraulic conductivity, porosity, and other parameters describing the characteristics governing GW flow and balance, and information on boundaries and internal fluxes such as rate of GW recharge, surface water table intersecting the GW at borders or within the system in the form of drains or canals. The following gives a brief description of the major input parameters used into GW modeling.

3.2.3.2 Hydro-geological constellation; characteristics governing the flow from cross-sections

A geomorphological-lithological map of the Khorezm region and Turtkul oasis (Pre-Aral Hydro-Geological Expedition 1982) covering Khorezm at a scale of 1:100,000 and hydro-geological cross sections close to the case study WCA were used to extract hydro-geological input data needed for FEFLOW. In the simulation area, the depth of the phreatic aquifer was 48 m, and the elevation of the bottom barrier (the impermeable layer) of the aquifer was 50 m above the sea level (asl). The thickness of the top layer (loam to sandy loam) was 1.5 m over a 31.5 m deep sandy layer. Between the sandy layer and the impermeable layer, the thickness of the sandstone layer was 15 m. Based on this layer information, the vertical discretization corresponded to 3 layers and 4 slices.

3.2.3.3 Boundary conditions

The domain of the model (Figure 3.2.3) was delineated by irrigation canals (Zey-Yop and Polvon) at the northern and eastern border and by collectors (Sapcha and South). The water levels in the canals and collectors were used as a boundary condition (Dirichlet-type).

Preliminary observations illustrated an impact of the drainage system on the GW flow, which was introduced in the domain of the model also. The digitized map of the drains and collectors was prepared in ArcGIS and imported into the FEFLOW model. The 1st kind of Dirichlet-type boundary condition was used with the same procedure as for the surrounding canals and collectors. The water table was derived from information on bottom elevation at characteristic points in the drainage system in combination with data regarding water depth on a monthly basis.

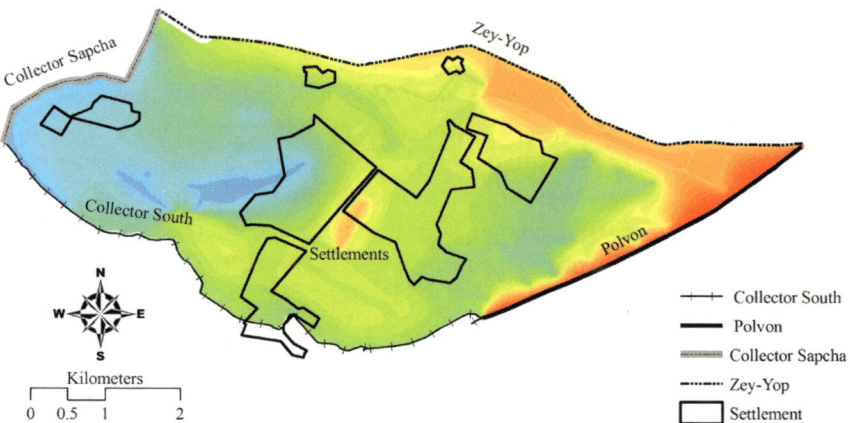

Figure 3.2.3: Simulated groundwater levels in the case study WCA delineated by canals and collectors (levels are given as absolute heights; dark red color indicates high levels with a maximum of 96.45 m asl and dark blue color represents low levels with a minimum of 90.17 m asl)

3.2.3.4. Water balancing approach and the irrigation efficiency concept for estimating the groundwater recharge

Recharge rates are major input required for the FEFLOW model. Recharge rates at field level depend on factors e.g., climatic conditions, soil texture, cropping pattern and GW tables. These factors can vary significantly in any irrigation canal command area. Thiessen polygons (on the available point data of GW wells and soil texture) were drawn in ArcGIS to capture the spatial variability of GW tables and soil texture, which determine the capillary rise and in turn, the water balance and the recharge. Areas of similar GW and soil texture in the WCA were defined as hydrological response units (HRUs). Satellite remote sensing data was used for the land-use classification in the area (Awan 2010).

The recharge rates were determined in a two-step procedure. In a first step, the water balance approach and the irrigation efficiency concept were applied to each HRU (Awan 2010). The water balance at field level was completed by considering evapotranspiration, capillary rise and effective precipitation from which the net irrigation amount and the irrigation timing could be derived. Based on the field application efficiency, network efficiency, the gross irrigation amount was estimated. Next the recharge was determined as the difference between gross and net irrigation amounts subtracting the evaporation component of the overall losses. In a second step, recharge rates were upscaled from field to HRU and then to WCA.

3.2.4 Results

3.2.4.1 Result 1: Spatial mapping of groundwater table by FEFLOW

FEFLOW has the capacity to capture the spatial variability of the GW tables which assists in planning of the water resources within an irrigation scheme. For example, the map of GW tables simulated by FEFLOW before the peak irrigation season is presented in Figure 3.2.3.

The map shows the usual trend of GW dynamics, i.e., GW tables are shallower around the main irrigation canals Polvon and Zey-Yop and deeper around the collectors. At the junction point of these canals, the effect of seepage is even higher. At the junction where the Sapcha collector falls into the South collector, the GW table is quite deep. Deep GW tables around these collectors are due to exfiltration from the GW to these collectors. In the settlement areas of the study WCA, GW tables were expected to be deep due to low recharge rates. However, in reality these levels equaled those in the fields caused by the poor drainage since a drainage infrastructure in the settlements was lacking, due to the intensive irrigation of gardens, and the very small GW slopes (0.00026–0.00033, Katz 1977). The general slope of the GW table is from east to south, which is in line with the overall slope of the Khorezm region (and the gradient from Amu Darya River as major water source towards the Ozerny main collector).

3.2.4.2 Result 2: Modeled temporal behavior of the groundwater

To assess the performance of the integrated model, monthly GW tables simulated by FEFLOW were compared with empirical values. Out of the 15 observation wells, 10 were selected for the evaluation whereas 5 wells were excluded due to their location close to canals or drains and therefore not being representative for the overall GW. The monthly averaged simulated (standard deviation = 0.37) and observed (standard deviation = 0.12) GW tables during the study period (April to September, 2007) matched quite well (Figure 3.2.4).

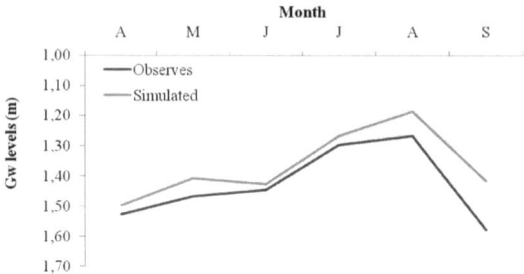

Figure 3.2.4: (Simulated and observed monthly averages of groundwater table: average from 10 wells over 6 months in 2008)

3.2.4.3 Results 3: Impact of raising irrigation efficiency on the groundwater table

The scenario simulations showed that improving the irrigation efficiency under the existing drainage system would lower the GW tables in August on average by 12, 38, and 44 cm in the scenarios respectively considering a theoretical overall irrigation efficiency of 36 %, 51 %, and 56 %.

3.2.5 Using groundwater information for water management

Detailed spatial information about GW tables and salinity levels (section 3.2.2.1) can be utilized to support water management towards site-specific solutions. The sites with shallow GW levels and lower salinity favor GW re-use or conjunctive use as a promising option for water saving. In contrast, shallow GW levels but with higher salinity indicate problematic areas requiring intervention. Analyzing the influencing factors on the spatio-temporal groundwater behavior revealed moderate and heavy texture (rather low hydraulic conductivity) as factors leading to shallow groundwater in the irrigation season. While considering interventions towards lowering the groundwater table either by improving the drainage system or reducing the groundwater recharge, respective areas could be used as starting-point.

Groundwater monitoring with high spatial and temporal resolution proved that heavy pre-seasonal leaching determines the groundwater table with a small delay only. In the irrigation season, rice cultivation has in addition a significant influence on the groundwater table of respective and neighboring fields. In such situations, often the dug-in irrigation canals perform then a drainage function.

Linking the GW model with an irrigation scheduling model can become an effective tool for water management. It would consider the capillary rise into irrigation schedules and describe the GW situation in response to the recharge by the current irrigation strategy. Furthermore the model allows assessing the impacts of raising irrigation efficiencies and on the lowering of the groundwater table (as an intervention by input-reduction).

References

Awan UK (2010) Coupling hydrological and irrigation schedule models for the management of surface and GW resources in Khorezm, Uzbekistan. Dissertation, University Bonn

Bobojonov I (2008) Modeling Crop and Water Allocation under Uncertainty in Irrigated Agriculture – A Case Study on the Khorezm Region, Uzbekistan. Dissertation, University Bonn

Clarke D, Smith M, El-Askari K (1998) CROPWAT for Windows: User Guide, University of Southampton

Diersch H-JG, Kolditz O (2002) Variable-density flow and transport in porous media: Approaches and challenges (Review paper). Adv. Wat. Res. 8(12): 899–944

Forkutsa I, Sommer R, Shirokova Y et al (2009) Modeling irrigated cotton with shallow GW in the Aral Sea Basin of Uzbekistan: I. Water dynamics. Irrigation Sci 27(4): 331–346

Ibrakhimov M, Khamzina A, Forkutsa I, Paluasheva G, Lamers JPA, Tischbein B, Vlek PLG, Martius C (2007) GW table and salinity: Spatial and temporal distribution and influence on soil salinization in Khorezm region (Uzbekistan, Aral Sea Basin). Irrig Drain Syst 21 (3–4): 219–236

Ibrakhimov M, Martius C, Lamers JPA, Paluasheva G, Lamers JPA, Tischbein B (2011) The dynamics of GW table and salinity over 17 years in Khorezm. Agr Water Management 101 (2011): 52–61

Katz D (1976) Influence of irrigation on groundwater (in Russian). M. Kolos

Rhoades JD, Kandiah A, Mashali AM (1992) The use of saline waters for crop production. FAO irrigation and Drainage paper 48, Rome

SIC-ICWC (2010) On-line data of the BWO Amudarya. Scientific Information Center – Interstate Commission for Water Coordination of Central Asia, (http://www.cawater-info.net)

Tischbein B, Awan UK, Abdullaev I, Bobojonov I, Conrad C, Forkutsa I, Ibrakhimov M, Poluasheva G (2012) Water management in Khorezm: current situation and options for improvement (hydrological perspective). In: Martius C, Rudenko I, Lamers JPA, Vlek PLG (Eds.) Cotton, water, salts and Soums – economic and ecological restructuring in Khorezm, Uzbekistan. Springer, Dordrecht pp 69–92

Akmal Akramkhanov, Bernhard Tischbein, Usman Khalid Awan

3.3 Effective management of soil salinity – revising leaching norms

Abstract

The effectiveness of the current soil salinity management, which mainly relies on pre-season leaching based on static norms, is limited despite the relatively high leaching water input (400–500 mm). The low leaching effectiveness is illustrated by the spatial distribution of soil salinity in a representative study area as monitored using the electromagnetic induction technique. Analyses of the soil salinity before and after leaching showed a shifting of salts from upper (20 cm) to lower layers rather than an effective removal of salts. To improve the leaching effectiveness, an approach is recommended that includes advanced monitoring tools, such as the non-destructive electromagnetic induction technique coupled with GPS and data logger and modeling of salt distribution within the soil profile. Due to a high working speed (70–80 ha per 1–2 days) and quick availability of results, the combination of monitoring and modeling tools provides high-resolution information that would allow timely and site-specific measures to cope with the temporally and spatially variable soil salinization.

Advanced leaching strategies demand an infrastructure for discharge dosage at the field level, which is not always and everywhere available. In addition, laser-guided land leveling facilitates a uniform application of appropriate amounts of leaching water as well as of irrigation water during the vegetation period. Overall, the simulation by the HYDRUS-1D model indicates that an improved management system has the potential to raise leaching effectiveness and to avoid salt accumulation in the root zone during the vegetation period.

Keywords: salinity monitoring, spatial distribution, leaching fraction, irrigation, interventions

3.3.1 Current situation

Soil salinization is a severe threat to sustainable crop production on irrigated lands. According to CISEAU (2006), approximately 10–15 % of the 280,000,000 ha irrigated worldwide is already impacted by salinization, and 0.5–1 % of the irrigated area is being lost each year. Irrigated agriculture in Central Asia in general and especially the schemes located at and fed by the lower reaches of the major rivers Amudarya and Syrdarya suffer from soil salinization. Khorezm region is one of the affected areas considered representative for the irrigated lowlands. Forkutsa et al. (2009) cited, for instance, the analyses by Ministry of Agriculture and Water Resources (MAWR, 2004) that classified 55 % of the irrigated lands in Khorezm, in the northwest of Central Asian Uzbekistan, as slightly saline (2–4 dSm^{-1}), 33 % as medium saline (4–8 dSm^{-1}) and 12 % as highly saline (8–16 dSm^{-1}) at the end of the vegetation period in 2004.

Besides the acknowledged spatial variability, soil salinization shows a distinct, often neglected, temporal behavior over the year as observed in Uzbekistan (Ibrahimov et al. 2007). Due to the arid climate, the field and even regional water dynamics in Khorezm are dominated by irrigation and drainage management activities. As a consequence, salt dynamics driven by water fluxes show a clear dependency on the water management. Forkutsa et al. (2009) simulated soil salinity dynamics with the HYDRUS-1D model (Simunek et al. 2005), which was parameterized using data from representative cotton fields in Khorezm on soils with sandy, sandy loam, and loamy textures, cultivated and managed by farmers. Fig. 3.3.1 depicts the soil salinity at characteristic times during the vegetation season in a 2-m layer divided into the root zone (0–0.8 m) and the layer below the root zone (0.8 to 2 m).

Typically, the pre-season leaching in March lowers the root zone salt contents, which increase again during the vegetation period. At the end of the vegetation period, the salt contents in the 2-m layer reach the same level as before leaching (fig. 3.3.1). The soil salinity in the layer below the root zone demonstrates the ineffectiveness of the present leaching activities: pre-season leaching more or less shifts the salts from the upper to the lower layer rather than removing salts from the entire 2-m layer. The salts pushed below the upper layer are later restored to the root zone due to the capillary rise. The reduction in soil salinity in the lower layer towards the end of the season is likely to be caused by the general lowering of groundwater tables from October onwards due to a cessation of the irrigation activities.

Input of salts into the root zone is caused by irrigation water and capillary rise from shallow groundwater. The salt content of the irrigation water provided by the Amudarya is in the range of 0.8–1 gl^{-1} (Ikramov 2004), whereas that of the groundwater averages 1.75 gl^{-1} (Ibrahimov et al. 2007). According to the FAO classification (Ayers and Westcot 1985), under such salinity levels a slight to moderate restriction on groundwater use should be exerted. Groundwater in Khorezm is shallow all the year round with a long-term annual average of 1.5 m (Ibrakhimov et al. 2007). The levels fluctuate between about 2.3 m below the ground at the end of the irrigation-free period in January, about 1.4 m after the pre-seasonal leaching in April, and about 1.2 m during the irrigation season in August. As a consequence, capillary rise from the shallow groundwater is a major driver of secondary soil salinization. Analyses from various locations showed that salt input from shallow groundwater with a salinity of 2.5 to 3.5 gl^{-1} was 40 % higher than that from irrigation water (Ibrakhimov et al. 2007).

Current soil salinity management strategies by farmers mainly include pre-season leaching. Depending on the coarse assessment of the soil salinity, which is classified as "low", "medium" or "high", the application of 3,000 – 7,500 m^3 ha^{-1} by 2 – 3 leaching events is recommended (Zaidelman 2003; SIC-ICWC 2004). On the regional scale, riverwater supply according to these static leaching norms accounts for around 25 % of the total annual water input of 4.5 km^3 in Khorezm (Tischbein et al. 2012). Considering such a large share of the leaching demand in the regional water consumption, the options for increasing water use efficiency need to include improved leaching practices that can be implemented by land managers and lead to water savings.

3.3.2 Shortcomings of current soil salinity management strategies

The high water input for pre-season leaching does not successfully deliver the desired effects given the limited decrease in soil salinity evident in a comparison of the soil salinity before and after leaching events. For this comparison, two approaches were used: (a) high spatial resolution monitoring with the electromagnetic induction meter EM (EM38[1], Geonics Limited, Canada) on a 30-ha land area divided into four fields (fig. 3.3.2), and (b) analysis of soil profiles.

(a) Monitoring the spatial distribution of soil salinity (represented as apparent electric conductivity EC_a) before and after leaching events was performed at the selected fields (numbered 21, 22, 23, 24) before and after leaching, and towards the end of the cotton vegetation period in 2008. The sensing depth of the EM in the vertical and horizontal mode allows monitoring salinity changes at 1.5 and 0.75 m soil depths. The EC_a measured by the EM is represented in mSm^{-1}. For loamy soils, values below 70 mSm^{-1} indicate non-saline soils, slightly saline soils range from 70 – 110 mSm^{-1}, and with values above 110 mSm^{-1} soils can be

---

[1] The mention of brand names does not include an endorsement of the authors and neither the institution

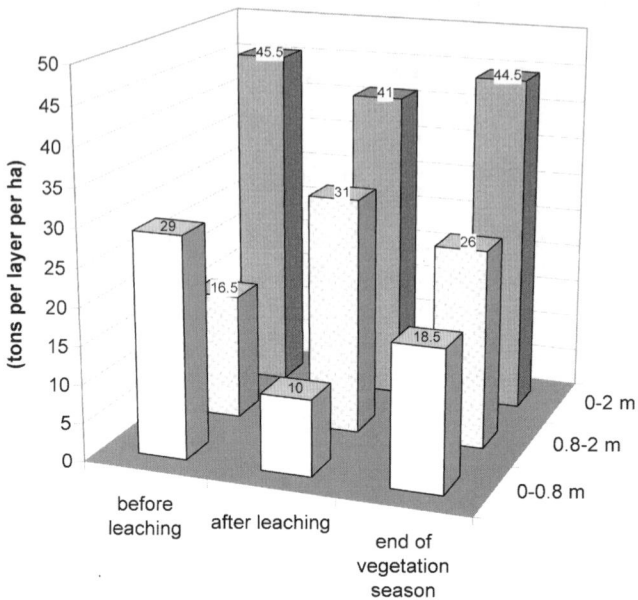

Figure 3.3.1: Typical annual course of soil salinity in Khorezm (modified data from Forkutsa et al. 2009)

considered as moderately saline. Readings above 150 mSm^{-1} indicate highly saline soils.

The monitoring results show that leaching events mainly diluted the salts in the 0.75 m soil layer. Moreover, under the present practices applied by land managers, soil salinity increased towards the end of the cotton vegetation period (August), indicating the movement of salts back to the upper layer. Field 22 was not leached, and showed an increased average soil salinity level in the 1.5 m layer after the leaching events on the neighboring fields. This could be attributed to a migration of dissolved salts from these fields and to the rise of the groundwater table.

(b) The impact of leaching on the vertical distribution of soil salinity was analysed by examining soil profiles in a farmer-managed field (about 7 ha). This field (No. 21; fig. 3.3.2) was selected because of its representativeness regarding size, soil, irrigation and drainage infrastructure. Soil salinity, expressed by the electrical conductivity of the saturation extract (EC_e), was monitored before and after leaching. A lowering of the EC_e (and thus of soil salinity) was observed in the upper soil layer (0–45 cm) only, whereas the EC_e did not change in the 45–75 cm layer, and even increased in the deeper soil layers (fig. 3.3.3).

Although a total 510 mm of leaching water had been applied during two

Effective management of soil salinity – revising leaching norms

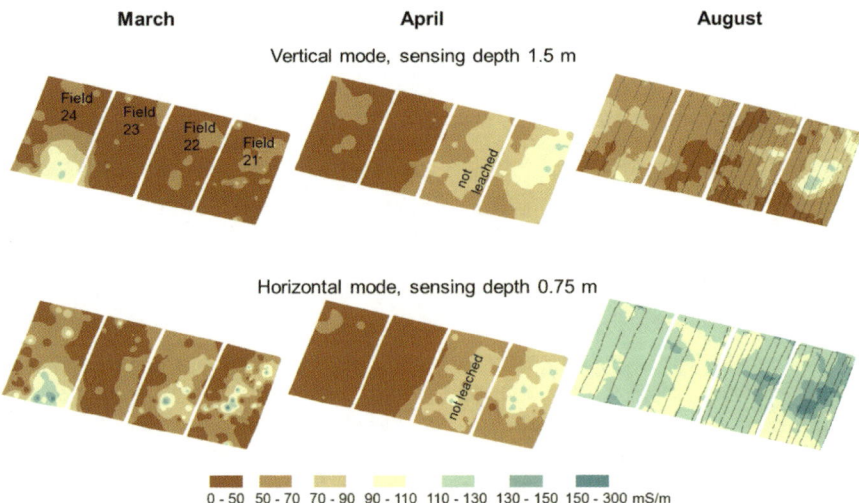

Figure 3.3.2: Soil salinity before (March) and after leaching (April), and towards the end of the cotton vegetation period on four selected fields (numbered 21, 22, 23, 24)

events by the farmer, leaching did not affect soil salt content in the entire crop root zone. This observation is in line with previous model simulations (Forkutsa et al. 2009), suggesting rather a relocation of salts from upper to lower layers than their removal from the profile. Furthermore, the observed shifting of salts within the soil profile supports the results of the EM monitoring, which also showed only a small change in soil salinity at a 1.5 m depth.

The low effectiveness of the current leaching practices also becomes obvious when using the FAO concept of leaching fractions (Ayers and Westcot 1985). Considering, for instance, 400 mm leaching water and assuming that 100 mm of this water is required to raise the actual soil moisture to field capacity before the percolation can be initiated for leaching, and taking into account the seasonal evapotranspiration of 800 mm (Conrad et al. 2007), the actual leaching fraction would be in the range of 35–40 %. This is far above the fraction of 7 % that is needed to avoid salt accumulation at levels exceeding the salt tolerance of cotton according to the FAO approach parameterized for the conditions in Khorezm (Awan et al. 2011).

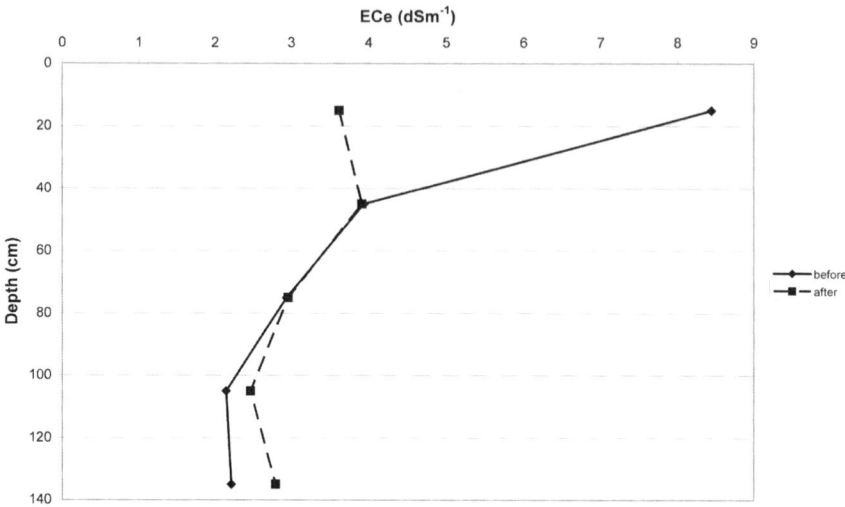

Figure 3.3.3: Vertical profile of soil salinity (expressed by ECe) before and after pre-season leaching on a farmer-managed field

3.3.3 Integrated management of soil salinity

Although the reasons for the currently low effectiveness of the salt management strategies are manifold and interdependent, four major causes can be identified: 1) the current strategies have a strong focus on the pre-season leaching, 2) the amount of leaching water is determined by static norms, 3) a correct dosing of leaching water is hindered due to the missing infrastructure for the discharge control at field level, and 4) the spatial distribution of leaching water within croplands is non-uniform mainly because of insufficient land leveling. Tackling these issues calls for an approach integrating the interventions in the leaching period with those in the salt accumulation (vegetation) period, including the leaching fractions to hinder re-salinization. Furthermore, combining advanced monitoring methods with modeling would allow site-specific and timely measures.

The suggested approach is based on different components:

(a) Framework: Considering the relevant processes in salt accumulation and adjusting the leaching periods in order to reach effective salt management, i.e., lowest possible input of leaching water, which avoids salt stress on crops.

(b) Tools: Capturing the relevant processes of salt leaching and salt accumulation by appropriate modeling of the hydrological processes including

parameters that represent the influence of management interventions.
Advanced monitoring with high spatial and temporal resolution to support modeling and mapping of the near-time situation, thus enabling timely and site-specific corrective interventions.
(c) Interventions: Deriving interventions with respect to leaching and accumulation processes based on model simulations and detailed monitoring.
(d) Prerequisites and context of overall water use (irrigation and leaching): Seeking for win-win-situations by introducing measures to improve leaching effectiveness and irrigation performance.

3.3.4 Advanced monitoring of soil salinity

At present, soil salinity assessment in Uzbekistan is based on laborious soil sampling in representative areas, with subsequent laboratory analyses – a process that is very resource and time consuming. Traditional soil salinity analysis determines the amount of total dissolved solids in a given soil sample. Consequently, maps of soil salinity are only available after long delays and are location specific (non-spatial), which limits their usefulness. Surveys conducted with devices such as the EM allow the identification of fine-scale spatial variation of salinity with nearly continuous measurements. Further advantages of the EM are: (i) rapid and non-destructive measurements, (ii) suitability for monitoring purposes, (iii) suitability for fine- and large-scale mapping, (iv) low cost per land area unit, (v) possibility of logging and geo-referencing when coupled with a GPS, (vi) can be mounted on a vehicle for mobile and fast mapping of large areas, and (vii) successful testing and adaptation for the conditions in Uzbekistan. Furthermore, the EM device can be managed by one operator, who can survey an area of 70–80 hectares in 1–2 days only, depending on the land cover and the density of transects within a field. The rapidity of the measurements depends on the walking speed of the operator or the speed of the mobile system.

A disadvantage of the EM device is its dependence on soil moisture; soils should not be very dry during measurements, ideally around 20 % moisture content, below which the effect is unclear according to Bennet et al. (1995); however, field studies by Hendrickx et al. (1992) indicate that the effect of soil moisture on EM readings is negligible after 3–8 days after irrigation. Since EM readings provide an average EC_a value of the soil profile, a calibration for soil salinity interpretation is needed for different soil textures.

3.3.5 Detailed modeling

The presently used static irrigation and leaching norms only allow an appropriate determination of leaching amounts if the situation is comparable to the settings for which the norms were established earlier (MAWR 1975). The disadvantage of these static norms is that they do not consider the high spatial and temporal variability of soil salinity. As argued above, monitoring of soil salinity by the EM can overcome these shortcomings. Furthermore, the static norms do not allow the simulation of a variable, complex environment, that considers the impact of interventions besides leaching, such as irrigation water management. In contrast, a modeling tool is capable of describing and simulating the interlinked processes of salt and water dynamics under various management interventions.

The HYDRUS-1D model (Simunek et al. 2005) is a tool to quantify water and solute fluxes with high temporal and spatial (vertical) resolution based on detailed deterministic approaches. Most importantly, the temporal course of soil salinity can be described by HYDRUS-1D depending on biophysical conditions of the site (i.e., soil characteristics, evapotranspiration, water and salt input by precipitation, groundwater levels, groundwater salinity) and under the influence of management interventions, e.g., of leaching/irrigation amount and timing, of crop selection on transpiration and surface residue on evaporation, and of drainage on groundwater level. The simulated course of soil salinity can be combined with crop-specific and time-depending salt tolerance and yield-response functions (Ayers and Westcot 1985). This allows working out appropriate salt management strategies to avoid salt stress or to minimize the impact of salt stress on the yield.

Forkutsa et al. (2009) parameterized, validated and applied the HYDRUS-1D model to three sites in Khorezm to assess the impact of alternative management options on soil salinity and amount of leaching water used. These simulation scenarios included application of a crop residue layer (to reduce soil evaporation), lowering the groundwater table to 2 m, re-scheduling leaching and irrigation (for saving the non-effective leaching water fractions and avoiding water stress in the vegetation period) and a combination thereof (improved system). Based on these simulation results, the impact of the current strategy and the improved system (marking the maximum of achievable improvement in the given situation) could be compared for the cotton root zone of 0–0.8 m (fig. 3.3.4a) and the soil layer down to 2 m (fig. 3.3.4b). Regarding the root zone, the improved approach has the potential to increase leaching effectiveness: soil salinity decreased from 28.5 to 8 tons per layer per ha compared to 29 to 10 tons per layer per ha by the current strategy. With the combined interventions (improved system), the currently common salt accumulation towards the end of the vegetation period (8–8.5 tons per layer per ha) could be almost completely avoided. There was an overall decrease in salt content under the improved system given the balanced situation in the entire 0–2 m layer (fig. 3.3.4b). Furthermore, the application of HYDRUS-1D to the typical setting in Khorezm on a sandy-loam field revealed a water saving potential of 25 % during the pre-season leaching by replacing the currently common three leaching events by two effective events, without compromising the overall effect of leaching (Forkutsa et al. 2009).

Currently, approximation of the salt leaching and accumulation driven by percolation and capillary rise, which are processes mainly in the vertical direction, using HYDRUS-1D is seen as a first step when moving from norms to more detailed approaches. A potential improvement of the applied modeling tool will be an incorporation of the lateral component using HYDRUS-2D, particularly when considering the system of furrow irrigation.

3.3.6 Recommendations for improving the current strategy

The results of the monitoring and modeling performed in Khorezm and related findings make it possible to derive options for action to improve the effectiveness of soil salt management in several steps.

When aiming to raise the effectiveness of leaching practices:
- The use of modeling for describing the soil water and salt dynamics allows compiling site-specific leaching schedules, which lead to a reduction in leaching water requirements to about 75 % of the presently used amounts.
- Introducing time-depending and crop-specific salt tolerance levels of cultivated crops after leaching allows simulating and conceiving site-specific leaching strategies (timing, amount), which in turn prevent over- as well as under-application of leaching water.
- Improved monitoring tools such as EM devices enable mapping of the spatial distribution of soil salinity with close time steps (e.g., before and after leaching), which in turn supports immediate, corrective responses to fields or basins in the vegetation season according to the detected impact of leaching. Similarly, an effective monitoring can be applied in the vegetation season (especially in critical periods, for sensitive crops and locations prone to salinization) and, in case needed, near-time reactions can be conceived (raising irrigation depth to achieve a leaching effect).

When aiming at lowering salt accumulation and improving irrigation performance:
- Through conservation agriculture practices (Egamberdiev 2007; Tursunov 2009; Devkota 2011; Pulatov et al. 2012) or their components, actual evapo-

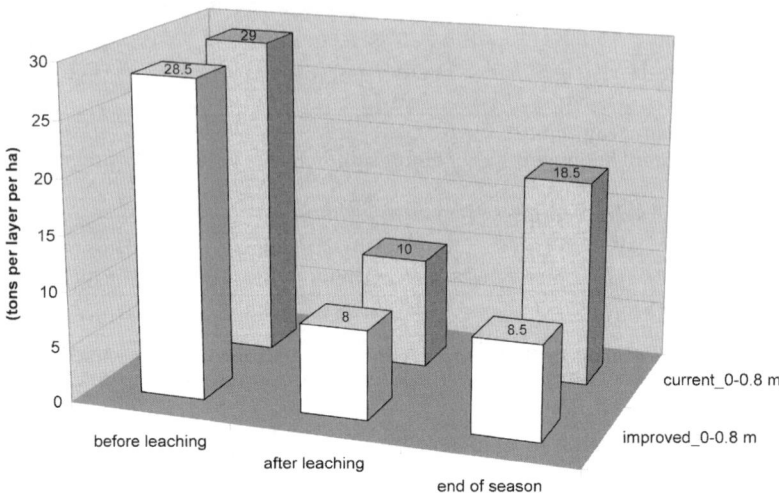

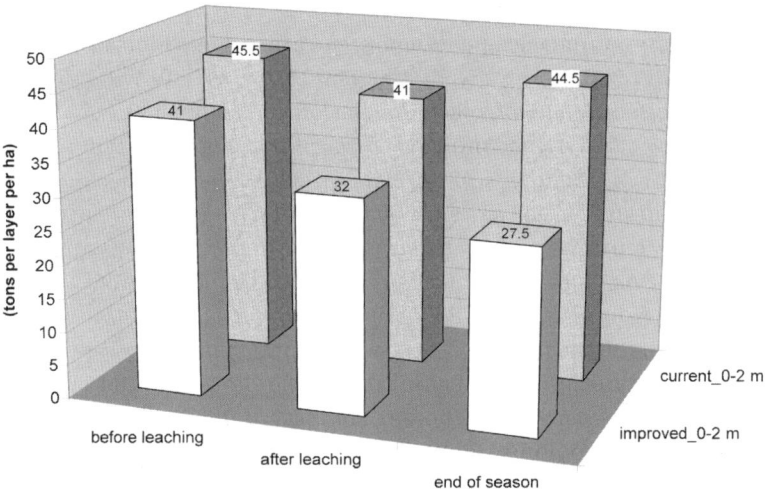

Figure 3.3.4a, b: Impact of improved management system (combined application of crop residues, improved drainage conditions, and re-scheduling of leaching and irrigation events) on the salt balance of the root zone 0 – 0.8 m (fig. 3.3.4a: top diagram) and the soil layer down to 2 m (fig. 3.3.4b: lower part) (modified data from Forkutsa et al. (2009))

ration could be reduced by 50 % of the potential level, which in turn lowers salt accumulation by 15 – 20 % during cotton growth under the typical conditions of Khorezm (according to the HYDRUS-1D simulation).
- Improving drainage to lower the groundwater table to 2.0 m can minimize capillary rise and hence reduce salt accumulation in the upper 2 m by 25 –

30 % (simulation by HYDRUS-1D); as a consequence, irrigation strategies can be modified to compensate for reduced water input via capillary rise into the root zone.
- When increasing the uniformity of within-field water applications (e.g., by effective land leveling), percolation losses can be utilized for leaching purposes.
- As the introduction of water dosage infrastructure at field level and the implementation of options to homogenize water application are also the prerequisites for the urgently needed improvement of irrigation efficiencies (Forkutsa et al. 2009; Tischbein et al. 2012), there is a win-win-situation for the realization of these combined measures.

The above-described combined approach of monitoring and modeling tools provides information that makes it possible to react to temporally and spatially variable soil salinization by implementing timely and site-specific measures. Yet, with respect to the implementation of the described improved leaching strategies, an adequate infrastructure for discharge dosage at field level is an important prerequisite. The establishment of sophisticated measures (e.g., flumes, weirs) does not seem achievable in the near future. However, the introduction of siphons or accounting for the hydraulic characteristics of the pipes currently in use at field level can be suitable to estimate discharges. As a typical win-win-situation, the discharge dosage infrastructure supports optimization of the irrigation water application in the vegetation period and volumetric water pricing (Djanibekov et al. 2012). However, the introduction of water fees as a stand-alone incentive for irrigation and leaching water savings is likely to increase costs and hence decrease income generation by producers, and should therefore be flanked by additional measures (Djanibekov et al. 2012). Akramkhanov et al. (2010) also suggested a system of boni and mali on taxes to support the implementation of measures to achieve both water saving and salinity control (Table 3.3.1).

Table 3.3.1: Potential scheme of incentives and disincentives for combined water and land management (Akramkhanov et al. 2010).

	Farmers without increase in water use efficiency	Farmers who increase water use efficiency
Farmers who increase soil salinity	Increased taxation	Reduced tax bonus
Farmers who maintain good land conditions or decrease soil salinity	Medium tax bonus	High tax bonus

3.3.7 Outlook

Introduction of advanced soil salinity monitoring (EM) and modeling (HYDRUS-1D) tools requires financial and human resources. For example, implementation costs of EM tools are estimated in the range of USD 15,000 – 20,000. Given the current farm sizes (approximately 100 ha), the lack of incentives for saving irrigation water, and the ongoing reforms in the agriculture sector, adoption by the farmers will be difficult due to the costs involved and the lack of human resources. Regional branches, Water Consumers Associations, and Farmer Councils, which cover large areas, are seen to be the organizations most suitable for handling the implementation of salinity monitoring tools, as this would mean improved services provision. Whereas modeling tools are suitable for research and education organizations, as they require data input, time to run and interpret results and, more importantly, a scientific background to derive recommendations, in the long run they could become more user friendly and accessible to a wider public.

However, despite the fact that the tools potentially have great benefits for the end user, the introduction of such innovations will require policy support in nationwide dissemination and financial instruments to promote the use of the tools among different stakeholder levels.

References

Akramkhanov A, Ibrakhimov M, Lamers JPA (2010) Managing soil salinity in the lower reaches of the Amudarya delta: How to break the vicious circle. In: Pinstrup-Andersen P and Cheng F (Eds.) Case Studies in Food Policy for Developing Countries, Volumes 1, 2, and 3. (Case Study 8 – 7). Cornell University Press, Ithaca, NY

Awan UK, Ibrakhimov M, Tischbein B, Kamalov P, Martius C, Lamers JPA (2011) Improving irrigation water operation in the lower reaches of the Amu Darya River – current status and suggestions. Irrigation and Drainage 60 (5): 600 – 612

Ayers RS, Westcot DW (1985) Water quality for agriculture. FAO Irrigation and Drainage Paper 29. Rome

Bennett DL, George RJ, Ryder A (1995) Soil Salinity Assessment Using the EM38: Field Operating Instructions and Data Interpretation. Miscellaneous Publication 4/95: Department of Agriculture, Western Australia

CISEAU (2006) Irrigation and salinization relationships. www.ciseau.org/servlet/CDSServlet?status=ND0xMDYzLjU0ODYmNj1lbiYzMz1yZXNvdXJjZSZzaG93Q2hpbG-RyZW49dHJ1ZSYzNz1pbmZv#koinfo. Cited 27 Sep 2010

Conrad C, Dech SW, Hafeez M, Lamers JPA, Martius C, Strunz G (2007) Mapping and assessing water use in a Central Asian irrigation system by utilizing MODIS remote sensing products. Irrigation and Drainage Systems 21 (3): 197 – 218

Devkota M (2011) Nitrogen management in irrigated cotton-based systems under conservation agriculture on salt-affected lands of Uzbekistan. PhD Dissertation, Bonn University

Djanibekov N, Bobojonov I, Djanibekov U (2012) Prospects of agricultural water service fees in the irrigated drylands, downstream of Amudarya. In: Martius C, Rudenko I, Lamers JPA and Vlek PLG (Eds.) Cotton, water, salts and Soums – economic and ecological restructuring in Khorezm, Uzbekistan. Springer, Dordrecht

Egamberdiev O (2007) Changes of soil characteristics under the influence of resource saving and soil protective technologies within the irrigated meadow alluvial soil of the Khorezm region. PhD dissertation, Tashkent

Forkutsa I, Sommer R, Shirokova YI, Lamers JPA, Kienzler K, Tischbein B, Martius C, Vlek PLG (2009) Modeling irrigated cotton with shallow groundwater in the Aral Sea Basin of Uzbekistan: II. Soil salinity dynamics. Irrigation Science 27 (4): 319–330

Hendrickx JMH, Baerends B, Raza ZI, Sadiq M, Chaudhry MA (1992) Soil salinity assessment by electromagnetic induction on irrigated land. Soil Sci Soc Am J 56: 1933–1941

Ibrakhimov M, Khamzina A, Forkutsa I, Paluasheva G, Lamers JPA, Tischbein B, Vlek PLG, Martius C (2007) Groundwater table and salinity: Spatial and temporal distribution and influence on soil salinization in Khorezm region (Uzbekistan, Aral Sea Basin). Irrigation and Drainage Systems 21 (3–4): 219–236

Ikramov RK (2004) Present salinity and drainage condition, and salinity control measures in Uzbekistan. Proceedings of the meeting on "Development of the salinity, land degradation and drainage-waste water reuse research program for Central Asia". IWMI, Tashkent, May 11–17, 2004

MAWR (2004) Annual report of Amu-Darya basin hydrogeologic melioration expedition, Khorezm region. Ministry of Agriculture and Water Resources (MAWR), Tashkent (in Russian)

MAWR (1975) Interim recommendations for leaching. Uzbekistan (in Russian). Ministry of Agriculture and Water Resources

Pulatov A, Egamberdiev O, Karimov A, Tursunov M, Kienzler S, Sayre K, Tursunov L, Lamers JPA, Martius C (2012) Introducing conservation agriculture on irrigated meadow alluvial soils (Arenosols) in Khorezm, Uzbekistan. In: Martius C, Rudenko I, Lamers JPA and Vlek PLG (Eds.) Cotton, water, salts and *Soums* – economic and ecological restructuring in Khorezm, Uzbekistan. Springer, Dordrecht

SIC–ICWC (2004) Drainage in the Aral Sea Basin: Towards a strategy of sustainable development. Scientific Information Center–Interstate Coordination Water Commission. Progress report. Tashkent, Uzbekistan

Simunek J, Van Genuchten MT, Sejna M (2005) The HYDRUS-1D software package for simulating the one-dimensional movement of water, heat, and multiple solutes in variably-saturated media. Version 3.0, HYDRUS Software

Tischbein B, Awan UK, Abdullaev I, Bobojonov I, Conrad C, Forkutsa I, Ibrakhimov M, Poluasheva G (2012) Water management in Khorezm: current situation and options for improvement (hydrological perspective). In: Martius C, Rudenko I, Lamers JPA and Vlek PLG (Eds.) Cotton, water, salts and *Soums* – economic and ecological restructuring in Khorezm, Uzbekistan. Springer, Dordrecht

Tursunov M (2009) Potential of conservation agriculture for irrigated cotton and winter wheat production in Khorezm, Aral Sea Basin. PhD dissertation, Bonn University

Zaidelman FR (2003) Reclamation of soils. Moscow University Press (in Russian)

Alexander Tupitsa, John P. A. Lamers, Asia Khamzina, Evgeniy Botman, Martin Worbes, Christopher Martius, Paul L. G. Vlek

3.4 Adaptation of photogrammetry for tree hedgerow and windbreak assessment in the irrigated croplands of the Khorezm region

Abstract

In the irrigated cropping areas in Central Asia, tree hedgerow systems were established during the Soviet era for protecting the bordering croplands from wind erosion and for improving their microclimate, thus increasing yields. Due to a lack of a suitable, cost-effective methodology in the current transition period of the Central Asian countries (CAC), little information is available about the condition of the tree hedgerows and their functioning as windbreaks. Using photogrammetry combined with geographic information systems and field surveys, the spatial distribution and windbreak function of the existing hedgerows were assessed with an overall accuracy of 85 % in Khorezm, Uzbekistan, a region representative for the irrigated lowlands in CAC. The share of the hedgerows within the cropland was lower than the nationwide recommended 1.5 %. The dominating single-species hedgerows consisted mainly of Morus alba L., Salix spp. and hybrid Populus spp., which were all concurrently used for production purposes. Therefore, the height, a primary windbreak criterion, was generally below 5 m, signifying an insufficient sheltering of the adjacent croplands. Up to 50 % of the hedgerows did not reach the edges of the adjacent fields. The orientation to the prevailing winds, crown closure, width and vitality of the hedgerows were, however, found to be satisfactory. The results indicate (i) the suitability of the photogrammetric procedure developed for assessing the tree hedgerows in the irrigated lowlands of CAC, and (ii) the necessity of extending the existing hedgerows based on appropriate design and maintenance and harvesting techniques to improve the windbreak function in these agroforestry systems in the irrigated croplands.

    Keywords: *windbreak functionality, Khorezm, aerial photographs, spatial distribution, Uzbekistan*

3.4.1 Introduction

Desertification, which has already impacted over 25 % of the earth's surface, is one of the greatest environmental concerns of this millennium (UNCCD 2005). Desertification caused by wind erosion is a persistent problem for agriculture worldwide, in particular in the arid and semi-arid regions. In Australia, for instance, more than a half of the cropland is at risk of sand storms (Hamblin,Williams 1996). In Uzbekistan, Central Asia, over an area of 2,000,000 ha of irrigated cropland, about 80 t ha^{-1} of topsoil is annually lost owing to wind erosion (FAO 2003). Furthermore, strong winds affect agriculture by decreasing the humidity of air and soil, scattering weed seeds, sandblasting the fields, and filling the irrigation canals and drainage collectors with sand and vegetation debris (Kayimov 1993). From 1960 to the 1980s, tree windbreak plantings in Uzbekistan were introduced on about 40,000 ha of the irrigated cropland to protect the cropped fields and reduce desertification at the margins of the irrigated area, and on 1,400,000 ha of steppe to protect the pasture land (Molchanova 1980). After the break-up of the Soviet Union in 1991 and transformation of the state agricultural institutions into private farms, the planting practices ceased, and windbreaks and shelterbelts have been largely cut down or have perished due to a lack of maintenance (MFD 2008).

Forest inventories, regularly conducted during Soviet times, focused mainly on natural forests and did not assess the functioning of the tree windbreaks. Since the break-up of the Soviet Union, no inventory has been conducted in Uzbekistan. Thus, the presently available information on tree windbreaks in Uzbekistan is out of date, precluding an objective decision making with regard to their maintenance and further extension. The lack of a suitable methodology still is a chief obstacle to obtaining reliable information on spatial extent and functionality of the existing tree windbreaks.

The use of high-resolution satellite images for a forest inventory in Uzbekistan would mean immense costs, which are presently unaffordable for the forestry administration. Photogrammetry in combination with geographic information systems (GIS) has shown its value and accuracy for forest inventories worldwide (Akça 2000). In Uzbekistan, aerial photographs are regularly taken for updating topographic maps, but they have not yet been applied in forestry assessments.

In this study, the windbreak function of tree hedgerows in the agricultural landscape was assessed with the photogrammetric technique in combination with GIS-based tools and field surveys in the Khorezm region of Uzbekistan, covering an area of 650,000 ha in the lower reaches of the Amudarya river. Due to the vegetation (crop) cover virtually throughout the growing season (Conrad 2006), and the prevailing wind speed in Khorezm of only 1 – 6 m s^{-1} (Glavgidromet 2003), the potential for wind-induced erosion is low. But in the extreme continental climate (Glazirin et al. 1999) even slight changes in the microclimate of the cropped fields, such as increasing air humidity and decreasing air and soil temperatures through tree windbreaks, can improve growing conditions for the adjacent crops and lead to crop yield increases and water saving (Botman 1986; Kort 1988; Abel et al. 1997).

The potential effectiveness of tree windbreaks was assessed based on generally acknowledged primary and secondary criteria, which include height, orientation towards prevailing winds, porosity, length, width, and species composition of the windbreaks (Botman 1986; Bolin, et al. 1987; Abel et al. 1997). The extent and distribution of tree windbreaks in the (irrigated) agriculture-dominated landscape as well as tree vitality and age are further recommended for an overall assessment (Kayimov 1993). Using these criteria, our study aimed: (i) to determine the feasibility of the photogrammetric methodology for assessing the tree hedgerows and windbreaks in the lower reaches of the Amudarya river in the Khorezm Region of Uzbekistan, (ii) to generate a calibration dataset for a rapid and inexpensive assessment of spatially distributed hedgerows in the irrigated croplands based on the example of Khorezm, and (iii) to assess the windbreak function of the identified hedgerows.

3.4.2 Materials and methods

3.4.2.1 Study area

As part of the Central Asian semi-desert zone, the Khorezm region (41 – 42°N latitude and 60 – 61°E longitude) is surrounded by the Kizylkum and Karakum deserts. The mean annual temperature is about 13 °C, the mean temperature during the coldest month January is approximately -2 °C with absolute daily minimums as low as -28 °C. Summers are hot, with mean monthly temperatures in July reaching about 30 °C and daily extremes as high as 47 °C. The long-term average annual precipitation is 100 mm, in some years amounting to only 30 – 40 mm (Glazirin et al. 1999). Most precipitation falls in November-March, which is outside the crop-growing season. The annual potential evapotranspiration amounts to about 1,500 mm. The relief of Khorezm is mostly flat with elevations varying between 122 and 128 m (Katz 1976). The soils are mainly loamy and clayey-loamy, but are highly stratified, mixed with sandy and sandy-loamy soils (Ibrakhimov 2005) and are potentially susceptible to wind erosion (Molchanova 1986). Threshold winds (3 – 6 m s^{-1}) and erosive winds (>6 m s^{-1}) make about 15 and 1 % of the annual wind activity, respectively, and can come from north-south and east directions throughout the year in the south of Khorezm and from different directions in the north (ZEF/UNESCO Khorezm Project 2006).

3.4.2.2 Photogrammetric assessment

Photogrammetry was selected for the inventory of the hedgerows in Khorezm using aerial photographs available from national governmental agencies. Two transects were delineated to cover all typical land-use and land-cover classes in the study region. These transects crossed the region in north-south (NS; 320 km^2) and west-east (WE; 230 km^2) directions and covered in total ca. 10 % of Khorezm.

Criteria and classes for assessing the tree windbreaks included species composition, orientation to the prevailing winds, crown closure (as a proxy for porosity), height, length (reaching the edges of the related field), width (number of rows), vitality (disturbance by pests and diseases) and age (Annex 1 and 2). Given the variability of the agricultural landscape in the study region, the extent of the hedgerows/windbreaks was assessed in three systematic land-use classes based on the crop pattern defined by Conrad (2006): flooded (rice fields), typical (other crops) and pre-desert (other crops at a distance of up to 1 km from a desert) (Annex 1).

The photogrammetric equipment consisted of a VISOPRET 10-DIG[1] analytical stereo plotter and AutoCAD software. Prior to the analyses, 70 stereo pairs of aerial photographs (black and white, scale 1:20,000, taken in November 2001) were prepared for the consecutive procedures of interior, relative and absolute orientation. ESRI ArcGIS 9.0 and ESRI ArcView 3.2 software and their extensions were used for the analyses.

The photogrammetric technique included ocular photo-interpretation with a plotter to identify hedgerows/windbreaks and tree species according to two developed tree-thematic classes (i) pollarded *Morus alba* L. and *Salix* spp., and (ii) hybrid *Populus* spp., *Salix* spp. and *Elaeagnus angustifolia* L. hedgerows (Table 3.4.1). These classes were developed from the photographic indicators in two training areas. The final species identification was completed in the field. Following the identification of each hedgerow/windbreak, its polygon was plotted and an output file (digitized polygons) was produced. With the GIS-based tools, the area of the digitized polygons and their orientation angle were computed to the nearest 0.01 ha and the closest 1°, respectively.

Ocular estimates of hedgerow/windbreak crown closure (to the nearest 0.05 coefficient) were carried out with the plotter. The stereoscopic photogrammetric measurements with the plotter established the mean hedgerow/windbreak height (to the closest 0.5 – 1.0 m), which was determined for three randomly selected representative points or trees in each of two 2.5 % segments based on

---

[1] The mention of any brand names does not indicate an endorsement by the authors or institution

Adaptation of photogrammetry for tree hedgerow

Table 3.4.1: Photographic indicators for tree hedgerows/windbreaks in Khorezm, Uzbekistan

Thematic class	Photographic indicator			
	Horizontal crown projection	Crown shadow profile	Site condition	Minimal area (ha)
Pollarded *Morus alba* and *Salix* spp.	Line-shaped and clumped (individual crowns or rows are unrecognizable) projection corresponding to single, double or triple rows within a line-shaped area	Poorly visible profile due to short length	Along irrigation network and bordering agricultural fields	≥0.1
Populus spp., *Salix* spp. and *Elaeagnus angustifolia*	Similar to above thematic class	Visible long and clumped profile (individual crowns/rows are unrecognizable)	Similar to above	Similar to above

"best" and "worst" appearance of crown closure in each hedgerow/windbreak. This procedure was preferred to a subjective selection of an "average" segment for each hedgerow/windbreak or to resource-consuming measurements of multiple sample points. Finally, the edges of each hedgerow/windbreak horizontal crown projection and the edges of the related field were compared during the photo-interpretation.

Photogrammetrically derived hedgerow/windbreak assessment data were verified on a randomly selected 10 % sample of hedgerows/windbreaks. Mean height was measured with a Suuto PM-5/1520 optical height meter and a telescopic aluminum rod for trees above and below 5 m height, respectively. The reaching of the hedgerow/windbreak to the edges of the related field was visually estimated in the field, comparing the edges of the hedgerow/windbreak and the edges of the related field.

Hedgerow/windbreak area and orientation to the prevailing winds were measured with GIS tools. The data on wind speed and direction for assessing the orientation of hedgerows/windbreaks to the prevailing winds were taken from secondary sources (database of the ZEF/UNESCO Khorezm Project 2006). Species composition, number of rows, vitality, and age classes were estimated with photogrammetry combined with field surveys. A 10 % random sample from a large number of the hedgerows/windbreaks was applied. Species composition and number of rows at a stand level were visually identified. Vitality was visually assessed at an individual tree level of all species within each of two 2.5 %

representative segments based on the "best" and "worst" appearance of crown closure of each hedgerow/windbreak. Tree age was determined by counting annual rings on stem samples collected with the Suunto 300 mm increment borer from the same sampled areas as used for the vitality assessment. More details on methodology are provided in Tupitsa (2009).

3.4.3 Results

3.4.3.1 Verification of photogrammetric technique

The overall accuracy of 85 % of the differentiation of hedgerows/windbreaks into the tree thematic classes by photo-interpretation was satisfactory, i.e., pollarded *M. alba* and *Salix* spp. versus *Populus* spp., *Salix* spp. and *E. angustifolia*. The photo-interpretation of species composition within the classes was not feasible due to minor or no differences within the photographic indicators, such as the clumped horizontal crown projections and crown shadow profiles of pollarded *M. alba* vs. pollarded *Salix* spp., and *Populus* spp. vs. *Salix* spp. or *E. angustifolia* (Annex 2). Another reason for the difficulty in distinguishing between species was that the aerial photographs available were from November when the trees had already shed their canopies. However, the horizontal crown projections and crown shadow profiles were visible through the appearance of branches within the line-shaped and clumped planting schemes applied for hedgerows/windbreaks (Annex 2). Also, the perspective view of the aerial photographs allowed an accuracy of ca. 90 % in the assessment of whether or not the hedgerow/windbreak reached the edges of the related field. Field verification of the hedgerow/windbreak assessment based on the area size and orientation angle to the prevailing winds was not required due the higher accuracy of the GIS-based measurements compared to that of field surveys.

The photogrammetrically derived mean hedgerow/windbreak height appeared to be overestimated by 11 % in comparison to the field-verified height (Table 3.4.2). In contrast, crown closure was underestimated by 11 %. The photo-interpretations in the "fair" height class and "fair" and "good" crown closure classes were particularly inconsistent with the results of the field surveys. Yet the differences in absolute values (m) between photogrammetrically-measured and field-verified data were in all cases relatively small (Table 3.4.2). The planting schemes of the hedgerow/windbreaks were characterized by a high stand density, which resulted in overshadowing and clumping of the horizontal crown projections, where rows of trees were not recognized in the given scale of 1:20,000. The limitations for identifying the lower parts of tree crowns and stems precluded the assessment of tree vitality. The photo-identification of age classes

Table 3.4.2: Verification of stereo-photogrammetric measurements for hedgerow/windbreak assessment criteria and classes

Criterion and class (class mark)* mean ±standard deviation	100 % photo-interpreted mean (n=2323)	n	10 % sample estimates (n=232)				Difference	
			Photo-interpreted mean ±standard deviation	CV (%)	Field verified mean ±standard deviation	CV (%)	absolute	%
Mean stand height (m)								
<5 (poor)	3.6	163	3.7±0.9	25	3.3±0.8	25	+0.4	+11
5–10 (fair)	8.3	36	8.1±1.1	14	6.9±1.1	16	+1.2	+15
>10 (good)	11.7	33	11.7±0.7	6	11.0±0.5	5	+0.7	+6
Overall	7.9	232	7.8±0.9	15	7.1±0.8	15	+0.8	+11
Crown closure (coefficient)								
<0.4 (poor)	0.28	67	0.29±0.09	13	0.32±0.09	13	-0.03	-4
0.4–0.6 (fair)	0.61	144	0.58±0.09	20	0.64±0.06	17	-0.06	-14
>0.6 (good)	0.81	21	0.80±0.05	24	0.83±0.04	24	-0.03	-15
Overall	0.57	232	0.56±0.08	19	0.60±0.06	18	-0.04	-11

* Selected classes and marks (Annex 1)

was limited in the case of pollarded hedgerows of *M. alba* and *Salix* spp., where crowns did not differ much in shape, size or shadow in the young and/or premature and mature age classes. Also, in the fast-growing hedgerows of *Populus* spp., horizontal crown projections and crown shadow profiles differed little between the young and premature age classes and thus could not be distinguished accurately.

3.4.3.2 Hedgerow/windbreak assessment

Within the two transects, over 2,300 hedgerows/windbreaks were identified, comprising about 1 % of the cropland area (Table 3.4.3).

The overall share of the hedgerows/windbreaks within both the NS and WE transects was similar. The lowest share of ca. 0.5 % occurred in the "pre-desert" and "flooded" classes, while the "typical" class had the highest share (ca. 1.4 %) within both the NS and WE transects (Table 3.4.4).

Table 3.4.3: Distribution of hedgerows/windbreaks within two transects in Khorezm

Transect statistics	NS transect	WE transect	Total
Transect area (ha)	32,020	23,020	55,040
Agricultural field area (ha)	22,292	17,059	39,351
Number of hedgerows/windbreaks	1,374	949	2,323
Hedgerows/windbreaks area (ha)	270	181	451
Share of agricultural fields assigned to hedgerows/windbreaks (%)	1.2*	1.1*	1.2*

* Corresponds to "poor" mark in "overall" class (Annex 1)

Table 3.4.4: Distribution of hedgerows/windbreaks by cropland class within two transects in Khorezm

Cropland class	Cropland area (ha)	Number of hedgerows/windbreaks (n)	Area of hedgerows/windbreaks (ha)	Share of cropped fields assigned to hedgerows/windbreaks (%)	Mark*
NS transect					
Flooded	1,790	17	10	0.6	Good
Typical	16,251	1,259	238	1.5	Good
Pre-desert	4,251	98	22	0.5	Poor
NS total	22,292	1,374	270	1.2	Poor
WE transect					
Flooded	3,026	17	3	0.1	Good
Typical	14,033	932	178	1.3	Poor
Pre-desert	-	-	-	-	-
WE total	17,059	949	181	1.1	Poor
Total	39,351	2323	451	1.2	Poor

* Annex 1

Multi-species hedgerows/windbreaks were not identified in the transects examined. The overall share of the four main species of the hedgerows/windbreaks is shown in Figure 3.4.1.

The share of other tree species such as *Populus euphratica* Oliv., *Ulmus pumila* L., *Fraxinus pennsylvanica* Marshall, *Gleditsia triacanthos* L., *Maclura pomifera* (Raf.) C. K. Schneid., *Robinia pseudoacacia* L. and *Acer ginnala* var. *semenovii* (Regel & Herder) Pax was ca. 2 %. The spatial distribution of the tree species was similar in both the NS and WE transects.

The analysis of the windbreak function of the existing hedgerows according to the selected criteria is summarized in Table 3.4.5.

Adaptation of photogrammetry for tree hedgerow

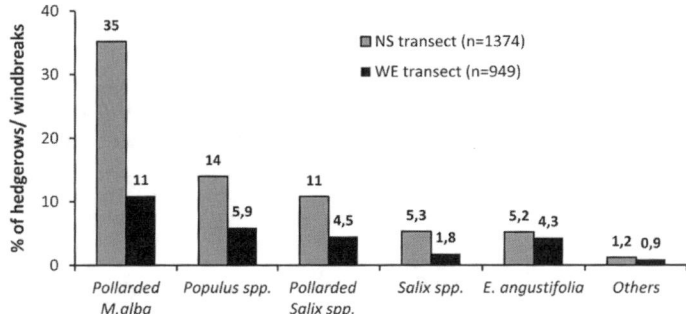

Figure 3.4.1: Dominant tree species in hedgerows/windbreaks within two transects in Khorezm

Table 3.4.5: Assessment of windbreak function of hedgerow systems within two transects in Khorezm

Number of hedgerows/ windbreaks by class mark*				Additional data
Poor	Fair	Good	Total	
Orientation to prevailing wind direction (primary criterion)				
488 Overall rank: +	256	1579	2323	SE+NW and N+S were the most suitable orientation in the south of the region to intercept the prevailing NS and E winds; any orientation was suitable in the north (Meteorological data, 2006)
Height (primary criterion)				
1696 Overall rank: −	418	209	2323	Mean stand height by class mark: 3.6 m (poor), 8.3 m (fair) and 11.7 m (good); mean stand height by tree species: 2.7 and 3.0 m (pollarded *M. alba* and *Salix* spp.), 5.7 m (*Salix* spp.), 7.8 m (*Populus* spp.) and 4.6 m (*E. angustifolia*)
Reaching to edges of the related field (primary criterion)				
1092	-	1231	2323	Overall rank: ±
Crown closure (secondary criterion)				
209	1440	674	2323	Overall rank: ±
Number of rows (secondary criterion)				
720	116	1487	2323	Overall rank: ±
Vitality (secondary criterion)				
209 Overall rank: +	511	1603	2323	Tree species by vitality class (poor, fair and good): *M. alba* (5, 10 and 85 %), *Salix* spp. (12, 18 and 70 %), *Populus* spp. (16, 58 and 26 %) and *E. angustifolia* (20, 19 and 61 %)
Age class (secondary criterion)				
1812 Overall rank: −	-	511	2323	Tree species by age class mark (poor = young/mature** and good = premature**): *M. alba* (80 and 20 %); *Salix* spp. (79 and 21 %); *Populus* spp. (85 and 15 %); *E. angustifolia* (53 and 47 %)

* Annex 1; + = optimal efficiency; ± = acceptable efficiency; − = poor efficiency;
** Table 3.4.6

Most of the hedgerows/windbreaks (ca. 80 %) were oriented towards the prevailing winds in the region. The crown closure was in general assessed as fair to good. Less than 50 % of the hedgerows/windbreaks extended to both edges of the bordering cropped field. Most hedgerows/ windbreaks had a double-row structure. The majority of the trees were in a good to fair health condition. Among the species, *Populus* spp. were relatively more affected by pests and diseases with ca. 60 % and 15 % of the trees in fair and poor condition, respectively. The mean stand height of the hedgerows/windbreaks did not exceed 4 m, whilst the height of *M. alba* and *Salix* spp. was below 3 m due to annual pollarding by farmers. The tallest tree stands were *Populus* spp. with a mean height of ca. 8 m. Young trees dominated in all hedgerows, i.e., ≤ 10 years for *Salix* spp., *Populus* spp. and *E. angustifolia*, and ≤ 20 years for *M. alba* (Table 3.4.6).

Table 3.4.6: Tree age classes by species in Khorezm and Uzbekistan

Principle species	Tree age class (yrs)		
	Young	Premature	Mature*
Hybrid *Populus* spp.	≤ 10	11 – 25	≥ 26
Elaeagnus angustifolia	≤ 10	11 – 25	≥ 26
Morus alba	≤ 20	21 – 50	≥ 51
Salix spp.	≤ 10	11 – 25	≥ 26

* Cutting age

3.4.4 Discussion

3.4.4.1 Verification of photogrammetric technique for windbreak assessment

According to Akça (2000), a standard error of the photogrammetrically-measured mean stand height is usually in the range of $\pm 0.7 - 0.8$ m, which is close to the values monitored in our study. Thus, the accuracy of the height measurements was in the range of typical method errors. The underestimation of crown closure was partly due to seasonal differences between the period of field verification in July-August, when trees had full canopies, and the November date of the aerial photographs after leaf fall. Also, the overshadowing of the understory trees affected the ocular estimation of crown closure, which led to underestimations, as frequently recorded previously (Akça 2000).

The available aerial photographs, although not ideal for assessing the windbreak function due to the intermediate scale of 1:20,000 and the time of collection, nevertheless allowed interpretation with an acceptable accuracy of most of the primary parameters defining the windbreak function. Larger-scale aerial photographs, taken for instance during the vegetative season, could improve the photo-interpretation of all windbreak parameters, but it is at present unlikely that the associated increase in costs can be covered by the low budgets of the forestry administration in Uzbekistan. Therefore, though the current approach is applicable, it should be combined with field surveys to assess the spatial extent and effectiveness of the hedgerows/windbreaks.

3.4.4.2 Windbreak function of the hedgerows

The tree hedgerows planted on the borders of the cropped fields in the Khorezm region represent the main type of agroforestry systems. They were introduced during the Soviet era for protecting the irrigated area from wind erosion and for improving crop yields (Molchanova 1980; Kayimov 1993). The findings show, however, that the extent of tree planting in all identified agricultural field classes was generally insufficient to fulfill the intended windbreak function effectively. The "flooded" fields (occupied by paddy rice) had the lowest share of trees, far below the recommended share of 1.5 %, as only a few tree species can withstand permanent water logging (Bouman,Tuong 2001). The risk of wind erosion in flooded rice fields is, however, obviously restricted. Consequently, the identified share of <1.5 % satisfies the requirements of wind shelter in the "flooded" field classes, irrespective of wind speed and soil conditions. In contrast, the "typical" agricultural fields are characterized by harsher microclimatic conditions, while "pre-desert" fields are at an even higher risk due to possible sandblasting of plants. Therefore, the share of hedgerows/windbreaks should be extended to 1.5 – 2 % and >2 %, respectively, in the "typical" and "pre-desert" fields using appropriate fast-growing, salt-tolerant species that could tolerate low irrigation water availability at the margins of the agricultural area (Khamzina, et al. 2006).

The prevalence of single-species hedgerows, consisting mostly of *M. alba* and *Salix* trees, which were both heavily pollarded for obtaining leaf fodder or wicker, adversely affected the primary windbreak parameter height (Abel et al. 1997). The tallest hedgerows/windbreaks with fast-growing *Populus* spp. were also frequently felled at an early age for construction purposes and other domestic uses (Kan, et al. 2008). This value of the tree products increased the motivation of land users to include and maintain trees on the cropland (Kan et al. 2008). Yet more sustainable harvesting techniques need to be adopted to preserve the prime windbreak function as has been shown elsewhere. For instance, every fourth tree (a 25 % harvest) or every tree in the interior rows (a 50 % harvest) could be harvested, leaving the outer rows intact to ensure permanent protection (Rocheleau, et al. 1988). Furthermore, the use of the windbreaks/hedgerows for production purposes requires continuous replacement of harvested trees. In this respect, a mix of fast- and slow-growing species planted in at least two rows is recommended (Bolin et al. 1987; Rocheleau et al. 1988) to support the completeness and longevity of the windbreaks. The present dominance of double-row hedgerows with "fair" to "good" crown closure values in Khorezm appears to be a suitable design, but enriching the species composition could also help diversifying the risks, e.g., those associated with pest or disease attacks that can result in a significant loss in a single-species plantation (Rocheleau et al. 1988). This particularly applies to the *Populus* spp. hedgerows, which were relatively more highly infested by pests and diseases.

During the growing season (April-October), the agricultural lands in the study region are less prone to wind erosion due to the virtually permanent vegetation cover (Conrad 2006). The erosion risk of the bare fields increases during the cold season when the windbreak function is reduced because of the deciduous nature of the species used. The coniferous trees, which have a crown throughout the year, do not grow well under the climatic conditions of Khorezm and suffer under the soil salinity there (Khamzina et al. 2006). Consequently, other techniques for combating wind erosion, including improved farm practices such as mulching of croplands (Hailu, Runge-Metzger 1992), minimal or zero tillage practices (Pulatov et al. 2012), or crop rotations might be considered as stand-alone measures or in combination with tree windbreaks for preventing soil loss and improving crop yields.

Ample research results underscore that the orientation of windbreaks to the prevailing problem winds, which could erode soils, must be perpendicular or at an angle $>30-45°$ (Minselkhoz 1972). From this perspective, the existing hedgerows/windbreaks in Khorezm provide an optimal to fair shelter efficiency. But the insufficient length of these hedgerows/windbreaks along the edges of the agricultural fields, frequently due to poor management, should be extended to prevent wind turbulences that enhance erosion at the field edges (Bolin et al. 1987).

At present, all the hedgerows/windbreaks are dominated by young trees, indicating that, unless trees are harvested prematurely, the shelter efficiency might increase along with canopy and height growth. Nevertheless, the presently low height of the hedgerows/windbreaks due to excessive harvesting remains the greatest hindrance with regard to an effective sheltering of the cropland against erosive winds in Khorezm.

3.4.5 Conclusions and recommendations

Even though the available aerial photographs were initially considered less suitable for assessing windbreak functions due to the intermediate scale of 1:20,000 and the time of their collection, they turned out to be suitable for an identification of hedgerows/windbreaks in terms of thematic classes, i.e., for estimating crown closure and mean stand height, orientation to the prevailing winds and whether the hedgerow/windbreaks reached the edges of the related fields. In contrast, species composition, windbreak width, tree vitality and age classes were assessed with less accuracy. For these parameters, complementary field surveys are needed to obtain a higher accuracy.

In the agricultural landscape of Khorezm, windbreak orientation to the prevailing winds, crown closure, length, width and vitality of the hedgerows were satisfactory. In contrast, the hedgerow systems generally failed to meet the primary criteria height, spatial extent and multi-species composition, which all determine an effective windbreak function. Therefore, implementation of proper design schemes and, in particular, appropriate tree harvesting techniques are needed to allow combining the productive and windbreak functions of the existing hedgerows. Enriching species composition in the present single-species hedgerow/windbreaks will also help diversifying the risks associated with disease and pest attacks and improve windbreak longevity. With respect to the share of trees within the cropland area, an additional 0.5–1.5 % of agricultural field area needs to be allotted to hedgerows/windbreaks in the "pre-desert" and "typical" agricultural classes to achieve the minimal recommended area under windbreaks given the moderate winds prevailing in the region.

References

Abel N, Baxter J, Cambell A, Cleugh H, Fargher J, Lambeck R, Prinsley R, Prosser M, Reid R, Revell G, Schmidt C, Stirzaker R, Thorburn P (1997) Design principles for farm forestry. A Guide to assist farmers to decide where to place trees and farm plantations on farms. Rural industries research and development corporation, Kingston, Australia

Akça A (2000) Forest Inventory. Lecture materials. Institute of Forest Management and Yield Sciences, Goettingen

Bolin M, Oliver R, Brady S (1987) Illinois Windbreak Manual. Soil Conservation Service, U.S. Department of Agriculture

Botman EA (1986) An impact of proper maintenance on shelterbelts efficiency in rain-fed lands. Afforestation in Central Asia, Tashkent, pp 88–94

Bouman BAM, Tuong TP (2001) Field water management to save water and increase its productivity in irrigated rice. Agric Water Manag 49: 11–30

Conrad C (2006) Remote sensing based modeling and hydrological measurements to assess the agricultural water use in the Khorezm region (Uzbekistan). PhD dissertation, University of Wuerzburg

FAO (2003) Gateway to Land and Water Information: Uzbekistan National Report. Available via http://www.fao.org/AG/aGL/swlwpnr/reports/y_nr/z_uz/uz.htm#soilerosion. Accessed 10 Jan 2009

Glavgidromet (2003) Main Department on Hydrometeorology of Uzbekistan under the Cabinet of Ministers. Meteorological database. Tashkent (in Russian)

Glazirin GE, Chanishev SG, Chub VE (1999) Brief outlines of climate in Uzbekistan. Chinor ENK – Galaba, Tashkent (in Russian)

Hailu Z, Runge-Metzger A (1992) Sustainability of land use systems. The potential of indigenous measures for the maintenance of soil productivity in Sub-Sahara African Agriculture. Tropical Agroecology 7, Weikersheim, Magraf

Hamblin A, Williams J (1996) Alarming erosion of Australia's soil and land base, The Australian Academy of Technological Sciences and Engineering Focus, No. 90, Jan/Feb 1996. Available via http://www.atse.org.au/index.php/index.php?sectionid=358. Accessed 20 May 2009

Ibrakhimov M (2005) Spatial and temporal dynamics of groundwater table and salinity in Khorezm (Aral Sea Basin), Uzbekistan. ZEF series Ecology and Development 23: 175

Kan E, Lamers JPA, Eschanov R, Khamzina A (2008) Small-scale farmers' perceptions and knowledge of tree intercropping systems in the Khorezm region of Uzbekistan. Forests, Trees and Livelihoods 18: 355–372

Katz DM (1976) The influence of irrigation on groundwater. Kolos, Moscow (in Russian)

Kayimov A (1993) Ecosystems of agroforestry landscapes in the irrigated lands. FAN, Tashkent, (in Russian)

Khamzina A, Lamers JPA, Worbes M, Botman E, Vlek PLG (2006) Assessing the potential of trees for afforestation of degraded landscapes in the Aral Sea Basin of Uzbekistan. Agrofor Syst 66: 129–141

Kort J (1988) Benefits of windbreaks to field and forage crops. Agriculture, Ecosystems and Environment 22/23: 165–190

MFD (2008) Main Forestry Department of Ministry for Agriculture and Water Resources of Uzbekistan. Available via http://forestry.uz. Accessed 15 Dec 2008

Minselkhoz (1972) Ministry for Agriculture of USSR. Shelterbelts and their arrangement. Department of Shelterbelt Systems in Collective and State Farms (in Russian)

Molchanova A (1980) History of shelterbelt plantings on irrigated and rain-fed lands in Central Asia. Afforestation Recommendations No.20. Central Asian Forest Research Institute, Tashkent, pp 11–22 (in Russian)

Molchanova A (1986) A role of windbreaks in irrigated lands. Protective forest planting 23: 102–109 (in Russian)

Pulatov A, Egamberdiev O, Karimov A, Tursunov M, Kienzler S, Sayre K, Tursunov L, Lamers JPA, Martius C (2012) Introducing conservation agriculture on irrigated meadow alluvial soils (Arenosols) in Khorezm, Uzbekistan. In: Martius C, Rudenko I, Lamers JPA, Vlek PLG (Eds.) Cotton, water, salts and soums – economic and ecological restructuring in Khorezm, Uzbekistan. Springer, Dordrecht pp 195–217

Rocheleau D, Weber F, Field-Juma A (1988) Agroforestry in Dryland Africa. Science and Practice of Agroforestry International Council for Research in Agroforestry (ICRAF) Nairobi, Kenya, pp 163–191

Tupitsa A (2009) Photogrammetric techniques for the functional assessment of tree and forest resources in Khorezm, Uzbekistan. PhD dissertation, Rheinischen Friedrich-Wilhelms-Universität Bonn

UNCCD (2005) Promotion of traditional knowledge. A compilation of UNCCD documents

478 and reports from 1997–2003. UNCCD. Available via http://www.unccd.int/publicinfo/publications/docs/traditional_knowledge.pdf. Accessed 15 May 2009

ZEF/UNESCO Khorezm Project (2006) Meteorological database of the project. Urgench, Khorezm

Annex 1: Acknowledged primary windbreak criteria and classes for assessing existing hedgerows in Khorezm

Windbreak criterion	Source	Windbreak criterion class (characteristic)	Mark
Share of agricultural fields assigned to windbreaks	Adopted from Kayimov (1993)	<1.5 % (flooded)	Good
		1.5 – 2.0 % (typical)	Good
		>2.0 % (pre-desert)	Good
		1.5 – 2.0 % (overall)	Good
Species composition	Bolin et al. (1987); Rocheleau et al. (1988)	Single-species (reduced range of functionality)	Poor
		Multiple-species (extended range of functionality)	Good
Orientation to prevailing wind direction	Abel et al. (1997); Bolin et al. (1987); Minselvodkhoz (1972)	>±45° of 90° (low efficiency of shelter)	Poor
		Range of ±31 – 45° of 90° (fair efficiency of shelter)	Fair
		Range of ±1 – 30° of 90° (optimal efficiency of shelter)	Good
Mean stand height	Abel et al. 1997; Bolin et al. 1987; Botman 1986	<5 m (little shelter)	Poor
		5 – 10 m (fair shelter)	Fair
		>10 m (good shelter)	Good
Reaching the edges of related field	Adopted from Bolin et al. (1987)	Not reaching (reduced extent of shelter)	Poor
		Reaching (optimal extent of shelter)	Good

Annex 2: Acknowledged secondary windbreak criteria and classes for assessing existing hedgerows in Khorezm

Windbreak criterion	Source	Windbreak criterion class (characteristic)	Mark
Crown closure	Adopted from Abel et al. (1997); Akça'(2000)	<0.4 (low efficiency of shelter)	Poor
		0.4–0.6 (fair efficiency of shelter)	Fair
		>0.6 (optimal efficiency of shelter)	Good
Number of rows	Adopted from Abel et al. (1997)	Single-row (less potential shelter if a tree is missing)	Poor
		Double and multiple row (more potential shelter)	Good
Vitality	MacDicken (1991)	Poor (dominance of severely affected trees = low efficiency)	Poor
		Fair (dominance of lightly affected trees = fair efficiency)	Fair
		Good (dominance of healthy trees = optimal efficiency)	Good
Age class	Adopted from Botman, *unpublished*	*Populus* spp., *Salix* spp., *E. angustifolia* ≤10 yrs. old (young) and ≥26 yrs. old (mature) and *M. alba* ≤20 yrs. old (young) and ≥51 yrs. old (mature) (reduced efficiency of shelter)	Poor
		Populus spp., *Salix* spp., *E. angustifolia* 11–25 yrs. old (premature) and *M. alba* 21–50 yrs. old (premature) (optimal efficiency of shelter)	Good

Dilfuza Djumaeva, John P. A. Lamers, Asia Khamzina,
Shirin Babajanova, Ruzumbay Eshchanov, Paul L. G. Vlek

3.5 Nitrogen fixation by trees with P-fertilization enhances growth and carbon sequestration in degraded irrigated croplands

Abstract

Environmental damage in the irrigated areas of Central Asia calls for the adoption of land-use alternatives. The findings from a field experiment conducted on highly salinized cropland in the Amu Darya lowlands between 2006 and 2008 illustrate numerous benefits of afforesting marginal cropland. In particular, nitrogen (N_2)-fixing species such as actinorhizal Eleagnus angustifolia L. and leguminous Robinia pseudoacacia L. thrived on degraded croplands, as they could satisfy their own N demand. In the early growth stage, they sequestered without Phosphorus (P) applications carbon (C) varying between 3.4 t C ha^{-1} by R. pseudoacacia and 20.5 t C ha^{-1} by E. angustifolia as opposed to 0.5 t C ha^{-1} by the non-N fixer Gleditsia triacanthos L. The application of 90 kg P ha^{-1} increased N_2 fixation and in turn C sequestration after three years by 20 % in E. angustifolia and up to 70 % in R. pseudoacacia. The findings reveal that the conversion of marginal agricultural areas to N_2-fixing tree plantation systems would mitigate land degradation and enhance the productive capacity of marginalized, irrigated croplands in Central Asian dryland and in particular when adding small amounts of P.

Keywords: Land use, N_2-fixing tree species, ecosystem services, carbon sequestration, saline soils, P-fertilization, Aral Sea Basin

3.5.1 Afforesting marginalized, irrigated croplands in Central Asia: state of the art

Seven decades of irrigated crop production during the Soviet era in Central Asia has led to widespread cropland degradation manifested in erosion, water-logging, soil compaction, nutrient depletion and soil salinization. The annual costs of land degradation are an estimated USD 31 million in entire Central Asia (Sutton et al. 2007). Mounting concerns over these environmental damages have triggered the exploration of land-use alternatives in Uzbekistan for degraded croplands[1] such as afforestation pioneered by Khamzina et al. (2012). The concept proposes setting aside the marginalized croplands for afforestation and the more effective use of the spared resources on productive land, thus combining the benefits to the ecology and livelihoods (Vlek et al. 2001). Afforestation of marginalized croplands is not only a means to increase their value but, since the agreement in Kyoto in 1992, it is also a recognized instrument to mitigate the impact of global warming through capturing and storing atmospheric CO_2.

Land reclamation through afforestation often includes leguminous species capable of symbiosis with *Rhizobium* bacteria. The actinorhizal species that benefit from a symbiotic relationship with N_2-fixing *actinomycetes* of the genus *Frankia* (Brewbaker 1989) have only occasionally been considered even though they are recognized as *"pioneers on nitrogen-poor soils"* (Baker 1990; Peoples and Crasswell 1992), also due to their ability to tolerate salinity, waterlogging, and mineral deficiencies (Brewbaker 1989). A decade of afforestation research on the degraded, nitrogen (N)-depleted croplands in northwest Uzbekistan (Khamzina et al. 2012) illustrates that the actinorhizal tree species *Eleagnus agustifolia* L. (Russian olive) was capable of biological N_2 fixation (BNF). This afforestation trial on salinized, abandoned croplands considered a minimum of site preparation to reduce the costs and therefore excluded the use, for example, of mineral and/or organic fertilizers (Khamzina et al. 2012). Later, Djumaeva et al. (2012) compared the performance of *E. agustifolia* with that of the tree legume *Robinia pseudoacacia* L. (black locust) under phosphorus (P) amendments. Whereas N-containing amendments are known to reduce or even inhibit N_2-fixation (Fried and Broeshart 1975) owing to the inhibitory effect on nodulation and/or nitrogenase activities (Streeter 1988), the use of P-containing fertilizers usually increases N_2-fixation rates (Balasubramanian and Joshaline 1996; Wheeler et al. 1996).

---

[1] Smit et al. (1991) postulated that the marginality of land *"relates fundamentally to the economic viability of land uses"*. They defined marginal land as *"land parcels or types for which there are few viable uses, or for which uses are either only just viable or not always viable"*.

Prior to the above-mentioned study (Djumaeva et al. 2012), the ability of P to enhance BNF of trees growing on impoverished soils was concluded for West Africa (Sanginga 2003), but not for other regions (e.g., Binkley et al. 2003; Gökkaya et al. 2006; Uliassi and Ruess 2002). In these studies, the impact of growth-enhancing treatments was judged by the absolute growth rates although relative growth rates (RGR) can be more effective in detecting the impacts (Hunt 1990). Furthermore, the BNF quantification methods for perennial vegetation were compared from a technical point of view only (Boddey et al. 2000; Peoples and Crasswell 1992). Financial considerations have hardly been included in such assessments, despite their being of importance for BNF research particularly in developing countries. Finally, the impact of P-enhanced BNF on carbon (C) sequestration by trees has not been quantified although it is of crucial interest in assessing the role of afforestation as a mitigation strategy against climate change impacts. Hence, the option of afforestation of degraded croplands in Central Asia was elaborated upon by assessing (i) the impact of P amendments on relative growth rates of *E. angustifolia* and *R. pseudoacaci*, (ii) N_2-fixation quantification methods from a financial perspective, and (iii) the amount of C capture and storage biomass following P applications.

3.5.2 Relative growth rates of trees

In the highly saline soil conditions at degraded cropping sites in Uzbekistan, actinorhizal *E. angustifolia* and leguminous *R. pseudoacacia* fixed considerable amounts of atmospheric N_2 (Khamzina et al. 2012). The annual amount of N derived from the atmosphere (Ndfa) by *E. angustifolia* varied from 200–500 kg N ha^{-1}, indicating an effective fixation (Khamzina et al. 2012). This was due to the combined effect of a high %Ndfa by the tree species (ca. 80 %) and the high stand density (about 5700 stems ha^{-1}) (Khamzina et al. 2012). This density is much higher than usually recommended for tree plantations, but it was nevertheless suggested as a means of generating a considerable bulk of biomass per land area unit in a short time for an early use as fuelwood and fodder (Khamzina et al. 2012). Furthermore, the BNF of both *E. angustifolia* and leguminous *R. pseudoacacia* was boosted through P additions as low as 45 and 90 kg P ha^{-1}, respectively very likely due to the satisfaction of the ATP demand of the energy-intensive N_2-fixation process. The increases in BNF went hand in hand with an increased biomass production of *E. angustifolia* (27 %) and *R. pseudoacacia* (57 %) compared to trees of the same species without P applications, although the differences were statistically not significant. In addition to the assessment of absolute biomass increments for growth analysis, Djumaeva et al. (2012) used more sensitive indicators such as RGR of height (RGR_H) and diameter (RGR_D)[2], complemented with N-productivity (NP) estimates, which is defined as the rate of weight increase per unit leaf N per time (Lambers et al. 1998)[3].

The analyses showed that during three successive years P applications significantly increased the NP of *R. pseudoacacia* (Table 3.5.1). This indicates that the increased *R. pseudoacacia* growth with P applications was related to the increased N_2 fixation caused by the added P, which was not evident from the analyses of biomass dynamics alone. Furthermore, applications of 90 kg P ha^{-1} increased the RGR_H of *E. angustifolia* (Table 3.5.2). The findings with respect to *R. pseudoacacia* were less consistent. In 2006, P applications increased the RGR_H and RGR_D of this species, but this impact was not significant in all years (Table 3.5.2). Hence, whilst the differences in absolute growth of both N_2-fixers were not statistically significant, the findings from the RGR analyses illustrate that various growth parameters were stimulated with P applications, with differing effects between the N_2-fixers. The findings confirm that RGR is a sensitive indicator of tree growth responses to treatments and therefore ought to be considered particularly for perennating tree biomass parts, as also suggested by Hunt (1990). Furthermore, the Ndfa (both in % and kg N ha^{-1}) increased with modest amounts of P already during the early growth of the trees.

From these analyses, the afforestation of salt-affected croplands in the irrigated areas of Central Asia could be a low-cost option to increase the land value. Based on a domestic price for Supersimple Phosphate of 400 Uzbek Soum kg^{-1}, an application of 90 kg P ha^{-1} would demand an additional investment of 36 000 Soum ha^{-1} (or about USD 15 ha^{-1}). This seems a reasonable investment given (i) the expected returns (Lamers et al. 2008) and (ii) that the purchase of only tree saplings would cost about 530 000 Soum ha^{-1} (or ca. USD 220 ha^{-1}). Although in-depth financial analyses on the use of P applications have not been included yet, the first indications appear favorable.

---

[2] The RGR of height and diameter. These are estimated as:
RGR_H *(in mm mm^{-1} d^{-1})= Relative height growth rate: (lnH$_2$ - lnH$_1$)/(t$_2$ - t$_1$)*
and RGR_D *(in mm mm^{-1} d^{-1}) = Relative diameter growth rate: (lnD$_2$ - lnD$_1$)/(t$_2$ - t$_1$)*
where H_1 and H_2 are the initial and subsequent height, and D1 and D2 the initial and subsequent diameter (diam$_{10}$) at the time of harvest at t_1 and t_2.

[3] The estimation of NP was completed according to Lambers et al. (1998): NP = RGR/PNC, where RGR (mg g^{-1} day^{-1}) is the relative growth rate and PNC is the tree N concentration, i.e., total tree N per total tree mass.

Table 3.5.1: Nitrogen productivity (NP) of three tree species following phosphorus (P) treatments in 2006, 2007 and 2008

Species	Treatment	2006	2007	2008
		NP (mg g^{-1} day^{-1})		
E. angustifolia	0-P	6.94 a	2.37 ab	1.92 a
	45 kg P ha^{-1}	6.91 a	2.32 b	2.03 a
	90 kg P ha^{-1}	6.93 a	2.41 a	1.95 a
Overall mean		6.93 A	2.37 A	1.97 A
G. triacanthos	0-P	5.88 b	2.21 a	0.95 b
	45 kg P ha^{-1}	6.50 a	1.40 b	1.58 a
	90 kg P ha^{-1}	6.63 a	1.89 a	0.90 b
Overall mean		6.34 B	1.83 B	1.14 C
R. pseudoacacia	0-P	6.37 b	2.33 b	1.01 b
	45 kg P ha^{-1}	6.11 c	2.45 a	1.79 a
	90 kg P ha^{-1}	6.76 a	2.45 a	1.64 a
Overall mean		6.55 B	2.41 A	1.48 B
		ANOVA, probability > F(=α)		
Year		<0.0001		
Species		<0.0001		
Phosphorus (P)		<0.0001		
Time*species		<0.0001		
Time*P		<0.0001		
Species*P		0.031		

Within a column and each tree species, values followed by the same letter are not significantly different at P <0.05. Species overall means with a column followed by the same capital letter are not significantly different at P <0.05

3.5.3 Financial considerations of BNF quantification methods

The quantification of the Ndfa of older trees under open-field conditions is reportedly challenging (e.g., Boddey et al. 1995). This is substantiated by findings derived with the use of various acknowledged N$_2$-quantification methods in Uzbekistan. Djumaeva et al. (2010) conducted a lysimeter experiment with 2- and 3-year-old tree saplings designed in a way to eliminate various uncertainties associated with the selection of a suitable reference plant and β value. The experimental design allowed quantifying BNF by E. angustifolia referenced against G. triacanthos and Ulmus pumila through four methods, i.e., ^{15}NA, ^{15}N enrichment (requiring no correction for isotopic fractionation), A-value, and total N difference. The combination of the ^{15}NA and ^{15}N enrichment methods permitted determining the "adjusted" β value by comparing ^{15}NA estimates to the best match of those based on the ^{15}N enrichment method. This value was, however, higher than the field-observed ^{15}N natural abundance values and hence not applicable for the field-grown plants (Khamzina et al. 2012). This stresses the challenge of accurately measuring field N_2-fixation in trees with the ^{15}N natural abundance method.

Table 3.5.2: Species means of relative growth rate in height (RGR_H) and diameter (RGR_D) following P treatments in 2006, 2007 and 2008.

Species	Treatment	RGR_H (mm mm^{-1} d^{-1})	RGR_D (mm mm^{-1} d^{-1})
2006			
E. angustifolia	0-P	0.029 b	0.002 b
	45 kg P ha^{-1}	0.040 ab	0.002 ab
	90 kg P ha^{-1}	0.044 a	0.003 a
G. triacanthos	0-P	0.006 b	0.0005 b
	45 kg P ha^{-1}	0.009 b	0.0008 ab
	90 kg P ha^{-1}	0.019 a	0.0010 a
R. pseudoacacia	0-P	0.012 b	0.002 b
	45 kg P ha^{-1}	0.019 ab	0.002 ab
	90 kg P ha^{-1}	0.025 a	0.003 a
2007			
E. angustifolia	0-P	0.024 b	0.004 a
	45 kg P ha^{-1}	0.037 a	0.004 a
	90 kg P ha^{-1}	0.041 a	0.005 a
G. triacanthos	0-P	0.011 c	0.002 b
	45 kg P ha^{-1}	0.028 b	0.002 b
	90 kg P ha^{-1}	0.053 a	0.004 a
R. pseudoacacia	0-P	0.043 a	0.004 a
	45 kg P ha^{-1}	0.042 a	0.004 a
	90 kg P ha^{-1}	0.051 a	0.005 a
2008			
E. angustifolia	0-P	0.028 b	0.003 a
	45 kg P ha^{-1}	0.028 b	0.004 a
	90 kg P ha^{-1}	0.038 a	0.005 a
G. triacanthos	0-P	0.028 a	0.001 b
	45 kg P ha^{-1}	0.029 a	0.002 a
	90 kg P ha^{-1}	0.028 a	0.002 a
R. pseudoacacia	0-P	0.028 a	0.003 a
	45 kg P ha^{-1}	0.038 a	0.003 a
	90 kg P ha^{-1}	0.039 a	0.003 a

Within a column and for each tree species, values followed by the same letter are not significantly different at $P < 0.05$ according to post hoc Tukey test

The %Ndfa in lysimeter-grown *E. angustifolia* trees narrowly resembled that of field-grown trees of the same age and measured against the same reference species (Khamzina et al. 2012). On the other hand, the amounts of Ndfa (kg ha^{-1}) in the lysimeters were, irrespective of the quantification method used, much lower than those estimated in the field (1 g tree^{-1} vs. 18 g tree^{-1} for 2-year-olds and 3 g tree^{-1} vs. 78 g tree^{-1} in 3-year-olds) (Djumaeva et al. 2010). Such discrepancies are not unusual, as lysimeter-grown trees develop differently from those grown in the field. Findings from the lysimeter experiment illustrate furthermore that, aside from the reference tree species used, the quantification method significantly influenced the Ndfa (kg N ha^{-1}) estimates.

The ^{15}N enrichment method has appeared out of reach for many researchers, because it demands large funds for the preparation and purchase of isotopic materials and subsequent analysis with a spectrometer (Table 3.5.3). Djumaeva et al. (2010) postulated, therefore, that if assuring that all tree biomass fractions in lysimeters can be comprehensively collected and weighted and consequently N contents accurately measured, the findings of the N-difference method resemble those of the isotopic method. The N-difference method therefore is an appropriate alternative for lysimeter studies with young trees in case the available funds do not permit using isotopic-based methods.

Table 3.5.3: Financial resources spent (in USD) for quantifying nitrogen fixation with the total nitrogen difference (TND), ^{15}N natural abundance (^{15}NA), ^{15}Nenrichment (^{15}NE) and A-value (AV) methods in 2008.

Method		Cost of enriched ^{15}N fertilizer	Chemical analyses				Total amount spent (in USD)
			No. of samples	Plant ^{15}N	Soil ^{15}N	Total N	
TND		-	3			9	27
^{15}NA		-	3	19			57
^{15}NE	4 g	40*	3	----19----			2280
	12 g	120*	3	----19----			6840
AV	4 g	40*	3	----19----			2280
	12 g	120*	3	----19----			6840

* The price of 1 g of 15N fertilizer (15NH4 15NO3, 35 % N, 5 atom% 15N excess) was USD 10 in the study year 2008

3.5.4 Carbon sequestration in trees following P applications

Afforestation of marginal croplands provides an important ecosystem service, i.e., capturing and storing C. The importance of afforestation and reforestation for mitigating climate change impacts is endorsed by the Kyoto Protocol (UNFCCC 1998). Capture and storage of C in tree wood is suggested as one of measures to counteract the increase in atmospheric CO_2 concentrations, which measure 3.4 Pg yr^{-1} (FAO 2000). The global assessments of the land suitability for re- and afforestation as a means to mitigate atmospheric CO_2 excluded *a priori* the irrigated dryland areas (Zomer et al. 2008). The authors argued that irrigated croplands were highly productive, thus rendering their conversion to tree plantation as cost ineffective. In contrast to this generalization, recent studies in Uzbekistan revealed cropland degradation of significant spatial extent in irrigated areas (cf. Fritsch et al., this book; Dubovyk et al., 2013) and the potential of afforestation to rehabilitate these degraded areas with little irrigation demand (Khamzina et al. 2012) and cost effectively (Djanibekov et al. 2012). The economic viability of afforestation in degraded cropland areas was based on the production of tradable commodities by the tree plantations, including the biomass C (Djanibekov et al. 2012). The information on the C-sequestration potential by afforestation species in highly salinized irrigated croplands is rare in Central Asia, particularly on the belowground C accrual because labor-intensive coarse root excavations usually are not performed in bio-sequestration studies in adult tree plantations.

The average C concentration in the perennial fractions (Table 3.5.4) of 4-year-old three species of *E. angustifolia*, *R. pseudoacacia* and the non-fixer *G. triacanthos* ranged from 41 to 46 %, with the minimum value observed in the coarse roots of *R. pseudoacacia*. This is below the 50 % value routinely used in C sequestration studies in the absence of field data, thus showing the importance of specific %C measurements.

Table 3.5.4: Carbon concentrations (%) in perennial tree fractions of 4-year-old tree species in 2008, Khorezm.

Fraction	*E. angustifolia*	*G. triacanthos*	*R. pseudoacacia*
	C concentration, %		
Stem/Stump	45 ± 0.3 b	44 ± 0.1 a	45 ± 0.4 a
Twigs	46 ± 0.2 a	44 ± 0.9 a	43 ± 1.2 a
Coarse roots	45 ±0.2 b	42 ± 0.6 a	41 ± 1.2 b

Values are means ± standard deviation. Letters indicate statistically significant differences within the columns according to the Post Hoc Tukey test.

The total C (t C ha^{-1}) stored in all woody fractions (twigs, stem/stump, and coarse roots), estimated through the weighted means of the C stored in these fractions of the 4-year old trees (Table 3.5.5) with P applications followed the order: *E. angustifolia* > *R. pseudoacacia* > *G. triacanthos*. With 90 kg P ha^{-1}, the total C increased when compared to 0-P by 20 % after three years (from 20.5 to 25.5 t C ha^{-1}) in *E. angustifolia*, by 20 % (0.5 to 0.6 t C ha^{-1}) in *G. triacanthos*, and by 70 % (2.8 to 4.7 t C ha^{-1}) in *R. pseudoacacia*. The findings for *E. angustifolia* are within the range of 11 – 23 t C ha^{-1} previously observed in 6-year-old tree plantations in Uzbekistan (Khamzina et al. 2012).

Most of the C was stored in the stem and twig fractions irrespective of species and level of P application (Table 3.5.5). The share of the C stored in the coarse root fraction differed among species and P application. It was highest in *R. pseudoacacia* (23 % with 0-P, 11 % with 45-P, 17 % with 90-P) and quite similar between *E. angustifolia* (9.5 % with 0-P, 7.6 % with 45-P, 6.1 % with 90-P) and *G. triacanthos* (10.7 %, 10.5 % and 10.5 % respectively with 0-P, 45-P, and 90-P). The share of C in the coarse roots is however considerable, showing the importance of integrating this fraction in C-accounting as well despite it being more difficult to monitor than the aboveground fractions.

Table 3.5.5: Carbon storage (in t C ha^{-1}) in woody fractions of 4-year-old tree species and following P application (0-P, 45-P and 90-P)

Fraction	*Eleagnus angustifolia*			*Robinia pseudoacacia*			*Gleditsia triacanthos*		
	0-P	45-P	90-P	0-P	45-P	90-P	0-P	45-P	90-P
Stem	8.2	8.9	11.1	0.6	1.0	1.4	0.1	0.3	0.1
Stump	2.8	2.1	3.3	0.6	0.8	1.0	0.2	0.2	0.3
Twigs	7.6	6.9	9.5	0.9	1.3	1.5	0.1	0.1	0.2
Coarse roots	2.0	1.5	1.6	0.7	0.4	0.8	0.1	0.1	0.1
Total	20.5	19.3	25.5	2.8	3.4	4.7	0.5	0.7	0.6

A change in land use towards tree-based systems may increase soil C as well. Based on repeated sampling of the top 20-cm soil layer during five years after afforesting a degraded cropping site, Khamzina et al. (2012) reported a 2–7 t C ha^{-1} increase in the soil-organic C stock, peaking in plots with *E. angustifolia*.

Results of a chrono-sequence study in hybrid poplar and elm plantations in Khorezm, following afforestation of the agricultural soils suggest that the soil-C sequestration rate averaged 0.15 t ha^{-1} year^{-1} in the first 20 years and 0.09 t ha^{-1} year^{-1} thereafter (Hbirkou et al. 2011), thus increasing the soil fertility status. Daily fluxes of N_2O and CH_4 (expressed in CO_2 equivalent (CO_2 eq)), in annual (N- and P-fertilized irrigated crops) and perennial (unfertilized poplar plantation and the natural *Tugai* floodplain forest) land-use systems followed the order of: paddy rice (10.1 kg CO_2 eq ha^{-1} day^{-1}) > irrigated cotton (8.8 kg CO_2 eq ha^{-1} day^{-1}) > poplar plantation (3.4 kg CO_2 eq ha^{-1} day^{-1}) > irrigated winter wheat (1.4 kg CO_2 eq ha^{-1} day^{-1}) > *Tugai* forests (0.2 kg CO_2 eq ha^{-1} day^{-1}). It therefore was postulated that the direct contribution of forest plantations to greenhouse gas emissions due to soil respiration is moderate compared to that from major crops (Scheer et al. 2008).

3.5.5 Overall discussion and conclusions

The various benefits of afforesting marginalized croplands in the irrigated drylands of Central Asia may still be insufficiently convincing to farmers and to motivate them to readily get engaged in such land-use practices. The present state policy for crop production may hamper the promotion of afforestation, because under this policy the farmers allocate defined croplands areas to cotton and comply with output targets that are preset based on the soil fertility levels of their lands. On the other hand, the local administrations have some leeway to exempt unproductive cropland parcels from the state plan in favor of income-generating afforestation that would provide benefits in terms of tree products. Furthermore, due to the low irrigation demand, the saved water could be used in productive areas (Djanibekov et al. 2012). Such decisions would also depend on the interest and the negotiation capacity of the land users.

One of the sources of motivation for farmers to opt for tree plantations particularly with N_2-fixing species is their suitability for high-quality fodder production, given the higher N contents in the foliage of N_2-fixers as opposed to non-fixing species. Therefore, enriching the feed diets of livestock with N-rich foliage has become a suggested means to enhance the roughage-based livestock nutrition (Djanibekov 2008; Djumaeva et al. 2009; Lamers and Khamzina 2010), however, this practice is not common in CA, yet. The findings in Uzbekistan confirm that an application of P has the potential to further enhance N contents and therewith the crude protein contents of foliage of both *E. angustifolia* and *R. pseudoacacia*. Hence, an admixture of tree leaves and conventional feed has the option to increase farmers' profits from dairy production by almost 50 % at the season onset, about 40 % at midseason, and ca. 35 % at the end of the season (Djumaeva 2011). Based on nutritive quality, the foliage of *E. angustifolia* is best collected in July and September but based on the biomass quantity, September (the end of the growing season) can be recommended. Thus, long-term investments in tree plantations on low-yielding or abandoned marginal cropland have substantial prospects for improving the income of farmers through improved livestock diets at lower costs than with the present diets. This is an important finding given the rapidly growing importance of livestock for sustaining rural livelihoods in CA, the vast stock of animals (Iñiguez et al. 2005), and the large demand-supply gap in high-quality feed (Djanibekov 2008). For an effective outcome, the awareness among livestock keepers of the nutritive quality of tree foliage needs to be improved (Kan et al. 2008).

Introducing tree plantations, especially when including N_2-fixing species, on degraded cropland in Central Asia where plant growth is often constrained by N deficiency, would generate biomass and soil organic matter (Djumaeva 2011; Khamzina et al. 2012). The enhanced production of foliage bears the dual potential to enrich soils through the leaf litter fall decomposition (Lamers et al. 2010) and to supplement livestock diets (e.g., Lamers and Khamzina 2010). This in turn could increase farm profits based on a systematic organization of trade-offs between the productive function of tree plantations and their ameliorative purpose.

References

Bailey R, Rayner A (2001) Carbon rights. Agriculture Notes. AG0904. Australia, State of Victoria, Department of Primary Industries, pp 2

Baker DD (1990) Optimizing actinorhizal nitrogen fixation and assessing actual contribution under field conditions, In: Werner D, Muekker P (Eds.) Fast growing trees and nitrogen fixing trees Stuttgart, Germany, Fisher-Verlag, pp 291–299

Balasubramanian V, Joshaline CM (1996) The effect of *Frankia* and VAM inoculation on growth of Casuarina seedlings. In: Wheeler CT, Narayanan R, Parthiban KT, Kesavan A, Surendran C (Eds), Proceedings of the national seminar on the root microbiology of tropical nitrogen fixing trees in relation to nitrogen and phosphorus nutrition: Coimbatore, Tamil Nadu Agricultural University, 55–59

Binkley D, Senock R, Kermit CJ (2003) Phosphorus limitation on nitrogen fixation by *Facaltaria* seedlings. For Ecol Manage 186: 171–176

Boddey RM, Octavio C de Oliveira, Alves BJR, Urquiaga S (1995) Field application of the ^{15}N isotope dilution technique for the reliable quantification of plant-associated biological nitrogen fixation. Fertil Res 42: 77–87

Boddey RM, Peoples MB, Palmer B, Dart PJ (2000) Use of the ^{15}N natural abundance technique to quantify biological nitrogen fixation by woody perennials. Nutr Cycl Agroecosyst 57: 235–270

Brewbaker JL (1989) Nitrogen fixing trees, In: Werner D and Mueller P (Eds.) Fast growing trees and nitrogen fixing trees, Gustav Fisher Verlag, Marburg, Germany, pp. 253–262

Djanibekov N (2008) A micro-economic analysis of farm restructuring in the Khorezm region, Uzbekistan. Dissertation, Center for Development Research (ZEF), University of Bonn

Djanibekov U, Khamzina A, Djanibekov N, Lamers JPA (2012) How attractive are short-term CDM afforestation projects in arid countries? The case of Uzbekistan. Forest policy and economics 21, 108–117. Elsevier B.V., London Academic Press

Djumaeva D (2011) The effect of phosphorus amendments on nitrogen fixation and growth of trees on salt-affected croplands in the lower reaches of Amu Darya, Uzbekistan. Dissertation, Center for Development Research (ZEF), University of Bonn

Djumaeva D, Djanibekov N, Vlek PLG, Martius C, Lamers JPA (2009) Options for optimizing dairy feed rations with foliage of trees grown in the irrigated drylands of Central Asia. Res J Agric Biol Sci 5: 698–708

Djumaeva D, Lamers JPA, Martius C, Khamzina A, Ibragimov N, Vlek PLG (2010) Quantification of symbiotic nitrogen fixation by *Elaeagnus angustifolia* L. on salt-affected irrigated croplands using two ^{15}N isotopic methods. Nutr Cycl Agroecosyst 88: 329–339

Djumaeva D, Lamers JPA, Martius C, Vlek PLG (2012) Chlorophyll meters for monitoring foliar nitrogen in three tree species from arid Central Asia. J Arid Environ 85: 41–45

Dubovyk O, Menz G, Conrad C, Lamers JPA, Lee A, Khamzina A (2013) Spatial targeting of land rehabilitation: a relational analysis of cropland productivity decline in arid Uzbekistan. Erdkunde (67):2 167–181

FAO (2000) Carbon sequestration options under the Clean Development Mechanism to address land degradation. World Soil Resources Reports, 92. Food and Agriculture organization of the United Nations, Rome. http://www.fao.org/clim/docs/998_carb%20seq%2011.pdf

Fried M, Broeshart H, (1975) An independent measurement of the amount of nitrogen fixed by legume crop. Plant Soil 43: 707–711

Gökkaya K, Hurd TM, Raynal DJ (2006) Symbiont nitrogenase, alder growth, and soil nitrate response to phosphorus addition in alder (*Alnus incana* ssp. *rugosa*) wetlands of the Adirondack Mountains, New York State, USA. Environ Exp Bot 55: 97–109

Hbirkou C, Martius C, Khamzina A, Lamers JPA, Welp G, Amelung W (2011) Reducing topsoil salinity and raising carbon stocks through afforestation in Khorezm, Uzbekistan. J Arid Environ 75: 146–155

Hunt R (1990) Basic growth analysis London, Unwin Hyman Ltd

Iñiguez L, Suleymanov M, Yusupov S, Ajibekov A, Kineev M, Kheremov S, Abdusattarov A, Thomas D, Musaeva M (2005). Livestock production in Central Asia – Constraints and opportunities, 22

Kan E, Lamers JPA, Eshchanov R, Khamzina A (2008) Small-scale farmers' perception and knowledge of tree intercropping systems in the Khorezm region of Uzbekistan For Trees Livelihoods 18: 355–372

Khamzina A, Lamers JPA, Vlek PLG (2012) Conversion of degraded cropland to tree plantations for ecosystem and livelihood benefits. In: Martius C, Rudenko I, Lamers JPA, Vlek PLG (Eds.) Cotton, water, salts and Soums – economic and ecological restructuring in Khorezm, Uzbekistan. Springer: Dordrecht, pp 235–248

Lambers H, Chapin III FS, Pons TL (1998) Plant Physiological Ecology. Springer Science, Utrecht University, The Netherlands, p 540

Lamers JPA, Bobojonov I, Khamzina A, Franz J (2008) Financial analysis of small-scale forests in the Amu Darya lowlands of rural Uzbekistan. For Trees Livelihoods 18: 373–386

Lamers JPA, Khamzina A (2010) Seasonal quality profile and production of foliage from trees grown on degraded cropland in arid Uzbekistan, Central Asia. J Anim Physiol Anim Nutr 94, e77-e85 Online at DOI: 10.1111/j.1439–0396.2009.00983.x.

Lamers JPA, Martius C, Khamzina A, Matkarimova M, Djumaeva DM, Eschanov R (2010) Green foliage decomposition in tree plantations on degraded, irrigated croplands in Uzbekistan, Central Asia. Nutr Cycl Agroecosyst 87: 249–260

Peoples MB, Crasswell ET (1992) Biological nitrogen fixation: Investments, expectations and actual contributions to agriculture. Plant Soil 141: 13–39

Quayle S (2001) Farm forestry: Its role in rural property management. Agriculture Notes, AG0989. Australia, State of Victoria, Department of Primary Industries, pp 4

Sanginga N (2003) Role of biological nitrogen fixation in legume based cropping systems; a case study of West Africa farming systems. Plant Soil 252: 25–39

Scheer C, Wassmann R, Kienzler K, Ibragimov N, Lamers JPA, Martius C (2008) Methane and nutrous oxide fluxes in annual and perennial land-use systems of the irrigated area in the Aral Sea Basin. Glob Change Biol 14: 1–15

Smit B, BrayJ, Keddie P (1991) Identification of marginal agricultural areas in Ontario, Canada. Geoforum 22, 3, 333–346

Streeter J (1988) Inhibition of legume nodule formation and N_2 fixation by nitrate. CRC Critical Reviews in Plant Science 7: 1–23

Sutton W, Whitford P, Stephens EM, Galinato SP, Nevel B, Plonka B, Karamete E (2007)

Integrating environment into agriculture and forestry. Progress and prospects in Eastern Europe and Central Asia. Washington, DC, The World Bank

Uliassi DD, Ruess RW (2002) Limitations to symbiotic nitrogen fixation in primary succession on the Tanana River floodplain. Ecology 83: 88–103

UNFCCC (1998) The Kyoto Protocol to the Convention on Climate Change. UNEP/IUC, France

Vlek PLG, Martius C, Wehrheim P, Schoeller-Schletter A, Lamers JPA (2001) Economic restructuring of land and water use in the Region Khorezm (Uzbekistan) (Project Proposal for Phase I). ZEF Work Papers for Sustainable Development in Central Asia, 1 [Online] http://www.khorezm.unibonn.de/downloads/WPs/ZEF-UZWP01_proposal.pdf. 75 pp

Wheeler CT, Narayanan R, Parthiban KT, Kesavan A, Surendran C (1996) Proceedings of the national seminar on the root microbiology of tropical nitrogen fixing trees in relation to nitrogen and phosphorus nutrition. Coimbatore, Tamil Nadu Agricultural University, 55–59

Zomer RJ, Trabucco A, Bossio DA, Verchot LV (2008) Climate change mitigation: A spatial analysis of global land suitability for clean development mechanism afforestation and reforestation. Agric Ecosyst Environ 126: 67–80

Sebastian Fritsch, Christopher Conrad, Teresa Dürbeck, Gunther Schorcht

3.6 Mapping marginal land in Khorezm using GIS and remote sensing techniques

Abstract

Marginalization of agricultural land in the Khorezm region, perceived as decline in crop production, is affecting livelihoods of the growing population. Unproductive areas could be used more efficiently through alternative land uses, such as afforestation, or via improved distribution of irrigation water. For enabling these options, essential information is needed on the spatial distribution of marginal land. To this end, we present a methodology for spatial modeling of marginal land based on GIS (Geographic Information System) and remote sensing data and techniques. The approach comprises a multi-criteria analysis of the weighted data including remote sensing-based, decadal cotton yields as an input with the highest weight. The output produced is a raster map with a resolution of 250 m that distinguishes between different classes of land marginality. The results show that 34 % of the agricultural land in Khorezm can be classified as marginal. The marginality increases with the distance to irrigation water intake points. The map allows the localization of marginal areas and the characterization of Water Consumers Associations with respect to marginality. Comparison with field data shows the plausibility of the results. It can be concluded that the Khorezm region is strongly affected by land marginalization. The generated map of marginal land provides the basis for further analysis and discussions among local stakeholders and scientists. From the perspective of land and water use planning, the map can support the adjustment of water distribution schemes and localization of sites for alternative, drought-tolerant crops.

Keywords: multi-criteria analysis, MODIS, time series, spatial analyses, marginality, Central Asia

3.6.1 Conceptualization of land marginality in irrigated crop production systems

Most agricultural areas in arid Central Asia can be classified as marginal in terms of prevailing natural conditions (Cassel-Gintz et al. 1997), and allow cropping practices only under irrigation, fertilizer, and pesticide inputs. Freshwater applications are also needed to cope with soil salinization, which lowers the suitability of land for agricultural production. Such as done in humid regions of Western Europe, the land marginality for agriculture can be further viewed with respect to the economic value that is gained from a certain piece of land (e. g. via crop yields). According to Smit et al. (1991, p.335), the marginality of land 'relates fundamentally to the economic viability of land uses'. Thus, the holistic definition of land marginality considers both environmental and economic factors (Deal 2006). Such definition is relevant for the Khorezm region of Uzbekistan, where the share of the rural population that depends on agriculture for livelihood and food security (Bekchanov et al. 2010) is high (78 %). Agriculture in Khorezm often operates at the margin of feasibility, for example because of high groundwater salinity and tables (Ibrakhimov et al. 2011) or the low water availability (Bekchanov et al. 2010). The risk of yield uncertainties is high due to the region's location in the downstream part of the Amudarya river, where the water availability is influenced by many external, climate-related and political factors (Manschadi et al. 2010). This is especially challenging in the light of the population growth, which reached 50 % between 1989 (1.014 million) and 2007 (1.504 million). Drought events such as those in 2000/2001 and 2008 had a significant impact on livelihoods in Khorezm and led to a decrease in both area under crop production and yields (FAO/WFP 2000). Moreover, the pressure on the region's water resources is likely to increase in future due to the climate change impact, water demands by upstream countries (Afghanistan), and the on-going energy-water nexus disputes between Uzbekistan and the upstream countries. In this light, sustainable land-use management in the Khorezm region is essential.

The challenge being addressed here is identification of marginal land, i.e., the cropland areas in Khorezm that could benefit from remedial measures or alternative land uses. The identification of marginal land is a prerequisite for implementing alternatives. In our analysis, we follow the definition of marginal lands formulated by Smit et al. (1991, p.336) as 'land parcels or types for which there are few viable uses, or for which uses are either only just viable or not always viable'. This definition corresponds to that of local farmers, who perceive marginal land as 'unproductive' (Dürbeck 2010), i.e., unable to support long-term cultivation of major crops such as cotton, wheat, rice, or maize. According to both, interviews with farmers and consultations with scientific experts, land constantly producing low yields is seen as an indicator of marginality, which is in line with the scientific perception (Deal 2006). This unproductive land is the result of multiple factors, including poor soil quality due to salinity, unsound management practices and a lack or untimely supply of irrigation water.

A general framework for land evaluation has been used extensively for a variety of purposes; its procedures are manifold and include, for example, consultation with stakeholders, land resource surveys, and spatial modeling and analysis (FAO 2007). The latter includes the use of GIS (Geographical Information System), and has often been applied in conjunction with a land suitability assessment, which allows the explicit evaluation of spatial problems concerning site search and site selection (Malczewski 1999). Besides GIS, remote sensing has furthermore offered the opportunity to include up-to-date information in such analyses. However, this has been mostly restricted to remote sensing-derived land-use or land-cover maps (e.g. Bandyopadhyay et al. 2009). Few studies have used higher-level information from remote sensing in the framework of a GIS-based land suitability assessment, and none have been targeted towards the identification of marginal land. This is a particular challenge given the wide number of factors determining it, as described earlier. Therefore, a framework for a land suitability assessment method based on remote sensing and GIS data for the detection of marginal land in Khorezm is proposed in the following, with a focus on the analysis of the resulting spatial distribution of marginal lands in the region.

3.6.2 Multi-criteria analysis for mapping marginal land

A multi-criteria analysis (MCA) based on GIS and remote sensing data was chosen for detecting marginal lands. This technique belongs to the suite of tools for land suitability assessment (Malczewski 2004). The land suitability assessment is generally defined as 'an appraisal and grouping, or the process of appraisal and grouping, of specific types of land in terms of their absolute or relative suitability for a specific kind of use' (FAO 2007, p.65) and thus can be used for detecting marginal land that has a low suitability for annual cropping.

3.6.2.1 Methodology overview

The main principle of a MCA is the evaluation of multiple variables within a weighing scheme that can vary in complexity. In terms of land suitability assessments, the variables can be understood as indicators of the suitability for a specific land use (Malczewski 2004). The weighted overlay concept, as implemented in the *weighted overlay* tool in ArcGIS (ESRI 2012), was therefore used for MCA. In technical terms, the concept of weighted overlay is to combine multiple (raster) datasets that use a common evaluation scale and weigh them according to their importance:

$$s = \sum_{i=1}^{n} (w_i \times v_i)$$

where S is the (land use) suitability, i is the layer number, w_i is the weight of layer i (between 0.1 – 1.0), and v_i is the coded value of layer i (here between 1 and 5). The weights (w_i) show the importance/relevance of a respective data layer (variable) as indicator for marginal land. To achieve a common evaluation scale, the data need to be classified in suitability categories (v_i), with, for instance, high values indicating a high probability of occurrence of marginal land. All weighted data layers are added up, and thus the final marginality of land is derived for each land unit (in the case of raster data: pixels). The resulting land-use suitability map depicts the distribution of areas of different marginality. According to Malczewski (1999), an important prerequisite of a successful MCA is to base all processing steps on the preference of the decision makers and on expert knowledge. This applies to the selection of the data, its evaluation, classification, and subsequent weighing. The detailed steps undertaken after expert interviews for land-use suitability evaluation based on marginality are described hereafter.

3.6.2.2 Input data selection and preparation

The selection of data for the identification of marginal land is decisive for the MCA, and often mirrors a compromise between data demand and data availability. A variety of approaches was used to concretize the MCA following the expert interviews, which indicated a combination of management-related and environmental factors. The available GIS data sets useful for the marginal land detection allowed elaboration of a comparatively coarse mapping scale only. Complementary information was derived from remotely sensed MODIS images with a resolution of 250 m pixel size. The coarse MODIS resolution impeded the precise assessment of single fields. Only parts of single fields and a mixture of different crop types and background is captured by the satellite pixels. The resulting map is therefore only an indication of marginal land at each pixel.

The common classification of cropland productivity in Uzbekistan is the so-called *bonitet*, which is the base for the governmental state order for lands of different productive potential. However, according to the opinion of experts, the existing soil *bonitet* data have to be treated carefully because they are not fully reproducible and often outdated. The information was used in the present study as a general indication of land suitability, but due to its limitations only a small weight was assigned to it.

Remote sensing-based geo-information could significantly improve the quality and amount of the input data, because it is the only reliable data source available at the regional scale. For the period 2000–2009, remote sensing-based annual land-use maps were selected based on Conrad et al. (2011). These maps comprise the spatial distribution of the following land-use classes: cotton, wheat-rice, wheat-fallow, wheat-other, rice and fallow/unused. Because the focus was on the identification of marginal lands within irrigated areas, those areas classified as unused for cropping throughout the past ten years were masked out.

The frequency of the occurrence of fallow/unused land during 2000–2009 served as an indicator of unproductive (i.e., marginal) land. As cotton is the dominant crop in the region, information on (average) cotton yield derived from the decadal (2000–2009) data was used as another input for the MCA. The light-use efficiency (LUE) approach was adapted to local environmental conditions to allow modeling the cotton yields (Shi et al. 2007; Fritsch et al. 2010). The derived crop yields were reclassified and used as the main input to the MCA (section 3.6.2.3 and Table 3.6.1).

The expert interviews suggested also including environmental factors in the MCA, such as soil or groundwater conditions as well as irrigation infrastructure. The capillary rise from the shallow, saline groundwater table enhanced by high rates of evapotranspiration is known to result in soil salinization. Whilst reliable details on soil information for the entire region were missing, data on groundwater depth and salinity were available for 1797 points throughout the region for the period from 1990 through 2004. The mean values of these point data were spatially interpolated using the Inverse Distance Weighted (IDW) method (Ibrakhimov et al. 2007) and used as additional input factors for the MCA to map the regional distribution of potentially marginal land. In the case of the groundwater table, this means that a low mean table during this period is indicative of insufficient irrigation water supply, because the measurements were generally taken in the months during which agricultural land is irrigated. This explains the classification in Table 3.6.1 (see below).

The access to irrigation infrastructure, represented in the density of the irrigation and drainage network (Tischbein et al. 2012) and weighed by its potential capacity (taken from the classification as magistral canals or canals of primary, secondary, or tertiary order) was assumed to reflect water availability for irrigation of crops and options to drain elevated water tables, and thus to reduce the soil salinization. Vector data on irrigation and drainage channels compiled from regional maps from 1960 were integrated. These data were updated using partly available cadastre maps from 2001–2006, and eventually refined using high resolution SPOT images obtained in 2006 (SPOT 2012). The channel capacity was described by its category (see above). Finally, all available vector data were converted to raster layers, adopting the pixel size of the remote sensing-based inputs. Table 3.6.1 summarizes the input data for the MCA.

3.6.2.3 Reclassification and weighing of data layers

The data compilation was followed by the classification of the data layer values according to suitability classes and a subsequent weighing of the data layers (Table 3.6.1). Reclassification was achieved by grouping the data layer values and assigning new values ranging between 1 and 5. In the following, the 1-rated conditions are referred to as 'favorable' and the 5-rated as 'unfavorable' for cropping, i.e., the higher the ranking value, the higher the land marginality. The subsequent weighing (in percentage) of the data layers was conducted in collaboration with local experts, and reflects opinions on the importance of any particular layer for the classification of marginal land levels (Dürbeck 2010). All weights summed up to the value 1, which is a prerequisite of the GIS tool. The processed data finally formed the input for the weighted overlay analysis in ArcGIS, and marginality was calculated according to (Malczewski 2004) (see section 3.6.2.1). This resulted in a floating value that ranged between 1 and 5. Next, the values were rounded to the nearest integer to finally obtain a five-class layer, which represented five different degrees of marginality: not marginal (class 1), slightly marginal (class 2), moderately marginal (class 3), marginal (class 4), and highly marginal (class 5).

3.6.2.4 Evaluation of results

Marginality of land was calculated within a pixel size of 250 m using the MCA, which rendered validation challenging due to the described limitations. Therefore, field surveys in 2010 were conducted in four Water Consumers Associations (WCAs) and consisted of interviews with local experts, farmers and land users, who pointed out locations that they perceived as marginal. These land areas were subsequently mapped and results compared with the MCA findings.

Table 3.6.1 Data layers, data ranges, classified values and weights used for the weighted overlay analysis in ArcGIS.

Data layer	Data range	Classification	Weight [%]
Cotton yield 2000–2009 [t/ha]	0–3	Natural breaks*	30
Occurrence of fallow/unused land during 2000–2009	0–10	2–0 = 1 4–3 = 2 6–5 = 3 8–7 = 4 10–9 = 5	20
Canal density [km/km²]	0–3.32	Natural breaks*	15
Collector density [km/km²]	0–1.79	Natural breaks*	10
Groundwater depth [m]	0–3	≤0.5 = 1 0.5–1 = 2 1–1.5 = 3 1,5–2 = 4 2–3 = 5	10
Groundwater salinity [g/l]	0 – >4	≤0.5 = 1 0.5–1.5 = 2 1.5–2 = 3 2–4 = 4 >4 = 5	10
Soil *bonitet*	0–100	81–100 = 1 61–80 = 2 41–60 = 4 <40 = 5	5

*Natural Breaks: This algorithm is implemented in ArcGIS and groups variables according to breaks inherent in the data by maximizing the difference between a given number of classes (ESRI 2012).

3.6.3 Spatial distribution of marginal land and policy implications

The comparison of the local perception and findings of the MCA show that the general tendencies of marginality were, with an accuracy of around 80 %, well captured by the MCA. However, the comparison between model results and farmer information also reveals that in addition to the marginal areas indicated by farmers, more areas were characterized as being marginal by the MCA. The reasons for the overestimation of the MCA have to be identified by further research. The overarching findings nevertheless show that the MCA approach can be efficient in a rapid assessment of land marginality for cropping practices over larger areas.

The following results correspond to a total irrigated area of 330,000 ha, and were calculated using the coarse resolution MODIS data. Due to the relatively small field sizes in Khorezm (as compared to the pixel size of MODIS), the results from MODIS tend to overestimate the cropped area. Yet, although the estimated size differs from the official size of 270,000 ha (Vlek et al. 2012), it still is useful for analyzing the regional distribution of marginal land and trends thereof because the assessment is based on repeatedly measured objective data.

Altogether, 41,563 ha in Khorezm (12.4 % of total irrigated land) were classified as 'slightly marginal'. The classes 'moderately marginal' and 'marginal' accounted for 55,346 ha (16.5 %) and 15,928 ha (4.8 %), respectively. The class 'highly marginal' covered 1,542 ha (0.46 %) of irrigated land in Khorezm. Within the irrigation system of the Khorezm region, 114,379 ha (34.2 %) of marginal land were identified, which was unevenly distributed throughout the region (Figure 3.6.1, left); hence the share of marginal lands differed according to district. Marginal conditions occurred mainly at the desert fringes, especially in the western and southern parts of Khorezm. However, smaller patches of marginality were also detected in the center of the irrigation system.

Figure 3.6.1 (right) shows the average marginality of agricultural land per WCA, which principally exhibits a similar spatial distribution to that mentioned above. This illustration can help in spatial planning, because WCAs with a high average marginality of agricultural land can be viewed as potential 'risk areas', particularly in water-scarce years, thus calling for adjustment of the current management practices. Consequently, the identified areas can be prioritized in further investigations by taking additional information into account such as the total irrigated area per WCA. For example, if a high marginality is observed over a large irrigated area, the corresponding WCA should be examined more closely. Thus, mapping the land marginality can serve as an input for more complex indicator systems when combined with additional data.

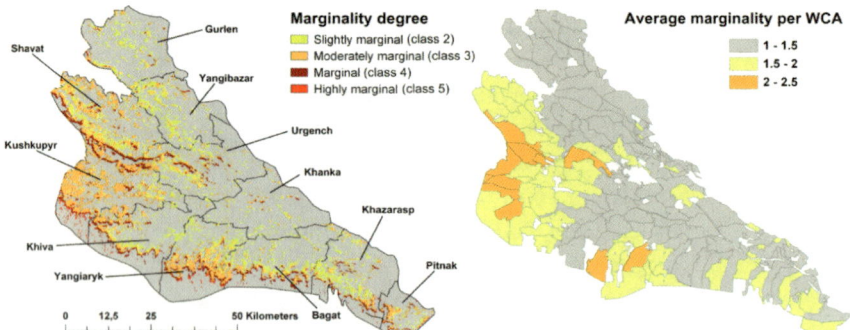

Figure 3.6.1: Spatial distribution of marginal land in the Khorezm region
Left: Marginality degree at the pixel scale, grey areas indicate land that is not marginal or does not represent agriculture (class 1). Right: average marginality per WCA

Occurrence patterns indicate a high correlation between the land marginality and the distance to river water intake points (Figure 3.6.2). The results of the spatial aggregation of the marginality per WCA, grouped according to the distance from canals to the water intake points from the Amudarya river (Conrad et al. 2007), suggest that marginality is determined by the location within the irrigation system of Khorezm, i.e., the further away from water intake points, the higher the general occurrence of marginal land. However, Figure 3.6.2 also shows that agricultural land with a better water access (often represented by a closer location to the water intakes) can also be affected by marginality: even in the class that represents the closest distance to intake points, total marginal area can reach up to 20 %. The presented methodology can consequently be used to locate such areas.

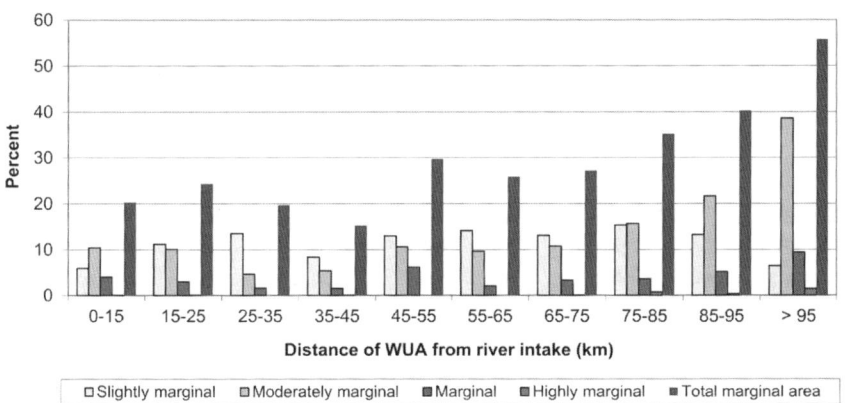

Figure 3.6.2: Distribution of marginal land depending on distance to points of water intake from the Amudarya river

The results presented in Figures 3.6.1 and 3.6.2 largely correspond to the trends identified by corresponding research in the Khorezm region. For example, the identified districts with the highest share of marginal land (Kushkupir, Shavat, Yangiarik) were also found to be amongst the ones with the lowest average water productivity in the period 2000 to 2007 (Bekchanov et al. 2010). Furthermore, Conrad et al. (2007) showed that those administrative districts were characterized by a lower seasonal evapotranspiration, which was in contrast to the districts of high crop productivity (e.g., Gurlen). Finally, Conrad et al. (2007) also postulated that evapotranspiration decreased with the distance to water intake points. This agrees with the land marginality trend revealed by the adapted MCA approach.

How can the above findings on the spatial distribution of marginal land be utilized? One possibility might be to establish tree plantations on marginalized croplands, which could help to support farmers' livelihoods (Lamers et al. 2008). Such plantings can meliorate the soil via bio-drainage, increase fertility and sequestrate carbon (Hbirkou et al. 2011; Khamzina et al. 2012). Although the combined measures could help to establish viable land uses on marginal land in Khorezm, the existing state order presently limits alternative land use options to smaller areas (Bobojonov et al. 2012); this should be taken into account.

3.6.4 Conclusions

We have presented an approach for mapping marginal land using remote sensing data and geo-information techniques. The developed methodology and the resulting maps are an objective basis for the development of alternative land-use options, like the introduction of alternative, more drought-tolerant crops or tree plantations, and for discussions amongst local stakeholders and scientists focusing on sustainable land management.

The present findings indicate that marginal land is mostly concentrated near desert areas, but it is also scattered as patches throughout the region. A GIS aggregation allowed identifying WCAs with a comparatively high marginality of agricultural land. Given a rising water scarcity in the region, this spatial information could help to allocate water resources more efficiently, particularly in drought years.

The results represent the scale of MODIS data and can only be an indication of the marginality of actual single fields. This scale represents parts of fields or a mixture of crops and background information, which can always lead to an over- or underestimation of the 'true' marginality. However, it helps in delineating risk areas that should be prioritized in designing remedial options, and further investigated using satellite data with a higher resolution.

References

Bandyopadhyay S, Jaiswal RK, Hegde VS, Jayaraman V (2009) Assessment of land suitability potentials for agriculture using a remote sensing and GIS based approach. International Journal of Remote Sensing 30 (4): 879–895

Bekchanov M, Karimov A, Lamers JPA (2010) Impact of water availability on land and water productivity: A temporal and spatial analysis of the case study region Khorezm, Uzbekistan. Water 2: 668–684

Bobojonov I, Lamers JPA, Djanibekov N, Ibragimov N, Begdullaeva T, Ergashev A,

Kienzler K, Eshchanov R, Rakhimov A, Ruzimov J, Martius C (2012) Crop diversification in support of sustainable agriculture in Khorezm. In: Martius C, Rudenko I, Lamers JPA and Vlek PLG (Eds.) Cotton, water, salts and soums – economic and ecological restructuring in Khorezm, Uzbekistan. Springer, Dordrecht, pp 219–233

Cassel-Gintz MA, Lüdeke MKB, Petschel-Held G, Reusswig F, Plöchl M, Lammel G, Schellnhuber HJ (1997) Fuzzy logic based global assessment of the marginality of agricultural land use. Climate Research 8: 135–150

Conrad C, Colditz R, Dech S, Klein D, Vlek PLG (2011) Temporal segmentation of MODIS time series for improved crop classification in Central Asian irrigation systems. International Journal of Remote Sensing 32 (23): 8763–8778

Conrad C, Dech SW, Hafeez M, Lamers JPA, Martius G, Strunz G (2007) Mapping and assessing water use in a Central Asian irrigation system by utilizing MODIS remote sensing products. Irrigation and Drainage Systems 21 (3–4): 197–218

Deal J (2006) The relationship between economically and environmentally marginal land. In: American Agricultural Economics Association Meeting, 34 p. Long Beach

Dürbeck T (2010) Land Suitability Analyse und marginales Land in der Bewässerungsregion Khorezm, Usbekistan. MSc dissertation, University of Würzburg

ESRI (2012) ArcGIS Resource Center. http://resources.arcgis.com/ (last accessed: 3.1.12)

FAO (2007) Land evaluation – towards a revised framework. Land and water discussion paper (6): 107. Rome

FAO/WFP (2000) Crop and food supply assessment mission to Karakalpakstan and Khorezm regions of Uzbekistan. Rome

Fritsch S, Conrad C, Dech S (2010) A MODIS-based approach for crop yield prediction in irrigation regions of Central Asia using light-use efficiency modeling. In: Proceedings of the ESA Living Planet Symposium, Bergen, Norway, 28th June – 2nd July 2010, ESA, SP-686 (CD-Rom)

Hbirkou C, Martius C, Khamzina A, Lamers JPA, Welp G, Amelung W (2011) Reducing topsoil salinity and raising carbon stocks through afforestation in Khorezm, Uzbekistan. Journal of Arid Environments 75: 146–155

Ibrakhimov M, Khamzina A, Forkutsa I, Paluasheva G, Lamers JPA, Tischbein B, Vlek PLG, Martius C (2007) Groundwater table and salinity: Spatial and temporal distribution and influence on soil salinization in Khorezm region (Uzbekistan, Aral Sea Basin). Irrigation and Drainage Systems 21 (3–4): 219–236

Ibrakhimov M, Martius C, Lamers JPA, Tischbein B (2011) The dynamics of groundwater table and salinity over 17 years in Khorezm. Agricultural Water Management 101: 52–61

Khamzina A, Lamers JPA, Vlek PLG (2012) Conversion of degraded cropland to tree plantations for ecosystem and livelihood benefits. In: Martius C, Rudenko I, Lamers JPA and Vlek PLG (Eds.) Cotton, water, salts and Soums – economic and ecological restructuring in Khorezm, Uzbekistan. Springer, Dordrecht, pp 235–248

Lamers JPA, Bobojonov I, Khamzina A, Franz J (2008) Financial analysis of small-scale forests in the Amu Darya lowlands of rural Uzbekistan. Forests, Trees and Livelihoods 18: 373–386

Malczewski J (1999) GIS and multicriteria decision analysis. John Wiley & Sons, New York

Malczewski J (2004) GIS-based land-use suitability analysis: a critical overview. Progress in Planning 62 (1): 3–65

Manschadi AM, Oberkircher L, Tischbein B, Conrad C, Hornidge A-K, Bhaduri A, Schorcht G, Lamers JPA, Vlek PLG (2010) White Gold' and Aral Sea disaster – towards more efficient use of water resources in the Khorezm region, Uzbekistan. Lohmann Information 45 (1): 34–47

Shi Z, Ruecker GR, Mueller M, Conrad C, Ibragimov N, Lamers JPA, Martius C, Strunz G, Dech S, Vlek PLG (2007) Modeling of cotton yields in the Amu Darya river floodplains of Uzbekistan integrating multi-temporal remote sensing and minimum field data. International Journal of Agronomy 99: 1317–1326

Smit B, Bray J, Keddie P (1991) Identification of marginal agricultural areas in Ontario, Canada. Geoforum 22 (3): 333–346

SPOT (2012) SPOT Image. www.spotimage.fr (last accessed: 3.1.12)

Tischbein B, Awan U, Abdullaev I, Bobojonov I, Conrad C, Jabborov H, Forkutsa I, Ibrakhimov M, Poluasheva G (2012) Water management in Khorezm: current situation and options for improvement (hydrological perspective). In: Martius C, Rudenko I, Lamers J and Vlek P (Eds.) Cotton, water, salts and soums – economic and ecological restructuring in Khorezm, Uzbekistan. Springer, Dordrecht, pp 69–92

Vlek P, Lamers JPA, Martius C, Rudenko I, Manschadi A, Eshchanov R (2012) Cotton, water, salts and Soums – research and capacity building for decision-making in Khorezm, Uzbekistan. In: Martius C, Rudenko I, Lamers JPA and Vlek P (Eds.) Cotton, Water, Salts and Soums – Economic and Ecological Restructuring in Khorezm, Uzbekistan. Springer, Dordrecht, pp 3–22

Elena N. Ginatullina, Laurel Saito, Lisa Atwell, Diana B. Shermetova, Dilorom Fayzieva, John P. A. Lamers, Sudeep Chandra, Margaret Shanafield

3.7 Water chemistry and zooplankton communities in drainage lakes in downstream Amu Darya, Central Asia

Abstract

In lakes, zooplankton are critical for energy transfer from primary production to upper trophic levels, and as a top-down control for phytoplankton abundance and composition. In addition, zooplankton themselves are strongly influenced by physical, chemical and nutrient characteristics of their habitat. We examined seasonal variability of zooplankton and lake characteristics in four shallow, saline lakes in northwestern Uzbekistan between June 2007 and November 2008. Twenty-nine zooplankton taxa were identified, including 13 rotifers, 7 cladocerans and 9 copepods. Because of shallow lake depths, salinity changed dramatically and appeared to negatively affect both zooplankton density/biomass and diversity when concentrations increased past ~3 g l^{-1}, and influence zooplankton community composition. For three slightly saline lakes, composition was dominated by Cyclops Thermocyclops vermifer *and* Cyclops vicinus, *cladoceran* Diaphanosoma mongolianum, *and rotifers* Brachionus plicatilis *and* Keratella quadratas. *The zooplankton community in the more saline fourth lake was dominated by the calanoid copepod* Arctodiaptomus salinus, *rotifer* Brachionus plicatilis, *copepods* Apocyclops dengizikus *and* Harpacticoida spp., *and cladoceran* Moina salina. *Generally low diversity may have been due to poor littoral habitat heterogeneity and the relatively young lake ages, as well as fluctuating salinity. While temperature corresponded with seasonal cycles of zooplankton abundance and community composition, salinity periodically appeared to override temperature as the dominant factor when salinity became high.*

Keywords: saline lakes, shallow lakes, zooplankton biodiversity, zooplankton abundance, Aral Sea Basin

3.7.1 Introduction

In water-scarce semi-arid regions such as Central Asia, understanding the balance of using limited water resources for crop production versus maintaining ecosystems for fish production or ecosystem services is an ongoing challenge (Oberkircher et al., 2011). Due to increased irrigation in Uzbekistan since the middle of the 20th century, delta lakes southeast of the Aral Sea that were once fed by the Central Asian river called the Amu Darya have disappeared. Simultaneously, other lakes in the irrigated regions have increased in volume and surface area due to intake from groundwater and recurrent irrigation water, forming a system of lakes in the Khorezm province in northwestern Uzbekistan. Most of these lakes have surface areas of several hectares (ha) or less, and they are connected to each other and to the Amu Darya by a collector-drainage network (Figure 3.7.1). A potential water resource for irrigation or aquaculture in the Khorezm region that remains largely unstudied is the more than 450 lakes (>1 ha) located in natural depressions in the landscape. In the Aral Sea basin, about 37 % of agricultural return flows are discharged to these types of natural depressions (UNEP 2005), which may become large enough to support fisheries.

The chemical composition of the river water has changed under the impact of drainage water discharges from irrigated fields and industries. Between 1960 and 1989, the Amu Darya's average salinity increased from 540 mg l^{-1} to 1161.2 mg l^{-1}. The chemical character of the Amu Darya water has changed from calcium-carbonate-dominated to sodium- or potassium chloride-dominated, with sulfate domination of anions (Pavlovskaya 1995). Since Amu Darya water is the main source of water input to the lakes (Scott et al. 2011), such changes in Amu Darya water chemistry directly affect the lakes. Furthermore, the lakes may be undergoing a process of gradual increase in water salinity due to inflows of drainage water and evaporation, which could lead to an environment unsuitable for many aquatic organisms. In the Khorezm region, some families are becoming increasingly interested in local fish production as a means to support the nutritional needs of their family members and increase their economic potential. While the broader Central Asia region including Uzbekistan was known for immense fish harvests in the past, little is known about the potential for fish production within the smaller lakes of this region despite supply of nutrient rich water from agricultural return water to these lakes. One of the first items to investigate would be the food supply available to fishes.

In this study, we examine zooplankton density and biomass in relation to water quality of representative lakes in Khorezm. Zooplankton are valuable study subjects because they are widely distributed geographically, fulfill many functional roles ecologically, have short generation times, and are subject to both top-down and bottom-up trophic influences in a lake ecosystem. The ratio

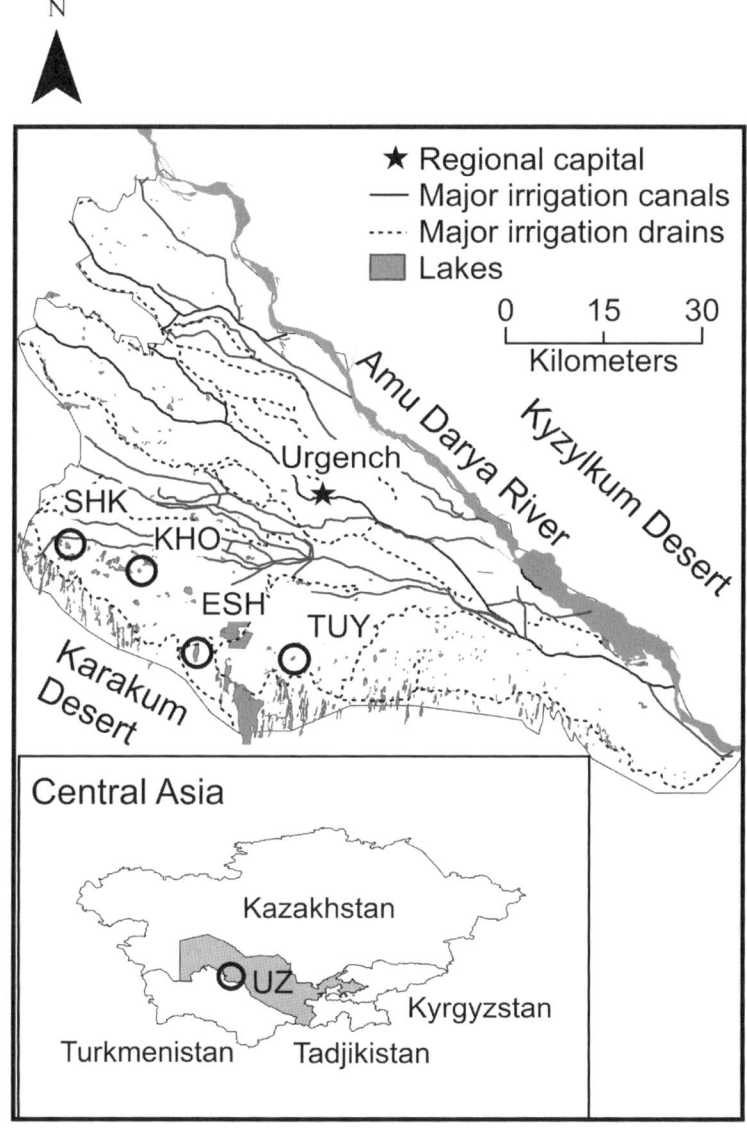

Figure 3.7.1 Khorezm province in northwestern Uzbekistan (UZ) showing location of study lakes Eshan Rabat (ESH), Khodjababa (KHO), Shurkul (Koshkopir) (SHK), and Tuyrek (TUY). Amu Darya flows from the southeast.

between zooplankton and phytoplankton biomass is sensitive to top-down control through cascading trophic interactions from fish to zooplankton and on to phytoplankton (Hessen et al. 2003). Conversely, zooplankton feed on phytoplankton and act as an important link in the lentic food chain by making phytoplankton energy available to upper trophic levels and supporting fisheries (Carpenter et al. 1985; Shurin et al. 2000). Zooplankton can be good indicators of community ecosystem health because they are easily identified and their generation time may be short enough to respond quickly to acute stress, but long enough to integrate the effect of chronic problems (Whitman et al. 2002).

Khorezm is a lowland plain region located between the Karakum and Kyzylkum deserts (elevation approximately 100 m). The average amount of precipitation in the Khorezm region is less than 100 mm per year, with most precipitation occurring from October to April. The region is characterized by hot dry summers and cold short winters, with average annual temperatures of about 13 °C (Conrad et al. 2007; Khamzina et al. 2008). In the Khorezm region, nitrate and total dissolved phosphorus in the lakes are low in spite of the heavy application of nutrient fertilizers on the surrounding agricultural lands, and the climate is variable (Saito et al. 2010; Shanafield et al. 2010). Since little work has been done to describe the aquatic ecology of the lakes in this region, we describe water characteristics and zooplankton communities in four shallow lakes characteristic of the Khorezm region. While these lakes are important to the local population to support fisheries, the small size of the lakes makes them sensitive to changes in water inflow and subsequently vulnerable to dramatic fluctuations in salinity. Such changes in salinity are likely to impact both the density and composition of the zooplankton community within the lakes. Because cladocerans generally are more sensitive to changes in water quality than are copepods, we hypothesized that the three less saline lakes would have: a) greater zooplankton diversity due to the presence of cladocerans and, b) greater zooplankton biomass due to the higher productivity rate of cladocerans compared to copepods. Our objectives in this study were to observe zooplankton and water chemistry changes throughout an 18–month period in 4 lakes of differing physical conditions to determine how zooplankton diversity and density/biomass were impacted by water quality in these shallow agricultural drainage lakes.

3.7.2 Materials and Methods

3.7.2.1 Study sites

The majority of lakes in Khorezm are shallow (typical depth is 1.5 – 2 m), ranging in area from about 2 to over 100 ha. We focused on four lakes that represent the size and volume typical of the region and that contain existing fish stocks that are of interest to local families for aquaculture (Table 3.7.1): Tuyrek (TUY), Khodjababa (KHO), Shurkul (Koshkopir) (SHK), and Eshan Rabat (ESH). The shorelines of TUY, SHK and KHO are overgrown by canes (*Phragmites vulgaris*), and the lake bottoms of SHK and ESH are covered by an aquatic macrophyte (*Myriophyllum aquaticus* for SHK, and *Chara* sp. for ESH). TUY and KHO lakes are turbid and dominated by phytoplankton, whereas the other two lakes (ESH and SHK) are clear and dominated by macrophytes.

Table 3.7.1: Characteristics of four agricultural drainage lakes sampled in the Khorezm region between June 2007 and November 2008. Surface area and volume are from Shanafield et al. (2010).

Lake name	Abbreviation	Latitude (°)	Longitude (°)	Surface Area (ha)	Volume (m^3)	Max. Depth (m)
EshanRabat	ESH	41.325	60.399	158.5	866,935	1.5
Khodjababa	KHO	41.437	60.288	23.6	265,298	2.0
Shurkul (Koshkopir)	SHK	41.471	60.156	48.7	770,952	3.0
Tuyrek	TUY	41.321	60.576	19.3	192,725	3.0

3.7.2.2 Sample collection

Quantitative collections of zooplankton were conducted approximately at the end of each month from June 2007 to November 2008. Eighty-six zooplankton samples were collected from the same region of each lake (mainly from pelagic zones in the deeper areas of the lakes) using a small conical net (diameter 18 cm, length 35 cm, mesh size No 76) for collecting the zooplankton. Collections were made by drawing the net from a depth of 1 – 1.5 m to the surface at 5 different locations in the pelagic regions of each lake. Counts and taxonomic identifications were conducted on 10 – 20 ml subsamples with a binocular microscope with a counting camera and then on an optical microscope at 50x and 100x magnification. Zooplankton species were identified according to taxonomic keys: rotifers (Kutikowa 1970), adult copepods (Borutskii 1952; Monchenko 1974) and cladocerans (Zalolichin 1995). Density by volume of zooplankton was expressed in thousand individuals m^{-3}, based on assumed net filtration of 100 %. Biomass of zooplankton was expressed in mg m^{-3} and mass was calculated as wet weight using a length-weight regression according to individual size of organisms (e.g. Salazkinet al. 1984; Ruttner-Kolislo 1977).

A portable YSI-85 instrument was used to measure the following parameters *in situ*, reported as the mean from multiple depths: temperature, dissolved oxygen (concentration and % saturation), specific conductivity, pH and salinity (Saito et al. 2010). A grab water sample was collected from the center of the lakes just below the surface approximately monthly for laboratory analysis of dissolved nitrogen and phosphorus species and major ions at the Hydrometeorological Research Institute (NIGMI) in Tashkent. Nitrogen species were analyzed according to modified United States Environmental Protection Agency (USEPA) 350.1 and 353.1 methods with method detection limits (MDL) of 0.001, 0.0005, and 0.005 mg l^{-1} for nitrite, nitrate and total ammonia nitrogen (TAN), respectively. Total nitrogen (TN) was calculated by summing nitrite, nitrate, and TAN. Total dissolved phosphorus (TDP) was analyzed at NIGMI based on USEPA 365.3 methods using photometric method with ascorbic acid and a MDL of 1 µg l^{-1}. To determine consistency of our field samples and precision of lab procedures, duplicates for nitrogen and phosphorus analyses were completed on 10 % of the samples and 72 % of the duplicates were within 15 % of each other. About 13 % of the samples were field duplicates, and 50 % of the duplicates were within 15 % of one another. Mean values of duplicates were used to represent measured values. Major ions were analyzed based on methods provided by Semenov (1977) (titrimetric method for calcium, chloride, magnesium and sulfate, back titrimetric method for hydrocarbonates, flame photometric method for silicon and sodium, and complexometric method for water hardness). Replicates for major ion analyses were done on 9 % of the samples; 77 % were within 5 % of one another, and 95 % were within 15 % of one another.

3.7.3 Results

3.7.3.1 Hydrochemistry

The seasonal dynamics of lake salinity increased from summer to winter (maximum salinity in January), with a reduction in salinity concentrations in February or March (Figure 3.7.2a). During 2007, salinity concentrations for three of the lakes (TUY, KHO and SHK) were within 2 – 4 g l^{-1}, however for ESH, salinity reached 15 g l^{-1} at the end of 2007. As a result of a dry spring in 2008, the salinity concentrations in all four lakes increased to 5 – 11 g l^{-1} for TUY, KHO, and SHK, and almost 50 g l^{-1} for ESH in July 2008. The predominant anions in all four lakes were chloride and sulfate, which were both measured at about 1 g l^{-1} for TUY, KHO, and SHK and about 6 g l^{-1} for ESH in 2007 (Saito et al., 2010). The pH in all four lakes generally ranged from 7.0 – 8.5 and often exceeded 8.0 if the salinity was higher than 4 – 5 g l^{-1}; pH in ESH dipped as low as 6.2 g l^{-1} in October 2008 (Figure 3.7.2b). Oxygen concentration fluctuated from <1 to over 13 mg l^{-1}, with maximum concentrations generally observed in winter in TUY and in summer in KHO and ESH (Figure 3.7.2c). Average concentrations in ESH, KHO, SHK, and TUY between June 2007 and November 2008 were 7.1, 8.0, 7.4, and

8.1 mg l^{-1}, respectively. Low dissolved oxygen was observed in TUY in summer 2007. During winter 2007–2008 all lakes were ice covered between mid-December and February, and nearly anoxic conditions were observed in ESH, KHO, and SHK. The winter anoxic conditions in SHK and ESH were likely related to decay of the previous season's aquatic plants under ice.

TDP varied between 0.01–0.09 mg l^{-1} during 2007–2008 (Figure 3.7.2d). Between June and November 2007, TN concentrations were low, except for a value in June 2007 at TUY that was 3.0 mg l^{-1}. At the beginning of winter, higher TN concentrations were observed, with highest concentrations in January-February 2008. Higher concentrations were also observed later in the year in April 2008. TN concentrations in 2008 were 0–2.6 mg l^{-1} for KHO, SHK, and TUY, and 0–1.5 mg l^{-1} for ESH.

3.7.3.2 Temperature conditions

The maximum measured water temperature of the studied lakes reached 28–30 °C in July-August (Figure 3.7.3). Temperature was lower beginning in October, decreasing to 0–4 °C in winter months; in ESH in winter 2008, temperature was observed at -1 °C (January-February 2008). By March 2008, lake water temperatures had increased to between 5 and 10 °C, and in September 2008 lake water temperatures were above 20 °C.

3.7.3.3 Zooplankton

Zooplankton species composition of the lakes was relatively non-diverse and consisted of twenty-nine forms: thirteen species of rotifers, seven species of cladocerans and nine species of copepods (Table 3.7.2). The maximum diversity of zooplankton species was observed in spring 2008 in three lakes (KHO, SHK, and TUY), where there were generally rotifers: *Keratella quadrata, Euchlanis dilatata, Brachionus leudigii, B. angularis, Filinia longiseta, Notholca angulata, Asplanchna priodonta and Anueraeopsis sp*. In June 2008 we discovered in these lakes additional rotifer species: *Lecane luna, Polyarthra sp., Keratella tropica* and *K. tropica reducta*. The zooplankton communities during the growth period 2007–2008 were dominated by two rotifer species: *B. plicatilis* for TUY and SHK, and *K. quadrata* for KHO. During 2007–2008, three lakes (KHO, SHK, and TUY) had high densities of crustaceans: ectang copepods *Thermocyclops vermifer* (from May to October) and *Cyclops vicinus* (from November to April). The cyclop copepod *Mesocyclops ogunnus* and harpacticoid copepod *Onychocamptus mohammed* were observed more rarely in these lakes. Cladoceran

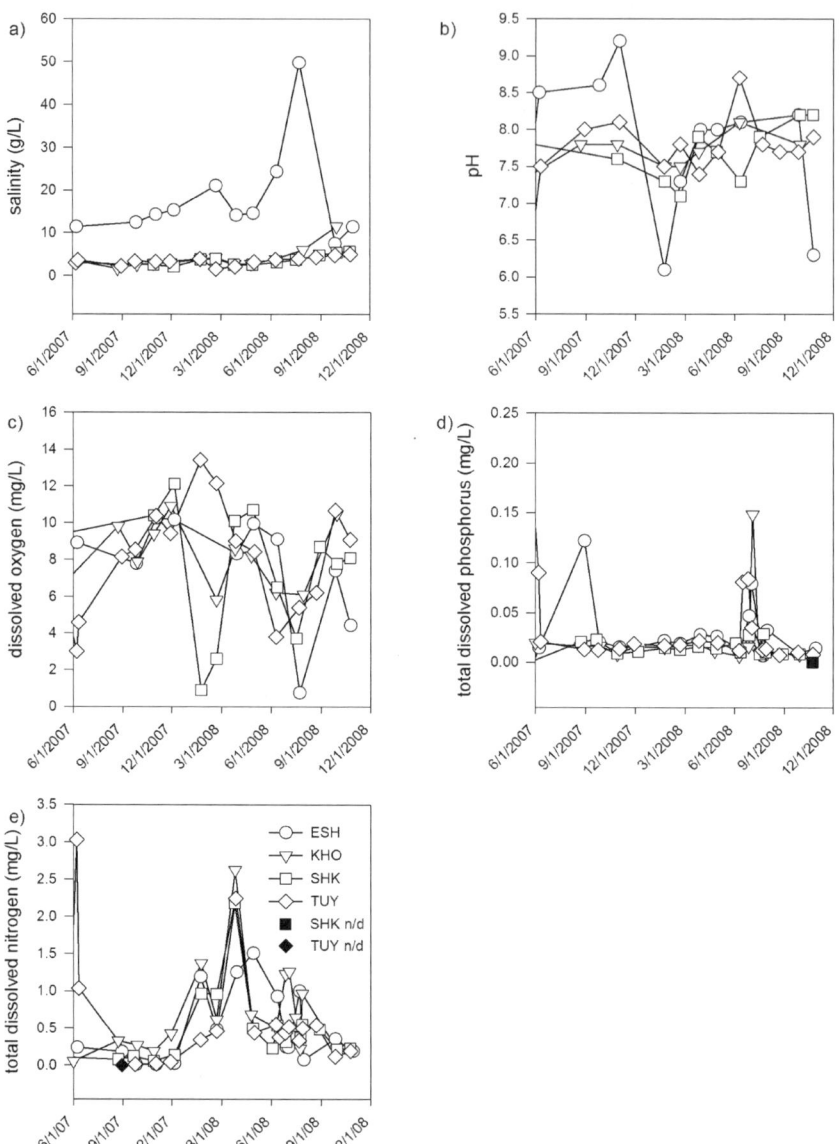

Figure 3.7.2 Measured a) lake water salinity, b) pH, c) dissolved oxygen, d) total dissolved phosphorus and e) total nitrogen for Eshan Rabat (ESH), Khodjababa (KHO), Shur (Koshkopir) (SHK), and Tuyrek (TUY) lakes. n/d indicates total dissolved phosphorus or total nitrogen was not detected at 0.001 mg l^{-1}.

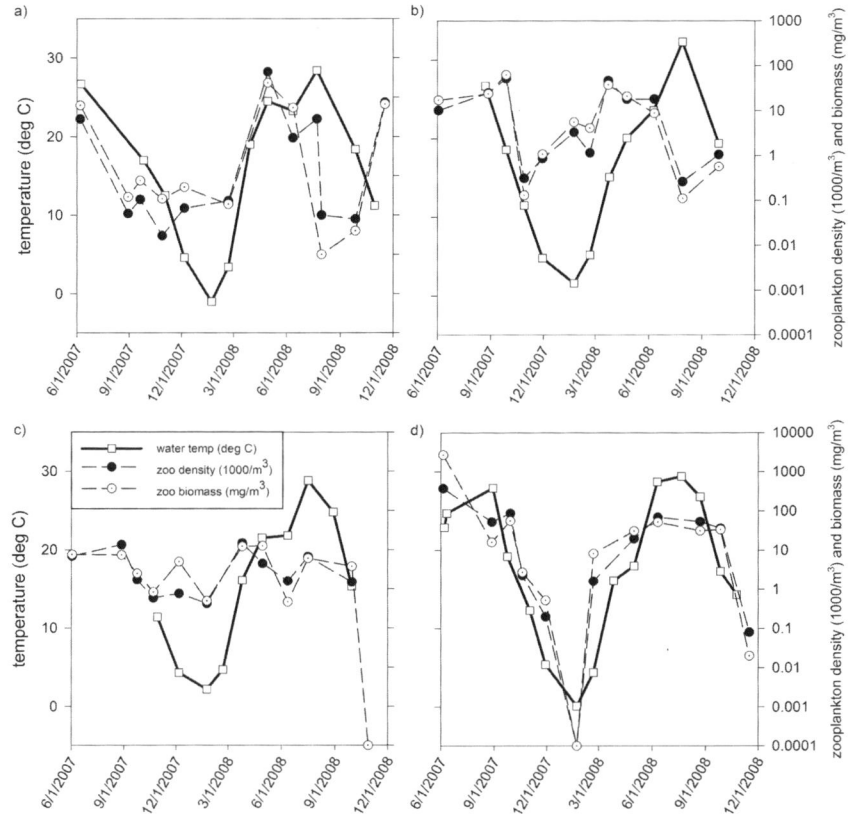

Figure 3.7.3 Water temperature versus zooplankton density and biomass for a) Eshan Rabat, b) Khodjababa, c) Shur (Koshkopir), and d) Tuyrek lakes.

Diaphanosoma mongolianum (May-October) and calanoid copepod *Arctodiaptomus salinus* played a considerable role in the zooplankton community in TUY, whereas cladocerans *Simocephalus vetulus* and *Chydorus sphaericus* were important for SHK and KHO, respectively. ESH had water salinity of more than 10 g l^{-1} higher than the other lakes except in September 2008 (Figures 3.7.1 and 3.7.4), and in turn the zooplankton community of this lake was different, consisting of widespread species *Thermocyclops vermifer*, *D. mongolianum*, *C. sphaericus* and *Alonarectangula*. The halophilic rotifer species *B. plicatilis*, calanoid copepods *Arctodiaptomus salinus*, harpacticoid copepods *Cletocamptus retrogressus*, *Onychocamptus bengalensis*, and *Schizopera sp.*, cyclopoid copepods *Apocyclops dengizicus*, and cladoceran *Moina salina* were also found in ESH in 2007. After salinity increased dramatically beginning in July 2008, species composition of ESH became even more sparse with only the rotifer *B.*

plicatilis and a small population of harpacticoid copepods remaining after salinity concentrations reached 49 g l^{-1}, though before salinity increased the zooplankton community was dominated by calanoid copepods *Arctodiaptomus salinus* (~78 % of community biomass in June 2008).

Table 3.7.2. Average abundance (thousands of individuals m^{-3}) of zooplankton species in Eshan Rabat (ESH), Khodjababa (KHO), Shurkul (Koshkopir) (SHK), and Tuyrek (TUY) lakes of the Khorezm region (Uzbekistan), during 2007 – 2008. White rows indicate species tolerant of 0 – 3 g l^{-1} salinity. Light gray shading indicates species tolerant of 3 – 10 g l^{-1} salinity. Dark gray shading indicates species tolerant of >10 g l^{-1} salinity. "ind" means only one individual was represented in the sample.

Species composition	ESH	KHO	SHK	TUY
Rotifera:				
Anueraeopsis sp.	-	-	0.07	-
Brachionus leudigii Cohn, 1862	-	-	-	ind
Bachionus angularis Plate, 1886	-	-	-	ind
Asplanchna priodonta Gosse, 1850	-	-	-	1.5
Euchlanis dilatata Ehrhg, 1832	-	-	0.1	-
Keratella quadrata Mueller, 1786	-	7.5	1.2	0.5
Keratella tropica Apstein, 1907	-	ind	-	-
Keratella tropica reducta Fadeev, 1927	-	ind	-	-
Notholca angulata Ehrhg, 1832	-	-	0.1	-
Filinia longiseta major Zacharias, 1863	-	-	-	ind
Lecane luna Mueller, 1776	-	-	ind	ind
Polyarthra sp.	-	-	-	ind
Brachionus plicatilis Muell, 1786	443.2	0.2	ind	118.8
Cladocera:				
Moina sp.	-	-	-	ind
Pleuroxus aduncus Jurine, 1820	-	-	-	ind
Simocephalus vetulus Muell, 1776	-	-	0.5	-
Alona rectangula Sars, 1862	ind	-	0.1	ind
Chydorus sphaericus Mueller, 1776	ind	3.7	0.5	0.3
Diaphanosoma mongolianum Ueno, 1939	1.0	4.2	0.1	208.8
Moina salina Daday, 1888	6.1	-	-	-
Copepoda:				
nauplii	3.5	28.2	13.0	11.5
Cyclops vicinus Uljanine, 1875	-	9.6	3.6	3.2
Mesocyclopso gunnus Onabamiro, 1957	-	1.1	0.1	ind
Onychocamptus mohammed Blet Rich, 1891	0.1	0.5	ind	0.3
Schizopera sp.	0.2	-	-	-
Thermosyclops vermifer Ulomskiy, 1963	0.3	71.2	0.2	101.6
Apocyclops dengizicus Lepeschkin, 1900	0.2	-	-	-
Arctodiaptomus salinus Daday, 1885	25.4	0.5	-	154.1
Cletocamptus retrogressus Schman, 1875	0.7	-	-	-
Onychocamptus bengalensis Sewell, 1934	0.2	-	ind	-
Species richness:	13	12	16	19

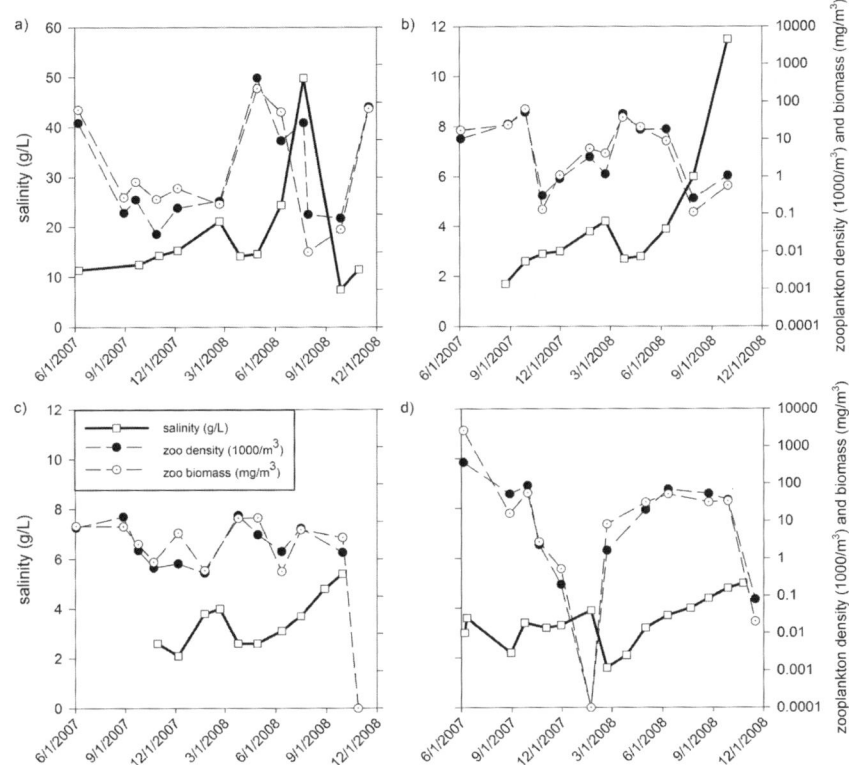

Figure 3.7.4 Salinity versus zooplankton density and biomass for a) Eshan Rabat, b) Khodjababa, c) Shur (Koshkopir), and d) Tuyrek lakes. Note the different scale for Eshan Rabat salinity.

Zooplankton biomass peaks were observed in TUY and KHO in March-April because of the development of cyclopoid copepods *Cyclops vicinus* (maximal biomass was 30–40 mg m^{-3}) and in June or August-September because of the development of four dominant species: *Brachionus plicatilis* (up to 17 mg m^{-3}), *Thermocyclops vermifer* (up to 60 mg m^{-3}, except in June 2007), *Arctodiaptomus salinus* (up to 621 mg m^{-3}) and *Diaphanosoma mongolianum* (up to 2021 mg m^{-3}). In ESH, biomass peaks were similarly observed in June 2007 (55 mg m^{-3}), April 2008 (230 mg m^{-3}), and November 2008 (69 mg m^{-3}), but 75 % of the biomass was formed by *Brachionus plicatilis*, and most of the rest (25 %) was *Arctodiaptomus salinus*. The highest biomass of the four lakes (2,640 mg m^{-3}) was observed in June 2007 in TUY (80 % cladocerans *Diaphanosoma mongolianum*, 15 % *Arctodiaptomus salinus* and 5 % *Thermocyclops vermifer*). Typical biomass observations in the four lakes varied from 0.1–60 mg m^{-3}. Of the four lakes, TUY had the highest average zooplankton biomass and the highest peak biomass. These values were also the highest mean and peak zooplankton biomass of shallow lakes in Uzbekistan since 2000 (Ginatullina, 2010). Biomass peaks were observed in ESH in June 2007 and 2008 for *Arctodiaptomus salinus* (biomass 62.2 and 54.7 mg m^{-3}, respectively) and because of the development of *Brachionus plicatilis* in April 2008 (232 mg m^{-3}) and in November 2008 (66.9 mg m^{-3}).

Generally, the factor that appeared to be most limiting *abundance* of zooplankton in the studied lakes was water temperature. From November to February, zooplankton biomass was reduced in all four lakes when water temperature was below 10 °C. During the growing season (from mid-February to mid-October) temperatures in excess of 25 °C also coincided with depressed zooplankton density/biomass (Figure 3.7.4). Temperature appeared to be related to generally predictable patterns of seasonal succession of zooplankton. Temperature is an important factor for zooplankton communities, influencing structure, distribution and seasonal sequence of species dominance (Tappa 1965; Geraldes and Boavida 2004). *Diaphanosoma mongolianum, Thermocyclops vermifer, Chydorus sphaericus, Moina salina* and Brachionus plicatilis are thermophilic (warm-season) species of these lakes and increase in their abundance requires sustained preheating of water (above 15 °C; Verbitsky 2012). These species disappeared from the studied lakes during winter when water temperatures were <10 °C. Cyclops vicinus, Simocephalus vetulus, Arctodiaptomus salinus and Keratella quadrata are rather thermolabile (cold-season) species. They were observed in the studied lakes during the fall-winter-spring months. However, temperatures above 25 °C in summer months coincided with a depression in abundance of the thermophilic species as well (Figure 3.7.4).

Salinity appeared to have an effect on zooplankton composition in the studied lakes. KHO, SHK, and TUY had salinity levels ranging from fresh to slightly saline (3 – 12 g l-1) during our study, but ESH typically ranged from saline (12 – 35 g l-1) to hypersaline (>35 g l-1, Brock and Lane 1983). In KHO, only a small amount of the rotifer Keratella tropica, harpacticoid copepod Onychocamptus mohammed and cyclopoid copepods Thermocyclops vermifer and Mesocyclops ogunnus remained when salinity reached 11 g l-1. In KHO, SHK, and TUY lakes, decreasing zooplankton density and biomass coincided with increasing salinity. In SHK, less salinity-tolerant cladocerans Simocephalus vetulus disappeared as salinity increased to more than 3 g l-1, while densities of more tolerant Chydorus sphaericus and Alona rectangula increased. The moderately halotolerant copepod Thermocyclops vermifer was observed in KHO, TUY and SHK as the dominant species in summer 2008 when salinity was increasing. In autumn-winter season 2007 – 2008 the species composition of KHO, TUY and SHK lakes was dominated by the moderately salinity-tolerant large cyclopoid copepod

Cyclops vicinus, which remained in the community until April 2008. After an increase in these lakes' salinity to more than 4 g l-1, these species were not found in November 2008.

3.7.4 Discussion

Over the 18-month long study period, zooplankton species compositions in the three less saline lakes had similar diversity (relatively non-diverse, 12 – 19 species in each lake; Table 3.7.2) to the saline lake (13 species), however the community in the less saline lakes included more rotifers, while the saline lake community was represented by more copepods. Density/biomass in the three less saline lakes was also generally indistinguishable from the more saline lake; at several times, density/biomass was slightly higher in ESH because of high densities of the halophilic rotifer *Brachionus plicatilis*. On a smaller time-scale, however, we found that the zooplankton communities did appear to respond to changes in lake water salinity, with reduced biomass and diversity in all lakes when salinity reached certain thresholds, but returning to previous levels of biomass and diversity when salinity decreased again.

In the small shallow lakes of the Khorezm region, zooplankton abundance and composition generally tracked seasonal temperature changes closely unless salinity rapidly increased to certain thresholds when it appeared to become the primary limiting factor. The first apparent threshold of salinity in our studied lakes was 3 – 4 g l^{-1} when widespread freshwater crustacean species, such as the large pelagic cyclops copepod *Cyclops vicinus* and plant-associated cladocerans *Simocephalus vetulus* and *Pleuroxus aduncus*, disappeared from zooplankton communities. A second salinity threshold appeared to occur around 10 – 12 g l^{-1}, when zooplankton species tolerant of slightly saline water disappeared from the community. Such species have been found to form the nucleus of typical communities in shallow, young drainage lakes in Uzbekistan in which inflowing water produces salinity between freshwater (<3 g l^{-1}) and slightly saline (<10 g l^{-1}) conditions (Ginatullina 2010). These species are halotolerant fresh water species such as *Thermocyclops vermifer*, *Diaphanosoma mongolianum*, *Ohycamptus mohammed*, *Mesocyclopso gunnus*, *Chydorus sphaericus* and *Alona rectangula* (Ginatullina 2010). Our findings indicated that when salinity increased to >12 g l^{-1}, a shift in the zooplankton community structure was observed that included more copepod species (mainly cyclops) and only one species of cladoceran (*Moina salina*).

All four of our study lakes showed a generally low diversity compared to other lakes in the Aral-Caspian basin such as those in the Karakalpakstan region of Uzbekistan (Ginatullina 2010). Findings from other brackish lake ecosystems around the world with varying salinities indicate that when salinity increases, zooplankton diversity decreases as the community shifts toward dominance by the few species capable of tolerating high salinity while sensitive species become extinct (Galat and Robinson 1983; Brock and Shiel 1983). Even relatively small increases in salinity can drive such systems to a state of depleted biodiversity and abundance (Galat and Robinson 1983; Brock and Shiel 1983), altering ecosystem function (e.g. Schallenberg et al. 2003). Fairly mild increases of salinity in our study lakes coincided with lower abundance of more vulnerable zooplankton groups such as the large crustacean *Simocephalus vetulus*. It is possible that the minimum of $2-3$ g l^{-1} salinity was enough to inhibit the most sensitive species and suppress diversity in even the three low-salinity lakes, KHO, SHK, and TUY. The zooplankton community of ESH was quite different than that of KHO, SHK, and TUY, likely due to its higher salinity. ESH supported a halophilic community of species that prefer hypersaline conditions until June 2008: *Arctodiaptomus salinus*, *Moina salina*, *Brachionus plicatilis*, *Cletocamptus retrogressus*, *Onychocamptus bengalensis* and *Apocyclops dengizicus*. After the very high salinity in July 2008 (49 g l^{-1}) in ESH, only *Brachionus plicatilis* remained, and other species that had disappeared from this composition did not re-appear before the end of the study. The high salinity also may have decreased overall biomass and density of zooplankton in ESH, which returned to the same order of magnitude as in June 2008 only after salinity dropped to below 10 g l^{-1} in November 2008.

We found low diversity in our study lakes despite the tendency of shallow lakes to typically support highly diverse biota due to a more extensive littoral zone. Maximum diversity of freshwater ecosystems tends to occur where wetland and littoral habitat heterogeneity interface with lake open waters (Wetzel 1999; 2001). Low species richness may have been influenced by the lack of varied macrophytes and poorly developed littoral zone of the four lakes, resulting in little spatial heterogeneity and habitat diversity. In contrast, the Sudochye wetland ecosystem in the Karakalpakstan region of Uzbekistan contains high littoral habitat heterogeneity and 66 species of zooplankton (38 rotifers, 12 cladocerans, 22 copepods) despite large salinity fluctuations between 3 and >35 g l^{-1} (Mirabdullaev et al. 2008); in comparison, we observed 29 species of zooplankton in our study lakes in the Khorezm region. Isolation is not likely a cause of low diversity because zooplankton species that were observed in the Sudochye wetland lakes could have colonized our study lakes as there is no geographical barrier to their dispersal to the Khorezm region. Cohen and Shurin (2003) showed that generally, local processes such as water chemistry or morphometric features of water bodies limit the richness and composition of local zooplankton communities, but because our study lakes are relatively young (around $80-150$ years old according to lake coring; Crootof 2011), it is possible that the littoral habitat has not had time to develop the potential complexity and heterogeneity that could support a more diverse zooplankton community.

General patterns of seasonal succession of zooplankton appeared to be temperature-related. When the temperature was below 10 °C (in winter season of 2008) all species of rotifers except individual examples of *Keratella quadrata* in KHO and SHK, and most crustaceans except nauplii and copepods of the large cyclop *Cyclops vicinus* in KHO, SHK, and TUY were not present. A few individuals of harpacticoids were found in ESH, a few individuals of cladocerans *Alona rectangula* and *Chydorus sphaericus* were found in SHK, and a few individuals of calanoid *Arctodiaptomus salinus* were found in TUY when water temperatures were <10 °C.

When temperature increased more than 23–25 °C in the studied lakes the mean size of zooplankton may have decreased because large species like the copepod *Cyclops vicinus* and cladoceran *Simocephalus vetulus* are less tolerant of high temperatures in contrast to smaller species such as the copepod *Thermocyclops vermifer* or the rotifer *Brachionus plicatilis*.

Because of pollution from irrigation, the quality of water available to the biota of the Khorezm lakes is likely as important as the quantity. Zooplankton biomass may be low due to nutrient limitation, but our limited nutrient data did not correspond with a strong change in zooplankton biomass or density when examined in comparison to either TN or TP, possibly due to overriding effects of temperature and salinity on zooplankton communities. While it is possible that monthly sampling did not provide enough resolution to observe a relationship between zooplankton abundance and N and P concentrations, it is worth noting several observations that suggest N- rather than P-limitation. We did observe that in all lakes and at virtually all times, copepods (as opposed to cladocerans or rotifers) were the most abundant zooplankton group (unpublished data, Ginatullina 2012). Copepods sequester nitrogen, removing it from the aquatic pool of nutrients, and are thus often associated with fresh water N-limited systems, particularly in arid regions (Johnson and Luecke 2012). Additionally, a notable increase in TN in all four lakes in April 2008 was coincident with observations in April/May 2008 of high or highest zooplankton biomass and density, even though TP did not notably increase at this time.

3.7.5 Summary and Conclusions

In the shallow lakes in Khorezm, water temperature and salinity appeared to interact to cause fluctuations in the abundance and composition of the zooplankton community. Temperature appeared to be the more predictable, dominant seasonal influence, with salinity contributing unpredictably as a dominant factor when high (i.e. at >3 g l^{-1}, and more markedly at >12 g l^{-1}).

Because of the unpredictable fluctuations in salinity and its potential influence on zooplankton biomass and seasonal declines in the zooplankton communities of these lakes, successful fisheries in the Khorezm region may be more likely to succeed when cultivating fishes that rely directly on benthic macro invertebrates than on zooplankton production. Vander Zanden and Vadeboncoeur (2002) reported that in diet data from 470 fish populations (15 species), direct consumption of zoobenthos contributed 50 % of total prey consumption and indirect consumption another 15 % for a total of 65 % reliance on benthic secondary production. We have studied food webs in several other lakes in Khorezm that have fish, but no zooplankton communities, which indicates that such benthic consumption to support fisheries exists in Khorezm.

Acknowledgments

This research was funded by the North Atlantic Treaty Organization (NATO)'s Science for Peace Program and the National Science Foundation (EAR-0838239). M. Shanafield received funding from a Fulbright Student Fellowship. The Center for Development Research (ZEF) provided valuable resources for field work and training. We also greatly appreciate the assistance of B. Nishonov for water quality analyses, M. Rosen for field training, and M. Bekchonova, A. Crootof, N. Mullabaev, A. Nigmadjanov, and J. Scott for field assistance during the project.

References

Borutskii EV (1952) Harpacticoida of fresh water bodies. Moscow-Leningrad: Academy of Sciences Publisher, V.3, issue4, 420 pp

Brock MA, Lane JA (1983) The aquatic macrophyte flora of saline wetlands in Western Australia in relation to salinity and permanence. Hydrobiologia 105, 63–76

Brock MA, Shiel RJ (1983)The composition of aquatic communities in saline wetlands in Western Australia. Hydrobiologia 105, 77–84

Brucet S, Boix D, Quintana XD, Jensen E, Natansen LW, Trochine C, Meerhof M, Gascon S, Jeppesen E (2010) Factors influencing zooplankton size structure at contrasting temperatures in coastal shallow lakes: implications for effects of climate change. Limnology and Oceanography 55, 1697–1711

Carpenter SC, Kitchell JF, Hodgson JR (1985) Cascading trophic interactions and lake productivity, BioScience 35, 634–639

Cohen GM, Shurin JB (2003) Scale-dependence and dispersal mechanisms in freshwater zooplankton. Oikos 103, 607–617

Coleski JA, Koch F, Marcoval MA, Wall CC, Johem FJ, Peterson BJ, Gobler CJ (2010) The role of zooplankton grazing and nutrient loading in the occurrence of harmful cyanobacterial blooms in Florida Bay, USA. Estuaries and Coast 33, 1202–1215

Conrad C, Dech SW, Hafeez M, Lamers JPA, Martius C, Strunz G (2007) Mapping and assessing water use in a Central Asian irrigation system by utilizing MODIS remote sensing products. Irrigation Drainage Systems 21, 197–218

Crootof A (2011) Assessing water resources in Khorezm, Uzbekistan for the development of aquaculture. M.S. Thesis. University of Nevada Reno, Reno. 129 p

Galat DL, Robinson R (1983) Predicted effects of increasing salinity in the crustacean zooplankton community of Pyramid Lake, Nevada Hydrobiologia 105, 115–131

Geraldes AM, Boavida MJ (2004) What factors affect the pelagic cladocerans of the mesoeutrophic Azibo reservoir? Annales de Limnologie – International Journal of Limnology 40, 101–111

Ginatullina EN (2010) Zooplankton of saline transformed drainage lakes in Uzbekistan. PhD Dissertation. Tashkent, Uzbekistan: Institute Zoology Uzbek Academy of Sciences. LAP LAMBERT Academic Publishing, AV Akademikerverlag GmbH & Co. KG.

Gyllstroum M, Hansson L-A, Jeppesen E, Garsia-Criado F, Gross E, Irvine K, Kairesalo T, Kornijow R, Maracle MR, Noges T, Romo S, Stephen D, Van Donk E, Moss B (2005) The role of climate in shaping zooplankton communities of shallow lakes. Limnology and Oceanography 50, 2008–2021

Hessen D, Faafeng B, Brettum P (2003) Autotroph: Herbivore biomass ratios: Carbon deficits judged from plankton data. Hydrobiologia 491, 167–175

Johnson CR, Luecke C (2012) Copepod dominance contributes to phytoplankton nitrogen deficiency in lakes during periods of low precipitation. Journal of Plankton Research 34, 345–355

Khamzina A, Lamers JPA, Vlek PLG (2008) Tree establishment under deficit irrigation on degraded agricultural land in the lower Amu Darya River region, Aral Sea Basin. Forest Ecology and Management 255,168–178

Kutikowa LA (1970) Rotifer's fauna of Soviet Union: Rotatoria. Class EurotatoriaOrders Ploimida, Monimotrochida, Paedotrochid. Leningrad: Science, 742 pp

Mirabdullaev IMEN Ginatullina E, Turemuradova GI (2008) Changes of zooplankton community in the ,Sudochye Lake as a result of increasing of mineralization.Uzbek Journal of Biology(Uzbekistan) 5, 45–48

Monchenko VI (1974) Free-living cyclopoid copepods Cyclopidae:Fauna of Ukraine in 40 volumes. Kiev: Haukova Dumka (Ukraina) 23, 12–230

Oberkircher L, Shanafield M, Ismailova B, Saito L (2011) Ecosystem and social construction: an interdisciplinary case study of the Shurkul Lake landscape in Khorezm, Uzbekistan. Ecology and Society 16, 20

Pavlovskaya LP (1995) Fishery in the lower Amu Darya under the impact of irrigated agriculture: FAO Fisheries Circula N 894, Rome: 42–57

Raven JA, Kubler JE (2002) New light on the scaling of metabolic rate with the size of algae. Journal of Phycology 38, 11–16

Ruttner-Kolislo A (1977) Suggestions for biomass calculations of plankton rotifers.ArchivfrHydrobiologia 8: 71–76

Saito L, Fayzieva D, Rosen MR, Nishonov B, Lamers JPA, Chandra S (2010) Final Report: Using Stable Isotopes, Passive Organic Samplers and Modeling to Assess Environmental Security in Khorezm, Uzbekistan. NATO Science for Peace Project No. 982159. Reno: University of Nevada Reno. 40 pp

Salazkin AA, Ivanova MB, Ogorodnikova VA (1984) Methodical recommendation atcollectingandprocessing materials byhydrobiologicalresearchesonfresh waterbodies: Zooplankton and its production. Moscow, Institute of Lakes and River Fishery of Soviet Union Republic. [in Russian]

Semenov AD (1977) Manual on chemical analysis of terrestrial surface water.-Gidrometeoizdat.Publisher, Leningrad 542 pp

Schallenberg M, Hal, CJ, Burns CW (2003) Consequences of climate-induced salinity increases on zooplankton abundance and diversity in coastal lakes. Marine Ecology Progress Series 251, 181–189

Scott J, Rosen MR, Saito L, Decker DL (2011) The influence of irrigation water on the hydrology and lake water budgets of two small arid-climate lakes in Khorezm, Uzbekistan. Journal of Hydrology 410, 114–125

Shanafield M, Rosen M, Saito L, Chandra S, Lamers JPA, Nishonov B (2010) Identification of nitrogen sources to four small lakes in the agricultural region of Khorezm, Uzbekistan. Biogeochemistry101, 357–368

Shurin JB, Havel JE, Leibold MA, Alloul BP (2000) Local and regional zooplankton species richness: a scale-independent test for saturation. Ecology 81, 3062–3073

Sosnovsky A, Rosso JJ, Quiros R (2010) Trophic interactions in shallow lakes of the Pampa plain (Argentina) and their effects on water transparency during two cold seasons of contrasting fish abundance. Limnetica 29, 233–246

Tappa DW (1965) The dynamics of the association of six limnetic species of Daphnia in Aziscoos Lake, Maine. Ecological Monographs, 35, 395–423

UNEP (2005) Aral Sea, Global International Waters Assessment Regional Assessment 24. Severskiy I, Chervanyov I, Ponomarenko Y, Novikova NM, Miagkov SV, Rautalahti E, Daler D. Kalmar, Sweden: United Nations Environment Programme, University of Kalmar

Vander Zanden MJ, Vadeboncoeur Y (2002) Fishes as integrators of benthic and pelagic food webs in lakes. Ecology 83, 2152–2161

Verbitsky VB (2012) Temperature optimum, preferendum and thermotolerance of freshwater crustaceans (Cladocera, Isopoda, Amphipoda). PhD dissertation. Jaroslavl region, Borok. Institute of Inland Water, Russian Academy of Sciences

Wetzel RG (1999) Plants and water in and adjacent to lakes. In Baird AJ, Wilby RL (Eds.), Eco-hydrology: Plants and Water in Terrestrial and Aquatic Environments. Routledge, London

Wetzel RG (2001) Limnology: Lake and River Ecosystem. Academy Press, San Diego

Whitman RL, Davis B, Goodrich ML (2002) Study of the application of limnetic zooplankton as a bioassessment tool for lakes of Sleeping Bear Dunes National Lakeshore, USGS Lake Michigan Ecological Research Station Porter, Indiana

Zalolichin SY (Ed.) (1995) Identifier of freshwater micro invertebrates in Russia and adjacent terrains. Institute of Zoology, Russian Academy of Sciences, St. Petersburg

Section 4: Production and Resource Economics

Maksud Bekchanov, John P. A. Lamers, Christopher Martius

4.1 Coping with water scarcity in the irrigated lowlands of the lower Amudarya basin, Central Asia

Abstract

Improving irrigation efficiency is of utmost importance in the irrigated lands of Central Asia, such as the Khorezm region of Uzbekistan, since water misuse and subsequent soil salinization threaten environment, economy, and livelihoods. To this end, several field-level 'water-wise' innovations were selected, which are classified into four groups that address crop pattern change, soil moisture maintenance, uniform water distribution, and furrow irrigation improvement. The potential of these innovations to raise irrigation water use efficiency from its current low level was analyzed from a socio-economic and technical point of view with a focus on short-term measures to cope with sudden water shortages. The overall water use reduction potential of these options was estimated considering their adoption feasibility within the time horizon of one year. To prioritize the examined innovations according to their contribution to overall water use reduction and water profitability, 'marginal water profitability curves' were developed. This integrated approach could serve as a simple but effective policy tool. The findings indicate that the option of replacing rice by maize contributes to more than 50 % of the total possible water use reduction. However, while all the other options increase the total revenue, reduced revenues will be unavoidable when paddy rice is replaced by maize. Manuring provides the highest additional profit per volume of reduced water use, but contributes less than 10 % of the total water use reduction potential. With water-wise options as an immediate and short-term measure to cope with sudden water shortages, the theoretical total estimated water reduction at the field level amounts to 183 – 376,000,000 m^3 or 9.0 – 18.5 % of the current total irrigation water requirement in the region. For coping with sudden shortages characterized by a water availability of only 60 % of the normal water supply, long-term planning and management of irrigation *activities focusing on a wider adoption of advanced irrigation technologies are necessary.*

Keywords: water-wise innovations, water use reduction rate, marginal water profitability, crop pattern change, manuring, Khorezm, Uzbekistan

4.1.1 Introduction

Coping with water scarcity is a major challenge in drylands due to an increasing demand for water in agriculture, industry and domestic consumption. By 2025, more than two thirds of the world's population is expected to be living under water stress (UN-WATER 2007). This threat is particularly alarming in the Aral Sea Basin (ASB), which covers 1,800,000 km² in seven nations: the five central Asian republics Uzbekistan, Turkmenistan, Kazakhstan, Kyrgyzstan, and Tajikistan, and parts of Afghanistan and Iran (Abdullaev 2005).

The combined effect of global warming, population growth and the need for further economic development are increasing water demands throughout the ASB. Furthermore, political disagreements frequently impede more efficient water distribution and use whilst the deteriorated irrigation and drainage networks prevent efficient conveyance and irrigation. Water users in downstream areas are more vulnerable to the continuous decline in water supply than their counterparts in the mid- and up-stream areas with respect to satisfying their demands for irrigation, drinking and industrial water. This is illustrated by the case of irrigated lowlands in the Khorezm region in Uzbekistan, located in the lower reaches of the Amudarya river in Central Asia.

Khorezm, located between 60.05 and 61.39 N and 41.13 and 42.02 E, has a total area of 670,000 ha, of which 270,000 ha are used for cultivating irrigated crops such as cotton, wheat, rice, vegetables and fruit trees (Vlek et al. 2012). The region is representative of about 8,000,000 ha of irrigated lowlands in Central Asia (Cai et al. 2003). Like in most of the other regions of the ASB, irrigated agriculture is pivotal for economic development, food security, and rural livelihoods. The agriculture sector contributes 46 % to the regional domestic product and 98 % to the total regional export revenues (Bekchanov et al. 2010b). The livelihoods of the rural population, which represents about 70 % of the total population, directly rely on irrigated agriculture.

Irrigated crop production and revenues in turn are driven by a high water consumption reaching 95 % of the total annual water withdrawal of 5 km³ (Tischbein et al. 2012). At the same time, a further decline in the probability of sufficient water supplies in the near future (Müller 2006) means enormous risks and would endanger sustainable development of the region. Water shortages are already frequent in the region due to decreased river runoff and increased water demand for hydropower generation and irrigation in upstream regions (Dukhovny, Sorokin 2007). Although in normal years sufficient water is provided, in dry years, such as in 2000, 2001 and 2008 (the worst droughts on record), the water users of Khorezm only had access to less than 60 % of the water required during the vegetation period (Statistics 2009). Such drought years cannot be foreseen and thus represent an unknown risk factor for farmers, particularly those downstream (Oberkircher et al. 2012). Nevertheless, despite the obvious decline in water sufficiency, water wasting is common, as indicated by the very low water use efficiency (about 30 % Tischbein et al. 2012) under the present furrow and basin irrigation methods and unlined canal systems (Martius et al. 2009). Since annual water shortages are temporary and characterized by a frequency of about 30 %, i.e., occurrence of droughts in three years out of ten (Müller 2006), and there is no direct payment for irrigation water (Bobojonov 2008), farmers are not encouraged to invest in efficient water use practices. Therefore, inadequate use of knowledge, technical and financial resources to adapt to the severe shortages causes substantial decreases in yield, overall agricultural production, and economic revenues (Bekchanov et al. 2010a). On the other hand, the current low efficiencies indicate an enormous potential for reducing total water consumption, e.g., by implementing what we call 'water-wise' innovations, i.e., "best water use practices" that raise water use efficiency by decreasing water losses while maintaining crop yields.

Against this background, we question in this paper: Which measures could be rapidly implemented by farmers to enable them to adapt to unexpected water shortages to reduce the risks of low yields? Are measures that can be feasibly, both financially and technically, implemented within a year adequate to cover the water supply and demand gap of about 40 %? This paper is a logical continuation of the study of Bekchanov et al. (2010a) where we addressed the technical and socio-economic efficiency of different water-wise options as well as incentives and threats to their wide-scale adoption. Here, we go one step further by estimating the total and individual overall water use reduction potential and economic value of measures such as alternative crops, hydrogel, manuring, laser-guided land leveling, drip irrigation, surge flow, alternate dry furrow, double flow, and short furrow. All measures implementable within one year to cope with unexpected water shortages.

4.1.2 Methodology

4.1.2.1 Selection of 'water-wise' innovations

The results of this study are based on a qualitative expert-knowledge survey to select the different water-wise options and on an in-depth quantitative analysis of the technological, economic, and adoption potentials of the selected options. First, experts were asked to rate the different parameters of water-wise options such as water use reduction rate, yield impact, adoptability with respect to different soil-climate conditions in the region, and adoption potential within a year (Bekchanov et al. 2010a). Then, the values rated by all experts corresponding to each criterion were averaged, and the top ten innovations with the highest technological and economic parameters were selected for detailed quantitative analysis. The selected water-wise options were grouped as follows: change in cropping pattern (Group A), which includes the option of replacing paddy rice by maize (A1) and aerobic rice (A2); preserving soil moisture (Group B), which consists of hydrogel application (B1) and manuring (B2); ensuring uniform distribution of water (Group C), which comprises laser-guided land leveling (C1) and drip irrigation (C2); and improved furrow irrigation (group D), which involves surge flow (D1), double flow (D2), alternate dry furrow (D3), and shorter furrows (D4).

4.1.2.2 Assumptions on the adoptability of 'water-wise' innovations

The study used an additive approach to evaluate the regional water use reduction potential by stacking the selected water-wise options since one single practice will unlikely be able to provide all the needed improvements due to the mixed initial conditions such as land size and quality, crops produced, location along the irrigation network, and farmer capitalization and willingness to pay for services rendered. The potential situations in which the different technical options were to be implemented were thus subject to a series of assumptions (Table 4.1.1). It should be remembered that all innovations had been suggested by experts as being able to contribute to the overall goal of water use reduction. Another underlying rationale was that the "implementation horizon" needed to introduce the innovations was estimated to realistically represent the technologically, economically or institutionally implementable change within the course of one year.

Table 4.1.1: Assumptions on potential adoption of water-wise options selected for Khorezm region based on expert opinion. Water-wise option codes: Group A – changes in cropping pattern; Group B – options to preserve soil moisture; Group C – options to ensure uniform distribution of water; Group D – options to improve furrow irrigation.

Code	Option	Assumption	Justification of assumption
A1	Maize vs. paddy rice	18 % and 45 % area of rice changed to maize	Based on the smallest areas of rice between 2000 and 2007; Statistics 2008a
A2	Aerobic vs. paddy rice	Aerobic rice replaced paddy rice on 10 % of the area	Lower yields of aerobic rice prevent the replacement of large areas under paddy rice
B1	Hydrogel	5 % of wheat, cotton, vegetables	Innovation with a high potential in non-saline and low-saline soils, but hydrogel is commercially not produced in Uzbekistan
B2	Manuring	22 % area of cotton, wheat, and potato	Maximal area that can be manured; estimation based on total amount and per-hectare use of manure
C1	Laser leveling	5 % of area of total cotton	Limited time frames each year for land leveling; limited availability of laser-guided land leveling equipment in the region
C2	Drip irrigation	7.5 % area of total vegetables	Applicable mainly to high cash crops mainly because of high investment costs
D1	Surge flow	2 % area of total cotton	Assumed to be implementable in the small fraction of cotton area because not well tested in Khorezm
D2	Double flow	10 % area of total cotton	Dependent on field slope
D3	Alternate dry furrow	5 % area of total cotton	Not common among farmers
D4	Shorter furrows (50 – 100 m)	20 % area of total cotton	More uniform distribution; easy and cheap to implement; does not require special technical skills

A change to less water demanding crops (Group A) would permit reducing water demand and in turn decrease water stress due to the anticipated lower amount of water resources in the region. Since paddy rice production is the most water-consuming practice in the Khorezm region, a decrease in rice area could contribute substantially to the reduction in irrigation water consumption. Considering the need to increase the crop portfolio with fodder crops to support the livestock sector (Iñiguez et al. 2005) given the recently issued presidential resolutions in Uzbekistan (#308 dated 23 March 2006 and #842 dated 21 April 2008), fodder maize could be a suitable crop to replace rice owing to its lower water requirement. Changing rice to fodder maize for decreasing water use requirements is also valid for regions beyond Uzbekistan due to the fact that the demand for feeding livestock is rising globally (Delgado et al. 1999). During the period between 2000 and 2007, the average share of rice in the total cropland area of Khorezm was 8.5 % decreasing to as low as 4.7 % in 2001 and 7 % in 2007. In other words, the rice area decreased by 45 % and 18 %, respectively, compared to the average level in these two years. These two values were accepted as the area of adopting the practice of replacing paddy rice by maize. For illustrating the option of dry-land rice production for increasing water use efficiency, it was assumed that 10 % of the present paddy rice area could be replaced by aerobic rice (Table 4.1.1).

Manuring is done to increase yields. It has also shown a reduction in crop irrigation water demand by maintaining higher soil moisture levels for longer periods (Ergashev 1993). Manuring is not only suitable for increasing the efficiency of "blue water" use (i. e., ground- and surface water used for production) in irrigated lands as in Khorezm, but also has a huge potential for enhancement of the efficiency of "green water" use (water from rain) in rainfed areas of the world, since green water use for crop production is globally 4–5 times higher than blue water use (Hoff et al. 2010). In this study, it is assumed that a suitable application in Uzbekistan amounts to 40 tons ha^{-1} of organic manure (Ergashev 1993). Based on the potentially available amount of manure, which is calculated using total livestock numbers (Statistics 2008a) and average annual manure production per head of cattle, it was estimated that the total amount of manure can potentially be applied on 22 % of the total cropping area. Cotton, wheat, and potato crops are considered as potential crops for manuring due to their high yield response to manuring (Karmanov 1978; Ergashev 1993) and significant share in total croplands (Statistics 2008a).

Hydrogel is a crystal polymer that can be applied to the rooting zone for absorbing and retaining water up to 200–400 times of its dry weight (Timirova, Salokhiddinov 2002). This water is slowly released and absorbed by the plants. Although the price of hydrogel is frequently quite low in other countries, it is not currently produced commercially in Uzbekistan. Moreover, it is suitable only on lands with low salinity. Therefore, it was assumed that hydrogel application is restricted to a limited area. We therefore assumed a 5 % change in the area of wheat, cotton and vegetables where hydrogel could be applied.

Laser-guided land leveling and drip irrigation both permit efficient irrigation due to a uniform distribution of irrigation water over the entire field (Egamberdiev et al. 2008; Mamatov 2009). Laser land leveling leads to an even field surface consequently improving water distribution, decreasing water losses, and maintaining uniform germination (Egamberdiev et al. 2008). Drip irrigation provides a high-frequency, low-volume supply of water directly to the plant roots, thus maintaining soil moisture at desirable levels throughout the growing season (Hillel 1997). But these innovations demand substantial initial capital investments. Given also the limited time frame each year where land leveling can take place (a maximum of 3 months), the potential area that can be laser-leveled per year is limited. In the scenario analyses, it was assumed that annually only 5 % of the total cotton area of 110,000 ha can be laser-leveled.

Considering the relative expensiveness of drip irrigation technologies, this option was assumed to be applied mainly on cash crops other than cotton, such as vegetables. It was assumed furthermore that in one year not more than 25 % of the entire tomato production, which equals about 7.5 % of the area allocated to vegetable production, could be put under drip irrigation.

Surge flow, double-sided furrow, alternate dry furrow and short furrow (Group D) are promoted for achieving a more uniform water application than conventional furrow irrigation while minimizing water losses (Paluasheva 2005; Abdullaev et al. 2007a). Surge flow irrigation uses devices such as gated pipe systems or buried pipelines that supply water in furrows intermittently following a cycle of flooding and dewatering (TWDB no date). It was assumed that the surge flow technique can be used on only 2 % of the cotton area, since it is an uncommon water application practice in the region.

As water is simultaneously applied from both sides of the irrigation furrow with the so-called double-sided furrow technology, a more uniform water distribution over the entire length of the furrow is achieved (Paluasheva 2005). Considering that this method can be applied only in fields with a slope of less than 0.001, it was assumed that 10 % of the cotton area could be irrigated with this technology.

During the application of the alternate dry furrow (ADF) technology, only every second furrow is irrigated (Devkota 2011). Therefore, 20 – 25 % less water is required than with traditional furrow irrigation (Kurbanbaev et al. 2006). Due to the well developed root systems of cotton plants with a depth of 1.5 – 1.9 m under ADF (Abdullaev et al. 2007a), cotton yields can increase by 0.4 – 0.5 tons ha^{-1} (Kurbanbaev et al. 2006). It was assumed that approximately 5 % of the total cotton area could be transferred to ADF.

Shorter furrows (length 50 – 100 m) are commonly associated with a higher uniformity of the applied irrigation water (Abdullaev et al. 2007a). Considering their cheap and easy implementation, it was assumed that 20 % of the cotton fields can be equipped with short furrows.

4.1.2.3 Technological and socio-economic efficiency of water-wise options

The water use reduction rate (WURR, in %) was considered as a technological efficiency parameter, and estimated as the share of the potential amount of water reduced by implementing each innovation in the amount of water used under conventional practices (furrow and flood/basin irrigation). For instance, a WURR of 8 % means an 8 % lower water use than under the conventional practices. The WURRs were either directly taken from various sources, or calculated on the basis of water use efficiency estimates, or assumed based on the discussion with the experts (Table 4.1.2). Since those estimates vary considerably in the secondary sources, minimum and maximum values for the WURR were determined. However, under the option of replacing paddy rice by maize, maximum and minimum WURRs are not different from each other, since the per-hectare water use requirement of rice is almost five times higher than that of maize.

Economic efficiency, or net income change (additional profit or loss) due to the implementation of an innovation, was estimated by partial budgets (Dalsted, Gutierrez 2000) considering the difference between additional revenues and costs after the implementation of an innovation (Bekchanov et al. 2010a). Additional revenues may occur due either to increased revenues because of increased yields or to decreased costs because of lower labor, machinery, and water requirements. In contrast, additional costs may occur due either to reduced revenues because of yield decrease or to increased costs because of additionally required labor, capital, machinery, and fertilizer inputs. Data on crop and agricultural input prices were obtained either from the market survey conducted in the ZEF/UNESCO Project or the Regional Statistics Department (Statistics 2008c). Data on crop yields and production area were taken from the regional statistical bulletins (Statistics 2008a, 2008b). Information on total water withdrawal was provided by the Regional Water Resources Management Department (Statistics 2009). Annualized investment, operation, and other costs related to technology adoption were estimated based on expert and literature survey (Hydrogel – Timirova, Salokhiddinov 2002; Manuring; Laser leveling – Abdullaev et al. 2007b; Drip irrigation costs – Mamatov 2009; Implementation costs and labor requirements of improved furrow irrigation methods were assumed low).

The partial budget analysis showed higher additional incomes for manuring of potato fields and drip irrigation of vegetables (Figure 4.1.1).

However, replacing paddy rice by maize or aerobic rice production would be accompanied by additional costs due to higher income from paddy rice.

4.1.2.4 Marginal water profitability curves

Prioritizing water-wise options with regard to their marginal water use profitability and overall water use reduction potential would improve investment allocation decisions. Therefore, marginal water profitability curves (comparable to the marginal water cost curves of McKinsey & Company 2009) were constructed under scenarios of maximum, minimum, and average water use re-

Table 4.1.2: Assumed or estimated minimum and maximum water use reduction rate (WURR) for selected water-wise options. Water-wise option codes: Group A – changes in cropping pattern; Group B – options to preserve soil moisture; Group C – options to ensure uniform distribution of water; and group D – options to improve furrow irrigation.

Code	Water-wise option	Water Use Reduction Rate Min	Water Use Reduction Rate Max	Source
A1	Maize vs. paddy rice	83 %	83 %	Müller, 2006
A2	Aerobic vs. paddy rice	30 %	50 %	Authors' assumption based on Sommer, unpublished, and personal communication with Dr. Krishna Devkota on 22.04.2010
B1	Hydrogel (cotton, wheat)	30 %	50 %	Timirova and Salokhiddinov, 2002
B2	Manuring	7 %	12 %	Authors' assumption based on personal communication with Dr. Jumanazar Ruzimov on 02.03.2009
C1	Laser leveling	20 %	30 %	Abdullaev et al. 2007b
C2	Drip irrigation vs. furrow irrigation	45 %	60 %	Mamatov, 2009
D1	Surge flow	15 %	20 %	TWDB, no date
D2	Double flow	10 %	20 %	Paluasheva, 2005
D3	Alternate dry furrow (ADF)	20 %	25 %	Kurbanbaev et al. 2006
D4	Shorter furrows (50–100 m)	5 %	8 %	Abdullaev et al. 2007a; Authors' assumption based on personal communication with Dr. Bernhard Tischbein on 14.05.2009

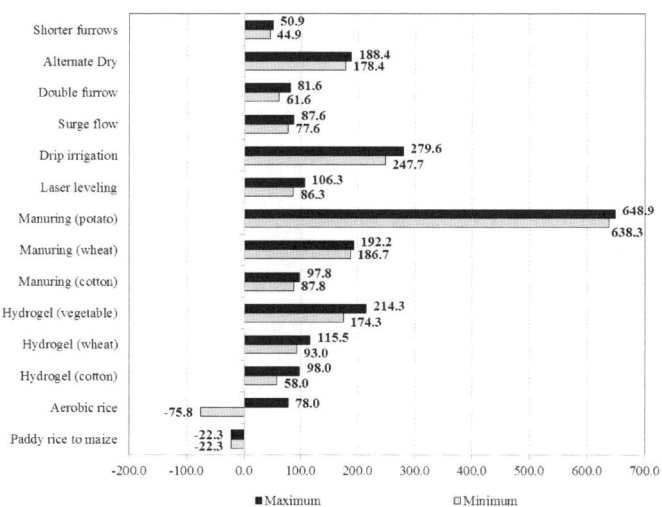

Figure 4.1.1: Net income change (economic efficiency) under different water-wise options

duction rate. Marginal water profitability was estimated as the ratio of net income changes (or additional economic profits that, in case of loss, may be negative) per water use reduction amount under each option. Concurrently, total water use reduction volumes were calculated based on potential area for adoption. Finally, marginal water profitability curves were obtained by contrasting ranked marginal profitability values to the cumulative water use reduction volume.

4.1.3 Results and discussion

4.1.3.1 Implementation area and total water use reduction potential of water-wise options

The overall estimated water use reduction potential when stacking various water-wise options in the additive approach was at least 9 %, and in the best case 18.5 % of the total amount of irrigation water applied in the Khorezm region (Table 4.1.3). The contribution of each option to potentially reduce water consumption depends on the WURR and the assumed potential area of adoption. Each option has its own (dis)advantages, and hence its use depends on the soil-climate conditions, risk of investment recovery, or limited amount of financial assets of the farmers and government (Bekchanov et al. 2010a).

A substantial part of the water use reduction comes from changing the paddy rice area to maize, as the latter consumes only 20 % of the water required for rice irrigation. But such change would reduce rice production and subsequently increase rice prices since rice is one of the most important staple food commodities in the region. Thus, it would be necessary to change the food consumption structure of the population by partly replacing consumer goods with a high virtual water content by those with a lower virtual water content (Bekchanov et al. 2012).

In spite of the highest water use reduction rate, the contribution of drip irrigation to the overall water use reduction is too low due to the assumption that high investment costs would limit its implementation to a small part of the vegetables area. However, drip irrigation of vegetables would be economically efficient. Since horticulture and orchards are the main activities in household plots of the rural population in Khorezm (Djanibekov 2008), and since fruits and vegetables marketing is not subject to any price regulations, producers can exploit this marketing niche thus taking advantage of the higher profits from the drip-irrigated crops for additional income generation. The introduction of cheaper drip irrigation technologies by national companies able to produce at lower costs or through the provision of credits at low interest rates could render

Table 4.1.3: Volume and percentage of overall water use reduction with selected water-wise options. Water-wise option codes: Group A – changes in cropping pattern; Group B – options to preserve soil moisture; Group C – options to ensure uniform distribution of water; and group D – options to improve furrow irrigation.

Code	Water-wise innovation	MIN efficiency		MAX efficiency	
		Reduction in water use, 10^3 m^3	Ratio to the total irrigation requirement at the field level, %	Reduction in water use, 10^3 m^3	Ratio to the total irrigation requirement at the field level, %
A1	Maize vs. paddy rice	89,175	4.4	223,421	11.0
A2	Aerobic vs. paddy rice	18,127	0.9	30,212	1.5
B1	Hydrogel	17,636	0.9	29,393	1.4
B2	Manuring	17,610	0.9	30,189	1.5
C1	Laser leveling	8,816	0.4	13,224	0.7
C2	Drip irrigation	2,698	0.13	3,598	0.18
D1	Surge flow	2,645	0.1	3,526	0.2
D2	Double flow	8,816	0.4	17,632	0.9
D3	Alternate dry furrow (ADF)	8,816	0.4	11,020	0.5
D4	Shorter furrows	8,816	0.4	14,105	0.7
Total:		183,155	9.0	376,320	18.5

this option suitable also for cotton production on non-saline and low-saline lands.

The laser-leveling technique can contribute 8,800,000 – 13,200,000 m³ to water use reduction on the assumed area. Sharing the purchasing costs of laser-leveling equipment through a joint investment instead of individual purchases by farmers seems to be the most attractive option (Egamberdiev et al. 2008; Bekchanov et al. 2010a). In view of the low prices for cotton and wheat mandated by the state order to these crops, government subsidies or credit facilities could incentivize the adoption of laser leveling (Abdullaev et al. 2007b).

ADF, short furrow and double furrow have the highest implementation potential for cotton producers because these practices require neither high investments nor much additional experience or skills of the farmers, and hence permit water use reduction with minor efforts. On the other hand, their total water use reduction volume is relatively low compared to the other options examined (Table 4.1.3).

4.1.3.2 Prioritizing water-wise options according to marginal water profitability and overall water use reduction potential

Marginal water profitability curves (Figure 4.1.2) allowed visualizing the results estimated and discussed above and ranking the different water-wise options in accordance with their economic benefits per unit of water reduced. The highest additional profits per reduced water amount (marginal profitability) under both maximum and minimum WURR are for manuring, reaching 0.12 US$ and 0.19 m^{-3}, respectively (Figure 4.1.2a and 4.1.2b). The highest additional benefits from manuring can be explained by the substantial yield and thus revenue increase, particularly for potato (Figure 4.1.1), in spite of high manure application costs. In contrast, replacing rice by maize reduces revenues but contributes more than 50 % to the overall water use reduction potential.

Since the order of water-wise options is different under minimum and maximum WURR, we also ranked the options considering their marginal profitability under average WURR (Figure 4.1.2c). The marginal water profitability curve shows the contribution of each option to the overall water use reduction and its economic profitability. This integrated approach can serve as a simple but effective analytical tool for land planners to easily compare, prioritize and select the relevant water-wise methods for efficient water resources management considering the levels of water scarcity. For instance, depending on the severity of the irrigation water scarcity, farmers could implement water-wise options starting from the one with the highest marginal water profitability and going to the option with the lowest marginal profitability until the water supply and demand gap is covered. Indeed, depending on assumptions on the time horizon of implementation, potential areas for adoption, and technical and economic efficiency parameters the shape of the curves changes. As result, the recommendations based on these curves also change. In our example, particularly fine-tuning the data and assumptions regarding technical and efficiency parameters of various water-wise options would greatly improve the applied value of the marginal water profitability curves.

Under the assumptions of our example, if the expected irrigated water scarcity at field level is less than 23,900,000 m^3 (average of minimum (17,600,000 m^3) and maximum (30,200,000 m^3) water use reduction due to manuring), the implementation of manuring is enough to cover the scarcity gap (Figure 4.1.2c). If the expected water scarcity level is between 23,900,000 and 33,800,000[1] m^3, the ADF technique can be implemented in addition to manuring and so forth (Figure 4.1.2c). Given the aforementioned assumptions, all options, except re-

---

1 =23.9 + 0.5 * (8.8+11) (8.8 and 11 million m^3 are minimum and maximum water use reduction, respectively, due to alternate dry furrow)

placing paddy rice by maize and the introduction of dryland rice practices under the maximum water use efficiency, are economically beneficial and thus recommended for implementation even in water-sufficient conditions because of substantial additional yields and incomes.

In general, the total estimated WURR potential of 183,200,000 – 376,300,000 m^3 (Table 4.1.3 and Figure 4.1.2) could considerably ease the present water stress in the Khorezm region, in particular in the short run. However, these operative measures recommended for short-term water stress recovery can help to avoid only 25 to 50 % of the observed water supply-demand gap in dry years[2]. Thus, for achieving sustainable agricultural production in the long run and completely closing the water demand gap, farmers and regional decision makers would need to follow in particular long-term strategies of adopting more far-reaching innovations and upgrading technologies more regularly. First, we should keep in mind that the technical and economic potentials of the options were calculated in this study using estimates of realistic annual implementation horizons. Implementing these measures permanently and increasing their area share by adding additional hectares in consecutive years could be the next step to raise the regional water use efficiency. Second, when aiming at a long-term reduction in water consumption, which is inevitable due to the predicted increase in future irrigation water demand and a concurrent lower supply, stimulating the use of more water-efficient practices such as drip irrigation and laser-guided land leveling has a much higher potential than options ADF, short furrow and double furrow. Thus, at the start of technology adoption, less water-efficient but cheaper options can be enhanced, and later more efficient but more expensive measures can be implemented.

In this context, it should be once more noted that innovations such as a laser-guided land leveler are examples of innovations not targeted to be owned by individual farmers. Instead, sharing costs of laser-guided land leveling equipment among several farmers through cooperative services or business units would facilitate a wider adoption of this technology (Bekchanov et al. 2010a). If drip irrigation and laser-guided leveling equipment were to be produced in Uzbekistan, this would increase the financial viability of these options due to lower transportation and taxing costs.

Although water use reduction rate, economic efficiency and financial viability of water-wise options are important criteria, additional aspects such as farmers' risk attitude, their willingness to learn and familiarize themselves with innovations, institutional environment and access to supporting services may impact farmers' decision making (Bekchanov et al. 2009; Bekchanov et

---

[2] 9 % to 16.5 % water use reduction potential (Table 4.1.3) is approximately 25 % to 50 % of 40 % water shortage (mentioned in Introduction)

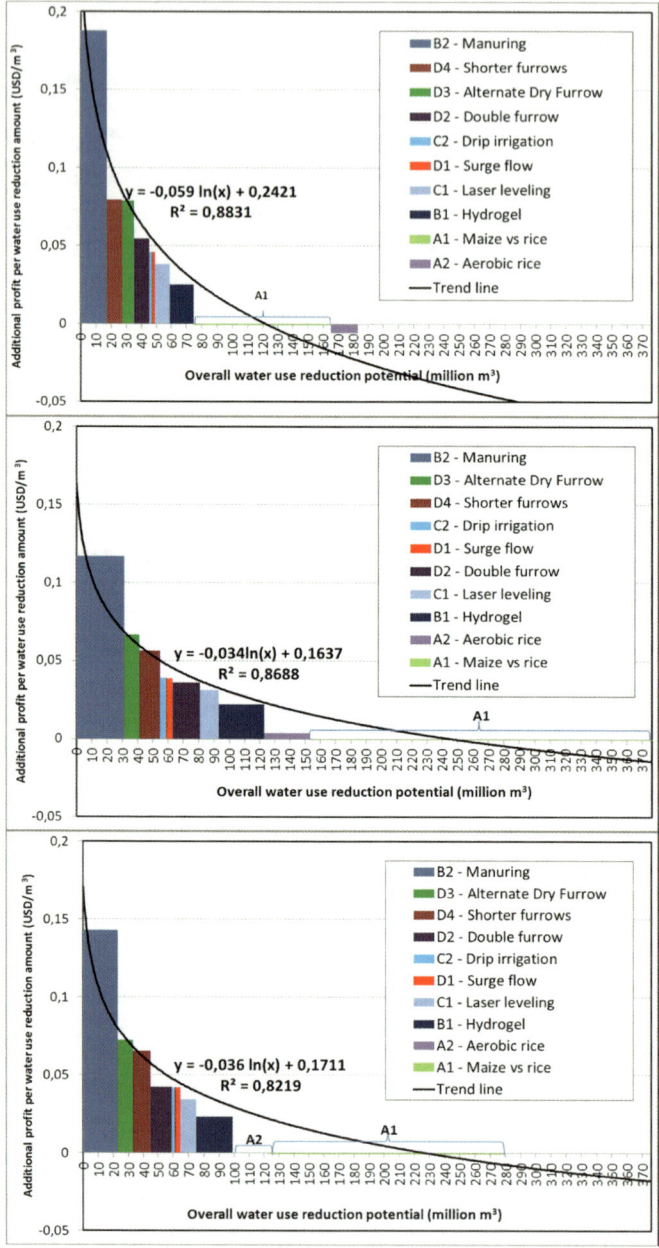

Figure 4.1.2: Marginal water profitability curves under minimum (a), maximum (b), and average (c) water use reduction rates

al. 2010a). Particularly, frequent changes in land and water use legislation and policies through redistributing land ownership rights and re-establishing water management structures with non-transparent rights and obligations prevent farmers from investing in water-efficient technology and production chain modernization, and thus from sharing the advantages of long-term planning and management. To this effect, strong cooperation and trust-building among farmers, governmental organizations, research institutions and extension services are of utmost importance for forming long-term management attitudes and thus effectively coping with water scarcity issues in Khorezm.

4.1.4 Conclusions

Considering the need for effective adaption to water-scarce conditions and for preventing future crop yield reduction, this study reviewed and assessed the adoption and overall water use reduction potential as well as the economic feasibility of a wide range of water-wise innovations in the Khorezm region, Aral Sea Basin, Uzbekistan. The established marginal water profitability curves proved to be an easy and effective tool for prioritizing water-wise measures and presenting their technical and economic efficiency potentials. However, their empirical usefulness and effectiveness can be improved further by fine-tuning various assumptions and data on technical and economic efficiency as well as adoption potentials of different water-wise innovations.

Irrigation methods such as double-sided flow, alternate dry furrow, and shorter furrow were found to have the highest implementation potential for irrigated cotton production in the short run, since little effort and low financial resources for immediate implementation would be needed from the farmers. In contrast, compared to all other options, manuring cotton, cereals, and potato had the highest water profitability potential. Concurrently, given the present high share of area cultivated with high-water-consuming crops such as paddy rice, replacing this crop by low-water-consuming crops has the highest potential to reduce the water demand, but with some revenues losses.

Despite the fact that the substantial overall water reduction potential of all water-wise options examined showed promising economic efficiency, the analyses indicated that the employment of such measures will not be able to cover the water supply-demand gaps in all years, which could be as severe as 40 % of the total water withdrawal in a normal year. Since the present projections for a region such as Khorezm show that the water supply in the near future will be even lower than the current supply, it is essential to create a more enabling institutional and legal environment that permits the introduction of water-wise options such as laser-guided land leveling and/or drip irrigation. Moreover, to better manage water risks through wider adoption of water-wise measures and hence obtain long-term sustainable gains from irrigated agriculture, improved ways of communication and awareness creation and learning by farmers and policy makers about the benefits of long-term management approaches will play a pivotal role.

References

Abdullaev I (2005) Addressing Central Asia's Water Problem. Retrieved 2 June, 2010, from http://www.cacianalyst.org/?q=node/2818

Abdullaev I, Kazbekov J, Molden D (2007a) Water Conservation Practices in the Syr Darya Basin of Central Asia: Water productivity impacts and alternatives. Journal of International Water and Irrigation 27: 14–17

Abdullaev I, Ul Hassan M, Jumaboev K (2007b) Water saving and economic impacts of land leveling: the case study of cotton production in Tajikistan. Irrigation and Drainage System 21: 251–263

Bekchanov M, Bhaduri A, Lenzen M, Lamers JPA (2012) The role of virtual water for sustainable economic restructuring: evidence from Uzbekistan, Central Asia. ZEF Discussion Papers on Development Policy 167

Bekchanov M, Kan E, Lamers JPA (2009) Options of agricultural extension provision for rural development in Central Asian transition economies: the case of Uzbekistan. In: "5th Annual Conference Proceedings of Research for Sustainable Development" held by Westminster International University, Tashkent

Bekchanov M, Lamers JPA, Martius C (2010) Pros and Cons of Adopting Water-Wise Approaches in the Lower Reaches of the Amu Darya: A Socio-Economic View. Water 2: 200–216

Bekchanov M, Karimov A, Lamers JPA (2010b) Impact of Water Availability on Land and Water Productivity: A Temporal and Spatial Analysis of the Case Study Region Khorezm, Uzbekistan. Water 2: 668–684

Bobojonov I (2008) Modeling Crop and Water Allocation under Uncertainty in Irrigated Agriculture: A Case Study on the Khorezm Region, Uzbekistan. Rheinische Friedrich-Wilhelms-Universität Bonn

Cai X, McKinney D, Rosegrant M (2003) Sustainability Analysis for Irrigation Water Management in the Aral Sea Region. Agricultural Systems 76: 1043–1066

Dalsted N, Gutierrez P (2000) Partial Budgeting. Retrieved 25 March, 2010, from http://agecon.uwyo.edu/RiskMgt/financialrisk/Partial%20Budgeting.pdf

Delgado C, Rosegrant M, Steinfeld H (1999) Livestock to 2020. The Next Food Revolution. Food, Agriculture, and the Environment. Discussion Paper No. 28

Devkota M (2011) Nitrogen management in irrigated cotton-based systems under conservation agriculture on salt-affected lands of Uzbekistan. University of Bonn, Bonn

Djanibekov N (2008) A Micro-Economic Analysis of Farm Restructuring in the Khorezm Region, Uzbekistan. Rheinische Friedrich-Wilhelms-Universität Bonn

Dukhovny BV, Sorokin A (2007) Assessing the Impact of Rogun Water-Reservoir on Amudarya River Flow Frequency. Tashkent, Uzbekistan

Egamberdiev O, Tischbein B, Franz J, Lamers JPA, Martius C (2008) Laser land leveling: more about water than about soil. ZUR no. 1. science brief. Retrieved 4 November, 2012, from http://www.zef.de/fileadmin/webfiles/downloads/projects/khorezm/downloads/Publications/ZUR/ZUR_No1.pdf

Ergashev A (1993) Agrochemical properties of the soils in irrigated areas of Uzbekistan and ways of increasing their productivity. Short summary of doctoral thesis. Tahskent, Uzbekistan (in Russian)

Hillel D (1997) Small – scale irrigation for arid zones: principles and options. FAO, Rome.

Hoff H, Falkenmark M, Gerten D, Gordon L, Karlberg L, Rockström J (2010) Greening the global water system. Journal of Hydrology 384 (3 – 4): 177 – 186

Iñiguez L, Suleymenov M, Yusupov S, Ajibekov A, Kineev M, Kheremov S, Abdusattarov A, Thomas D, Musaeva M (2005) Livestock production in Central Asia – Constraints and opportunities. ICARDA Caravan 22

Karmanov S (1978) Manual of potato farming. Moscow, Rosselhozizdat (in Russian)

Kurbanbaev E, Novikova A, Shirokova Y, Forkutsa I, Poluashova G (2006) Ways of water conservation at the field level in the lower reaches of Amudarya. In: Problems of irrigated lands reclamation: water availability and efficient water use, Shymkent, Kazakhstan

Mamatov S (2009) Drip irrigation system. SANIIRI (in Uzbek), Tashkent

Martius C, Froebrich J, Nuppenau E (2009) Water Resource Management for Improving Environmental Security and Rural Livelihoods in the Irrigated Amu Darya Lowlands. In: Brauch H, Oswald Spring U, Grin J, Mesjasz C, Kameri-Mbote P, Chadha Behera N, Chourou B and Krummenacher H (Eds.) Facing Global Environmental Change: Environmental, Human, Energy, Food, Health and Water Security Concepts. Springer, Berlin

McKinsey & Company, 2009. Charting Our Water Future – Economic Frameworks to Inform Decision-making. New York

Müller M (2006) A General Equilibrium Approach to Modeling Water and Land Use Reforms in Uzbekistan. Rheinische Friedrich-Wilhelms-Universität Bonn

Oberkircher L, Haubold A, Martius C, Buttschardt T (2012) Patterns in the Landscape of Khorezm, Uzbekistan: A GIS Approach to Socio-Physical Research. In: Martius C, Rudenko I, Lamers JPA, Vlek PLG (Eds.) Cotton, water, salts and Soums – economic and ecological restructuring in Khorezm, Uzbekistan. Springer, Dordrecht pp 285 – 307

Paluasheva G (2005) Dynamics of soil saline regime depending on irrigation technology in conditions of Khorezm oasis. In: Scientific Support as a Factor for Sustainable Development of Water Management, Taraz, Khazakhstan (in Russian)

Statistics KRDo (2008a) Agricultural indicators for the Khorezm region in 2000 to 2007. In: Urgench

Statistics KRDo (2008b) Socio-Economic Indicators for the Khorezm Region in 2007. In: Urgench

Statistics KRDo (2008c) Prices of agricultural commodities at the central markets of the Khorezm region in 2005 to 2007. In: Urgench

Statistics KRDo (2009) Irrigation water use in the Khorezm region in 2000 to 2009. In: Urgench

Texas Water Development Board (TWDB) (no date) Agricultural Water Conservation Practices. Retrieved 2 March, 2008, from http://www.twdb.state.tx.us/assistance/conservation/agricons.htm

Timirova M, Salokhiddinov A (2002) Non-traditional method of Water Conservation in irrigated Agriculture (Draft report)

Tischbein B, Awan U, Abdullaev I, Bobojonov I, Conrad C, Jabborov H, Forkutsa I, Ibragimov M, Paluashova G (2012) Water management in Khorezm: Current situation and options for improvement (Hydrological perspective). In: Martius C, Rudenko I, Lamers JPA and Vlek PLG (Eds.) Cotton, Water, Salts and Soums: Economic and Ecological Restructuring in Khorezm, Uzbekistan. Springer, Dordrecht pp 69–92

UN-WATER (2007) World Water Day 2007: Coping With Water Scarcity. Retrieved 10 August, 2010, from http://www.unwater.org

Vlek PLG, Lamers JPA, Martius C, Rudenko I, Manschadi A, Eshchanov R (2012) Cotton, water, salts and Soums – research and capacity building for decision-making in Khorezm, Uzbekistan. In: Martius C, Rudenko I, Lamers JPA and Vlek PLG (Eds.) Cotton, Water, Salts and Soums – Economic and Ecological Restructuring in Khorezm, Uzbekistan. Springer, Dordrecht pp 3–22

Anik Bhaduri, Nodir Djanibekov

4.2 Potential Water Price Flexibility, Tenure Uncertainty and Cotton Restrictions on Adoption of Efficient Irrigation Technology in Uzbekistan

Abstract

The study investigates different institutional and economic factors that can induce farmers to adopt water-efficient technologies in the irrigated agriculture of Uzbekistan. To investigate the effect of different policies on the investment of efficient irrigation technology, we simulate different scenarios of institutional design of the cotton policy and land tenure security using a farm-based optimization model that maximizes the total net benefit of a farmer. The results indicate that a flexible water price can increase the adoption rate of efficient irrigation techniques by 20 % compared to fixed water price levels. However, the high water supply variability may nevertheless delay the adoption of an efficient technology. Furthermore, the level of farmers' flexibility in decision making influences aggregate adoption of such technology. Likewise, the removal of the cotton yield target could increase the adoption rate of efficient irrigation techniques. Finally, the level of tenure security is a decisive factor in facilitating the adoption of water-efficient technologies. Farmers are likely to respond much stronger to flexible water pricing, when the risk of losing land is lower.

Keywords: water pricing, agricultural institutions, water-use efficiency, farm model

4.2.1 Introduction

Irrigation water is becoming an increasingly scarce resource for the agricultural sector in Uzbekistan, Central Asia. The water scarcity at the user end is due not only to a poor irrigation infrastructure, but also to the wasteful irrigation practices resulting from poor implementation of water pricing mechanisms, lack of incentives to adopt advanced irrigation technology, and lack of water-use control and measurement (Wegerich 2010; Oberkircher and Hornidge 2011; Djanibekov et al. 2012b).

Recent studies showed that the water-use efficiency of the users in the Khorezm region of Uzbekistan is not more than 30 % (Tischbein et al. 2012). The use of innovative irrigation technologies has been proposed as one of several possible solutions to increase the efficient use of the scarce water resources and reduce environmental degradation. Seckler et al. (1998) estimated that improvements in irrigation efficiency alone may meet one half of the increase in worldwide water demand until 2025. Similarly, through better and efficient use Uzbek farmers can save water that can then be re-allocated to products and services of higher value return in the water-scarce regions. This can take place without substantial investment in irrigation infrastructure. However, the relevant question is: What water-related policies in Uzbekistan can induce the increase in the efficiency of water use and maintenance of irrigated agricultural production while arresting and even reducing resource degradation in the irrigated areas?

It is often argued that water pricing is one way for inducing a higher adoption of efficient irrigation technology (Caswell et al. 1990; Sunding and Zilberman 2001). It could promote water-use flexibility and establish a recognized water value, thus providing incentives for more efficient water use (Gardner 1985; Shaliba and Bush 1987). Water pricing that takes the increasing value of scarce supplies of irrigation water over time into account can be an option also for inducing a wider adoption of more efficient irrigation technologies and improving water-use efficiency. There is evidence that the introduction of water pricing is effective in increasing water-use efficiency in Uzbekistan (FAO 1997).

In most cases, water charges are only a small fraction of the farmers' net income. Low collection rates for water charges and services in Uzbekistan are attributed to a lack of willingness to pay rather than by an inability to pay (cf. Saravanan et al. this book). Worldwide, the level of water fees remains low in order to motivate farmers to pay. But low water pricing often fails to meet the budget for cost recovery, and thus the incentive to pay more for water remains weak. This acts as a vicious circle of poorly managed water resources. Part of the failure can be attributed to a fixed water price. In Uzbekistan, water prices are fixed at a low level and determined administratively (Djanibekov et al. 2012b), thus they reflect neither supply cost nor scarcity value. Moreover, when the scarcity value of water increases, it insulates the water economy from the market forces of demand and supply of water.

Bhaduri and Manna (2014) finds that a flexible water price, which depends on stochastic water and aggregate demand, can increase the adoption rate of efficient irrigation techniques by more than 20 % if the farmer is risk averse.

However, the implementation of a flexible water price also requires supporting institutional arrangements. It is often argued that volumetric water fees (together with incentives such as additional services that increase willingness to pay) may induce farmers to adopt efficient irrigation technology. Uzbekistan has shown considerable interest in implementing volumetric pricing, and even initiated it in some places on a trial basis (Veldwisch 2008).

Other than water pricing, policies not related to water also influence the adoption of irrigation technologies in Uzbekistan. Pender et al. (2009) argue that despite their benefits on the experimental field level, in reality these technologies often fail to attract the farmers' attention, which is also due to the design of the cotton target policy.

Since the break-up of the Soviet Union, cotton production in Uzbekistan has remained strongly linked to the interests of the national export earnings (Djanibekov et al. 2013). The state purchases the entire cotton harvest from farms at prices below the potential border price (Pomfret 2008). To ensure a certain amount of cotton production, the state determines that farms should allocate about half of their cropland to cotton cultivation and produce raw cotton at amounts assigned according to the soil-fertility levels of their fields. Failure to fulfill the cotton production target can constitute the grounds for losing a land lease (Djanibekov et al. 2012a). In addition, input supply to cotton production is prioritized by the state. For example, irrigation water is diverted to cotton fields before it is delivered to other crops. This water distribution policy has often neglected the water demand for production of other crops, especially rice, which is an important staple food holding a special position in the national cuisine of Uzbekistan, and which is among the most economically attractive crops for farmers in Khorezm (Veldwisch and Spoor 2008). As a result, the effectiveness of water pricing mechanisms to induce the adoption of low-volume irrigation technologies may be limited because of the constraints imposed by the cotton policy and tenure insecurity (Djanibekov et al. 2012c).

Through scenario analysis, we investigate the opportunities for higher flexibility of farmers' production decisions created through the modification of the cotton policy and improved tenure security. Three interrelated questions are addressed: (1) Can flexible water pricing alone guarantee the adoption of efficient irrigation technology under higher variance in water flow? (2) Can tenure security support improvements in water-use efficiency? (3) Can a relaxation of the current cotton policy influence farmers' behaviour regarding efficient water use?

4.2.2 Model

A farm-based model is structured where a farmer is assumed to maximize an instantaneous profit function by choosing the amount of water to apply to his crops and the area of land to be irrigated with improved irrigation technologies. The model considers that farmers cultivate j crops where j= cotton (c) and rice (r). Cotton and rice are considered in this analysis since these crops are the main competitors for users of irrigation water in Khorezm. Furthermore, one crop (cotton) represents the state-imposed policies, while the other (rice) is commercially the most attractive crop for local farmers. Each farmer is endowed with A^j amount of land, which can be used to cultivate both crops. A^c represents the area used to cultivate cotton and A^r represents the rice area, thus the total agricultural area per farmer over t years is defined as:

$$A_t^r + A_t^c = A_t$$

Consider the total applied water (gross) as w^j for jth crop. Suppose there are two kinds of irrigation technologies: the inefficient, conventional furrow irrigation (F), and an efficient irrigation technology (H). For each irrigation technology k, irrigation effectiveness is denoted by e_k, where k=F, H and $e_H > e_F$.

Suppose the area covered under effective technology H at time t is H(t).

The rate of change in area covered with efficient irrigation technology can be shown as:

$$\dot{H} = h - \delta_H H \qquad (1)$$

where $H(t = 0) = H_0$ and h are the amount of new area brought under efficient irrigation technology; δ_H represents the depreciation rate of the new technology. We assume that the per-unit cost of installing efficient irrigation technology is constant, and is denoted by c^H.

The net effective water applied can be represented as:

$$w_e^j = w^j[e_F(A^j - H^j) + e_H H^j] \qquad (2)$$

The yields of the crops $y^j(w_e^j)$ are functions of respective effective water usage w_e.

We assume that the yield functions of both crops are concave in w_e with $y_{w_e} > 0$ and $y_{w_e w_e} < 0$. Given the institutional setup of agricultural production in Uzbekistan, the model considers the governmental assignment of a mandatory target yield and area for cotton, while rice is cultivated according to the farmer's decision.

The target yield of cotton is defined by $\hat{y}^c$, while the target cotton area is defined by $\hat{A}^c$. Thus, the total cotton production target for a given farmer is defined as $\hat{y}^c \cdot \hat{A}^c$.

We assume that the farmer can produce more than $\hat{y}^c$, and that this surplus can be realized at higher prices. If $\hat{y}^c \geq y^c$, then the surplus production is represented by $[\hat{y}^c \geq y^c] \cdot \hat{A}^c$. To introduce the tenure uncertainty we assume that if a farmer fails to meet the cotton yield target for three consecutive years, his land tenure contract will be terminated.

Assuming uncertainty in water supply and other factors, yields may vary. In this respect, farmers have knowledge about the probability distribution of yields. The cumulative distribution of cotton yield is defined by $F(\hat{y}^c) = P[y^c < \hat{y}^c]$. It is also assumed that yield y_t^j, is a function of water usage per hectare w^j. Therefore, two situations are included, one depicts the probability of fulfillment of the cotton target yield and thus defines the cotton produce surplus, and the second one is the cotton production target failure:

$$y^c \geq \hat{y}^c, \text{where } P[y^c \geq \hat{y}^c] = 1 - F(\hat{y}^c) \text{ and}$$
$$\hat{y} < \hat{y}^c, \text{where } P[\hat{y} < \hat{y}^c] = F(\hat{y}^c) \quad (3)$$

The model considers water supply W to be stochastic and follow a log-normal distribution with mean μ and standard deviation σ.

The price of water is denoted by p. The price of the water evolves over time and is a function of the aggregate excess demand of water:

$$dp/dt = \lambda \left[\sum_{l}^{N} w - W \right] \quad (4)$$

where λ is a constant and reflects the price sensitivity to excess demand function. It is assumed that there are N homogeneous farmers who can influence water demand and the price of water.

The expected net benefit of the farmer for a given period can be expressed as:

$$V = E(NB) = [1 - F(\hat{y}^c)] \cdot P^c * [E(y^c) - \hat{y}^c] * \hat{A}^c + (P^r - C^r) \cdot E(y^r) \cdot (A - \hat{A}^c) - c^H h - pw \quad (5)$$

where P^j and C^j are the price and marginal cost of jth crop j= c, r.

Considering the probability of fulfilling the cotton quota in three consecutive years as $\pi = 3.P[y^c < \hat{y}^c] = 3.F(\hat{y}^c)$, discount factor as δ, and terminal time period as T, the present discounted value of net benefits can be represented as:

$$PDV = V + \delta V + \delta^2 V + (1-\pi)\delta^3 V + (1-\pi)^2 \delta^4 V + \ldots + (1-\pi)^{T-2} \delta^T V \qquad (6)$$

Using a numerical analysis, the net benefit is maximized as expressed in eq. (6) with respect to gross water usage w as well as investment in efficient irrigation techniques in terms of area.

The @ Risk software package was used to simulate the model employing the Latin Hypercube sampling technique[1]. Table 4.2.1 illustrates the scenario definitions and Table 4.2.2 the parameter values of the model, distribution function and the crop-water response functions. The key variable used in the simulation is the water flow W. Using Best Fit Software and empirical data, the distribution function of the water flow of the Khorezm region was determined. Simulation results suggest that the stochastic factor ε_t is best fitted with a log-normal distribution with mean 15,000 m3 per hectare of cultivated area and a constant variance $\sigma = 8,000$ m3 ha^{-1} of cultivated area.

Table 4.2.1: Model scenarios

Attribute	Scenario 1 Water price fixed	Scenario 2 Water price flexible	Scenario 3 Secure tenure and water price fixed	Scenario 4 Secure tenure and water price flexible	Scenario 5 No cotton restriction and water price fixed	Scenario 6 No cotton restriction and water price flexible
Water price	Fixed	Flexible	Fixed	Flexible	Fixed	Flexible
Land tenure based on performance	Yes	Yes	No	No	No	No
Cotton yield and area restriction	Yes	Yes	Yes	Yes	No	No

Different scenarios were formulated based on water pricing, institutional design of the cotton policy and the land tenure insecurity to investigate the effect of different policies on the investment in efficient irrigation technology. Scenario 1 represents the baseline scenario under which the investment path of efficient irrigation tech-

---

1 The Latin Hypercube technique requires fewer model iterations to approximate the desired variable distribution than the simple Monte Carlo method, and ensures that the entire range of each variable is sampled. In the simulation, more than 10,000 iterations are used.

nology was simulated. In this scenario, the price of water used for rice production is fixed at US$ 5 per 1000 m³ of water according to the average values of the shadow price of water presented in Djanibekov (2008). Furthermore, the farmer produces cotton under the so-called cotton target system according to which the cotton area annually occupies on average about 55 % of the land. The cotton target yield is determined by the government at 2.5 t ha⁻¹ (Djanibekov 2008; Farm Survey 2010). Under this scenario, if the farmer fails to fulfill cotton production targets for three consecutive years, the land use contract is terminated, which will act as a disincentive. Scenario 2 pursues a similar situation to that in Scenario 1 but assumes a flexible water pricing scheme where the water price depends on the excess demand of water. Scenarios 3 and 4 simulate conditions with land tenure uncertainty. The latter two scenarios describe the situation where the farmer enjoys tenure certainty even if he does not fulfill the cotton target. Scenario 5 and 6 depict situations where there is neither a cotton yield target nor a restriction on the area for cotton production under fixed and flexible water price regimes, respectively.

Table 4.2.2: Parameters and values used in numerical analysis

Parameter (unit)	Value	Source
Depreciation rate (%)	10	Bhaduri and Manna (2014)
Cost of investment in irrigation technology (sprinkler) (US dollar per ha, c^H)	973	
Initial price of water (US$ per 1000 m³)	10	Djanibekov (2008)
Discount rate (in %)	10	
Efficiency of furrow irrigation e_F (%)	55	Bhaduri and Manna (2014)
Efficiency of sprinkle irrigation e_H (%)	80	
Price of water (US$ per 1000 m³)	5	Djanibekov (2008)
Water price sensitivity to excess demand (coefficient)	0.1	Bhaduri and Manna (2014)
Price of cotton (US$ per ton)	234	Farm survey 2010
Price of rice (US$ per ton)	1000	Farm survey 2010
Per unit cost of cotton (US$ per ton)	180	Farm survey 2010
Per unit cost of rice (US$ per ton)	600	Farm survey 2010
Average cotton target yield (tons per ha)	2.5	Farm survey 2010
Share of farmland under cotton target area (%)	55	Farm survey 2010
Water supply mean (1000 m³ per ha)	15	CAWATERinfo (2011)
Water supply variance (1000 m³ per ha)	8	CAWATERinfo (2011)
Yield function – cotton	$2.135 * \hat{w_e}(1/4)$	Estimated based on MAWR (2001)
Yield function – cotton	$1.723 * \hat{w_e}(1/3)$	Estimated based on MAWR (2001)

Note: Exchange rate of 1 US$ = 1,319 UZS as of 2009 according to Sommer et al. (2012).

4.2.3 Results and discussion

Figure 4.2.1 shows the optimal time path of investment in efficient irrigation technology under different scenarios[2]. According to the results of Scenarios 1 and 2, even when the price of water is fixed at a higher value, compared to the cases with flexible water price schemes it can be seen that (i) the rate of adoption will be much slower, and (ii) the aggregate level of adoption of efficient irrigation technologies will be lower. The results of Scenarios 3 and 4 imply that tenure security also determines the adoption rate of irrigation technologies. If the farmer's land tenure is secure and not related to the performance of cotton yield, then the difference between the adoption rates of efficient irrigation technology under flexible and fixed water price regimes will be more pronounced during the early periods. In such case, as far as adoption is concerned, a farmer is more likely to response to a flexible water price regime compared to a fixed water price. Hence, to motivate farmers to early adoption of efficient irrigation technologies, a series of frame conditions could be conducive such as flexible water pricing combined with tenure security for agricultural farmland. The results of Scenarios 5 and 6 also indicate that removal of the cotton policy regarding the fulfillment of yield targets and the area of production could encourage farmers to adopt efficient irrigation technologies. However, in the initial years the rate of adoption is likely to be lower, as a farmer has to account for costs of land and other fixed variables. However, recurring higher profits from cotton without a rigid cotton policy would provide greater opportunities for adopting irrigation technology in the following period and at a faster pace.

The model results show that the price of water may play a significant role in inducing a farmer to adopt efficient irrigation techniques. This in line with the findings of Oberkircher (2011), which show that when the cotton production targets are deregulated and the duration of land lease becomes more certain farmers' investments in efficient irrigation technology will increase. On the other hand, although the famers may invest in efficient irrigation technology (e.g., sprinkler), in all cases the timing of the adoption of the technology is likely to depend on the design of institutional arrangements (e.g., cotton production targets and/or land tenure security). In fact, farmers who face a short planning horizon because they risk losing their farmland do not consider sustainability as a priority (Oberkircher 2011). In this respect, the recent farm consolidation process, which involved the risk of losing farmland, has also reduced the farmers' interest in any agricultural technology (Djanibekov et al. 2012a).

---

[2] Here the aggregate adoption of efficient irrigation technology will decrease if the existing irrigation technology is depreciating and the representative farmer is no longer making new investments in irrigation technology.

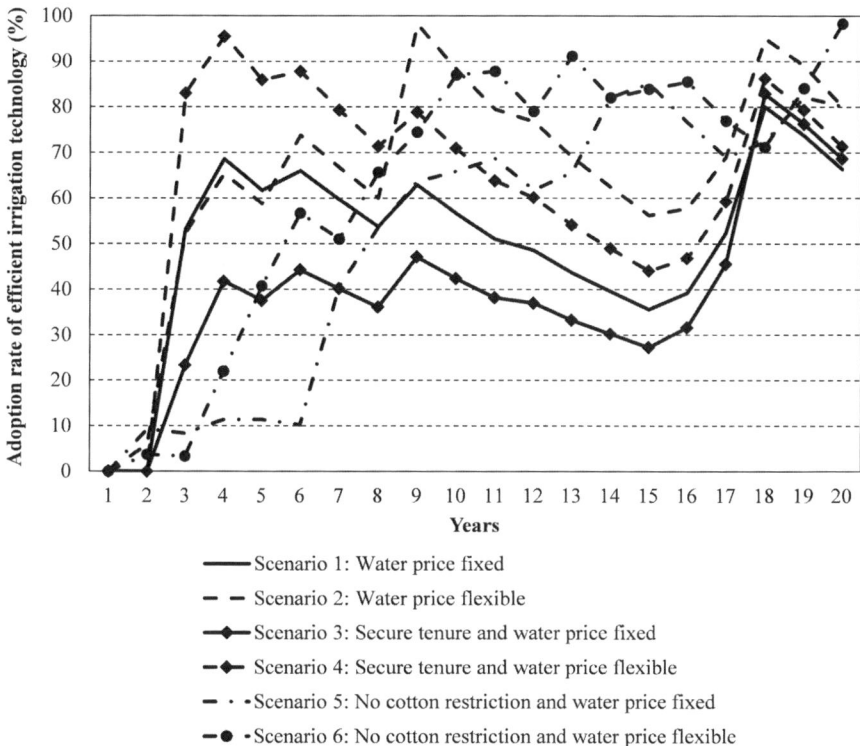

Figure 4.2.1: Adoption of efficient irrigation technology under different scenarios

Figure 4.2.2 shows how optimal investment amounts could respond to different levels of water supply variability. In the numerical analysis, it is assumed that the coefficient of variation changes from 0.55 to 0.80 indicating that the magnitude of the irrigation water supply increases over the years. The findings indicate further that the aggregate investment in efficient irrigation technology would be lower under higher variability in water supply. This result may hold if the decline in optimal water consumption has a lower impact on crop yield and the farmer adjusts his activities to the variability of the water supply.

Contrary to the prediction of the economic theory that resource use efficiency improves with its scarcity, irrigation water use is becoming less and less efficient while water is becoming scarcer (Saleth and Dinar 2004). This is particularly relevant for the case of Uzbekistan, where water scarcity conditions do not influence farmers directly to use water more efficiently. In 2010, farmers in Khorezm paid only US$ 2.5 per 1000 m^3 in form of water service fees, which barely covered the expenses of the Water User Association (Farm Survey 2010). Bobojonov (2008) has shown that farmers in the study region would need to be charged at least US$ 5 per 1000 m³ to cover operation and management expenses for the existing water management organizations and to achieve a water use efficiency of about 67 %. The findings of this study, however, indicate that water prices of US$ 5 per 1000 m³ of water can induce the farmer to irrigate the majority (80 %) of his land with efficient irrigation technology. However, if water prices are fixed, for example determined administratively and thus ignoring the water supply level, they reflect neither the supply cost nor the scarcity value of water, and the adoption rate of efficient irrigation technology is likely to be lower.

The duration and frequency of irrigation water scarcity in Khorezm has increased since 1980 (Müller 2006). It is expected that it will continue to negatively impact agricultural production in the Khorezm region. Together with climate change, this will have even greater negative impacts on agricultural incomes particularly in the downstream areas of the main river basins (Siegfried et al. 2012). The results indicate that water variability will have a stronger negative effect on the farmers' decision to invest in efficient irrigation technology during the early stages of adoption, and particularly under a fixed rather than under a flexible water price regime. Combined with high variability in the water supply, a fixed water pricing is likely to delay the adoption of efficient irrigation technology during the initial period. In such a case, the farmer may even prefer to buy water at a fixed price rather than to invest in irrigation technology. However, a consistently recurring variability in the water supply will in the end lead the farmers to invest in irrigation technology.

The findings of the study are consistent with the general view that water pricing is a valid tool for increasing the efficiency of water use. The theoretical validity of flexible water pricing, however, has been constrained by several factors. First, the effective application of this economic instrument requires that a strong institution is established, which owns and manages the physical infrastructure, monitors the allocation of water, assures water delivery, and provides the forum for conflict resolution, and thus alleviates the inefficiencies of public water management. In many instances, the absence of strong institutional arrangements often makes it difficult to implement water pricing. Second, the majority of the farmers in Uzbekistan may find it difficult to adopt efficient irrigation technology, especially when induced by water pricing. Under the existing state targets, the introduction of water pricing for creating water saving incentives can be limited due to the financial constraints of agricultural producers (Bobojonov et al. 2010). Another issue is that opportunities and constraints with respect to adopting efficient technologies vary between different groups of farmers who are different with regard to land characteristics and form of production (Veldwisch 2008; Oberkircher 2011). Hence any water pricing

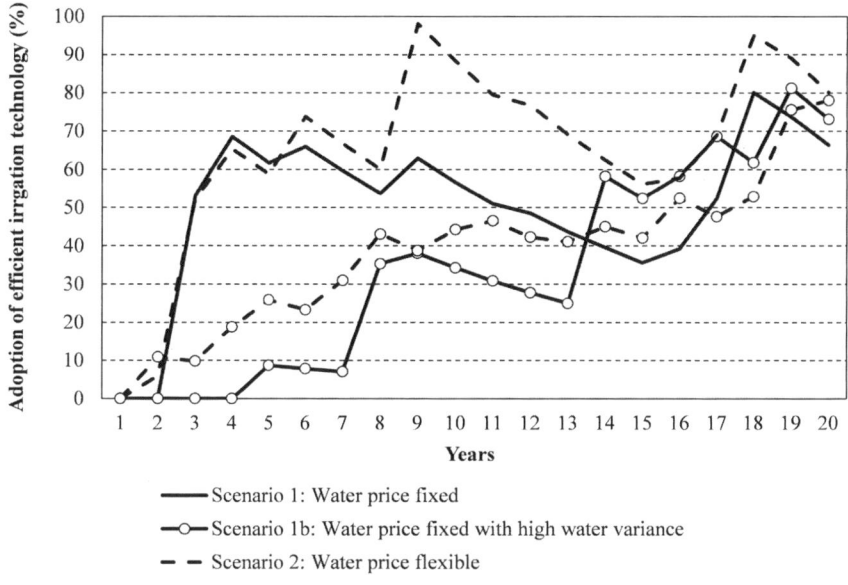

Figure 4.2.2: Adoption of efficient irrigation technology under different water price regimes and supply variability

approach must be sensitive to match the farmers' unique characteristics of low capital availability.

Water use improvement potential and financial feasibility of efficient irrigation technologies have an inverse relationship: The more efficient technology is also more capital intensive (Bekchanov et al. 2010). In this respect, to fit the low capital availability among farmers, cheaper but less water efficient options such as double flow, short and alternate dry furrow techniques may have the highest potential for adoption (Bekchanov et al. 2010). Among the technologies, the use of laser leveling has proven to increase water use efficiency and crop yields in many countries (Humphreys et al. 2005). Laser leveling could be extremely suitable for the region, as 70 % of the total cropped area is only poorly (unevenly) leveled in Uzbekistan (UNDP 2007). Proper leveling would allow a more equal distribution of water in flood and furrow irrigation and, therefore, decrease the amount of water needed for irrigation (Bobojonov et al. 2010). Laser-leveled fields have proven to reduce water demand by as much as 20 % compared to traditionally leveled fields. Laser-leveled fields are, however, 10 % less effective than sprinkler irrigation. The expense of leveling one hectare of land was estimated to be only 8 % of the cost of sprinkler irrigation.

The deregulation of the cotton production targets at the current stage may increase the pressure on irrigation water resources, as farmers would opt to cultivate more profitable, albeit more water-intensive, crops such as rice (Djanibekov et al. 2013). Additional policy measures will be required for promoting crops that require less irrigation water. On the other hand, the increasing variability of water supply stemming from climate change may also induce farmers to cultivate less water-intensive crops such as potatoes, vegetables and maize instead of cotton and rice as an alternative to introducing water pricing, and to adopting the efficient irrigation technologies for these crops (Bobojonov et al. 2010).

4.2.4 Conclusions

The study attempts to answer the question whether or not water pricing can induce farmers to adopt efficient irrigation technologies. The results show that a flexible water price, which depends on stochastic as well as aggregate water demand, can increase the adoption rate of efficient irrigation technologies (exemplified here by sprinklers) by more than 20 %. Overall, the results indicate that the negative effect of water variability on investment in efficient irrigation technology would be more prominent during the early stages of adoption, and particularly under a fixed rather than flexible water price regime. Fixed water price levels in combination with considerable variability in water supply are likely to delay the adoption of efficient irrigation technology during the initial periods. However, while even a flexible water price cannot guarantee its earlier adoption, under the increasing water supply variability as a result of climate change farmers are likely to decide to adopt efficient irrigation technology in the later stages.

Based on the current cotton target policy and tenure insecurity of farmland, according to which farmers bear the risk of losing their land lease if they do not achieve the production target, the results show that these aspects of Uzbekistan's agriculture are important factors influencing farmers' decisions to adopt efficient irrigation technology. With secure land tenure, farmers are likely to react more strongly to flexible water price regimes compared to fixed water regimes with regard to investment in irrigation technology. The farmers' aggregate adoption of efficient irrigation technology would be also greater if the minimum cotton yield targets were to be abolished. However, in such case, farmers would have to bear the full costs of cultivation of cotton, which in turn would restrict investment capacities during the initial periods. As a consequence, the adoption of efficient irrigation technology can be delayed.

References

Bekchanov M, Lamers JPA, Martius C (2010) Pros and cons of adopting water-wise approaches in the lower reaches of the Amu Darya: A socio-economic view. Water, 2: 200–216

Bhaduri A, Manna U (2014) Impacts of water supply uncertainty and storage on efficient irrigation technology adoption. Natural Resource Modeling, 27 (1): 1–24

Bobojonov I (2008) Modelling crop and water allocation under uncertainty in irrigated agriculture: a case study on the Khorezm region, Uzbekistan. Ph.D. diss., ZEF, Bonn University, Germany, 141 p

Bobojonov I, Franz J, Berg E, Lamers JPA, Martius C (2010) Improved policy making for sustainable farming: a case study on irrigated dryland agriculture in Western Uzbekistan. Journal of Sustainable Agriculture, 34 (7): 800–817

CAWATERinfo (2011) Amudarya River Basin. Online data of the BWO Amudarya. Available at: http://www.cawater-info.net/amudarya/index_e.htm#

Caswell M, Lichtenberg E, Zilberman D (1990) The effects of pricing policies on water conservation and drainage. American Journal of Agricultural Economics 72 (4): 883–890

Djanibekov N (2008) A Micro-economic Analysis of Farm Restructuring in the Khorezm Region, Uzbekistan. Ph.D. thesis, ZEF, University of Bonn, Bonn, Germany, 180 p

Djanibekov N, Van Assche K, Bobojonov I, Lamers JPA (2012a): Farm restructuring and land consolidation in Uzbekistan: new farms with old barriers. Europe-Asia Studies, 64 (6): 1101–1126

Djanibekov N, Hornidge A-K, Ul-Hassan M (2012b) From joint experimentation to laissez-faire: transdisciplinary innovation research for the institutional strengthening of a water users association in Khorezm, Uzbekistan. The Journal of Agricultural Education and Extension, 18 (4): 409–423

Djanibekov N, Bobojonov I, Djanibekov U (2012c) Prospects of agricultural water service fees in the irrigated drylands, downstream of Amudarya. In: Martius C, Rudenko I, Lamers JPA, Vlek PLG (Eds.) Cotton, water, salts and soums – economic and ecological restructuring in Khorezm, Uzbekistan. Springer: Dordrecht, pp 389–411

Djanibekov N, Sommer R, Djanibekov U (2013) Evaluation of effects of cotton policy changes on land and water use in Uzbekistan: Application of a bio-economic farm model at the level of a water users association. Agricultural Systems, 118: 1–13

Farm Survey (2010) Socio-economic survey of commercial farms in Khorezm province, June-August 2010

FAO (1997) Uzbekistan. FAO's Information System on Water and Agriculture URL: http://www.fao.org/nr/water/aquastat/countries/uzbekistan/index.stm

Gardner RL (1985) The potential for water markets In Idaho. Idaho Economics forecast, 7: 27–34

Humphreys E, Meisner C, Gupta R, Timsina J, Beecher HG, Lu Tang Yong, Singh Yadvinder, Gill MA, Masih I, Jia G, Thompson JA (2005) Water Saving in Rice-Wheat Systems. Irrigated Rice-Based Systems, Special Issue Proceedings of The Fifth Asian Crop Science Conference

Ministry of Agriculture and Water Resources of Uzbekistan (2001) Guide for Water Engineers in Shirkats and WUAs. Scientific production association Mirob-A, Tashkent

Müller M (2006) A general equilibrium approach to modelling water and land use reforms in Uzbekistan. Ph.D. thesis, ZEF, University of Bonn, Bonn, Germany, 170 p

Oberkircher L (2011) On pumps and paradigms: Water scarcity and technology adoption in Uzbekistan. Society & Natural Resources: An International Journal, 24 (12): 1270 – 1285

Oberkircher L, Hornidge AK (2011) 'Water is life' – farmer rationales and water saving in Khorezm, Uzbekistan: A lifeworld analysis. Rural Sociology, 76 (3): 394 – 421

Pender J, Mirzabaev A, Kato E, (2009) Economic analysis of sustainable land management options in Central Asia. IFPRI/ICARDA Final Report for ADB, 146 pp

Pomfret R (2008) Tajikistan, Turkmenistan and Uzbekistan. In: Anderson K, Swinnen J (Eds.), Distortions to Agricultural Incentives in Europe's Transition Economies. World Bank, Washington DC

Saleth M, Dinar A (2004) The Institutional Economics of water: A cross country analysis of institution and performance, Edward Elgar, Cheltenham, UK

Sekler D, Amarsinghe U, Molden D, de Silva R, Barker R (1998) World water demand and supply, 1990 to 2025: Scenarios and Issues. IWMI Research Report 19, Colombo

Shaliba BC, Bush DB (1987) Water markets in theory and practice: Market Transfers, Water Values, and Public Policy. Westview Press, Boulder, Colorado

Siegfried T, Bernauer T, Guiennet R, Sellars S, Robertson AW, Mankin J, Bauer-Gottwein P, Yakovlev A (2012) Will climate change exacerbate water stress in Central Asia? Climatic Change, 112 (3 – 4): 881 – 899

Sommer R, Djanibekov N, Müller M, Salaev O (2012): Economic-ecological optimization model of land and resource use at farm-aggregated level. In: Martius C, Rudenko I, Lamers JPA, Vlek PLG (Eds.), Cotton, Water, Salts and Soums – Economic and Ecological Restructuring in Khorezm, Uzbekistan. Springer: Dordrecht, pp 267 – 283

Sunding D, Zilberman D (2001) The agricultural innovation process: Research and technology adoption in a changing agricultural sector. In: Bruce LG, Rausser GC (Eds.), Handbook of Agricultural Economics Volume 1, Part A, Elsevier, pp. 207 – 261

UNDP (United Nations Development Program) (2007): Water – Critical Resource for Uzbekistan's future. UNDP Tashkent, pp. 38 – 60

Veldwisch GJ (2008) Cotton, rice and water: Transformation of agrarian relations, irrigation technology and water distribution in Khorezm, Uzbekistan. Ph.D. thesis, ZEF, University of Bonn, Bonn, Germany, 219 p

Veldwisch GJ, Spoor M (2008) Contesting rural resources: emerging 'forms' of agrarian production in Uzbekistan. Journal of Peasant Studies, 35 (3): 424 – 451

Wegerich K (2010) Handing over the sunset – External factors influencing the establishment of water user associations in Uzbekistan: Evidence from Khorezm province. Cuvillier Verlag, Göttingen, 169 p

Aziz A. Karimov, Miguel Niño-Zarazúa

4.3 Assessing Efficiency of input Utilization in Wheat Production in Uzbekistan

Abstract

Increased technical and scale efficiency in the production of wheat has been of major interest for farmers and administrators alike in Uzbekistan particularly since wheat became a strategic crop to achieve the goal of food self-sufficiency soon after the country's independence in 1991. A pioneer approach was adopted to estimate technical and scale efficiencies among wheat producing farms in the Khorezm and Fergana regions of Uzbekistan. A method was developed that consists of extending a nonparametric, output-based Data Envelopment Analysis (DEA) in two stages to allow the use of double bootstrapping techniques to produce bias-corrected estimates. The findings show that while most farmers have achieved scale efficiency under the current state of agricultural technology, there is room for increasing wheat production via enhanced technical efficiency. Interestingly, findings also show that the higher efficiency estimated for arable land with lower bonitet (soil fertility) scores indicates that farmers with better land quality use their resources less efficiently. It is argued that this in turn implies that under non-competitive market conditions, farmers have little incentives to use resources more efficiently.

Keywords: wheat production, technical efficiency, data envelopment analysis, Central Asia

4.3.1 Background

The efficient use of resources plays a fundamental role in maximizing grain productivity. In an era marked by major vulnerabilities associated with food insecurity, the efficient use of resources becomes critical for present and future agricultural policy. The resource-use efficiency of wheat production was examined in the context of Uzbekistan, where grain crops became central after the country gained sovereignty from the Soviet Union in 1991. Crucial changes in cropping patterns began with the Decree of the Cabinet of Ministries (*CMD*) No. 400 "*On measures of increasing grain production in irrigated lands starting from 1995*". It promoted significant increases in wheat production for achieving self-sufficiency in grain consumption (Kienzler et al. 2011). The allocation of arable land for wheat production was achieved largely at the expense of reducing land designated to other crops, including cotton[1], the main agricultural export commodity in the country. Irrigated land for wheat increased consequently from 626,900 hectares (ha) in 1992 to approximately 1.4 million ha in 2010, pushing the overall wheat production from 964,000 tons in 1992 to 6.7 million tons in 2010 (FAO, 2011).

Following independence, irrigated winter wheat became part of the state quota system next to cotton (Rudenko et al. 2012). The state quota system for wheat requires individual farms[2] to sell half of the produced wheat to the state at fixed prices in exchange for loans with below-market rates of interest, and subsidized agricultural inputs (Abdullaev et al. 2009). The remaining wheat is either home-consumed, traded at local markets, or in some cases, sold to the state at negotiated prices (Rudenko et al. 2012). Individual farms are presently the main wheat producers, producing nearly 82 % of *total wheat* in 2007 (State Statistical Committee of Uzbekistan, 2008).

Despite recent improvements in productivity, wheat yields have remained low in Uzbekistan *vis-à-vis* other countries with similar climatic conditions (ADB 2009). Studies conducted in developing countries by, *inter alia*, Hopper (1965), Getu et al. (1998), and Coelli et al. (2002), suggest that grain yields can be increased by a more efficient use of production factors. These studies explored whether or not there is a need for the adoption of new technologies, or if it is still possible to achieve higher yields with the current technologies. This is a major challenge to Uzbekistan given its only recent involvement in domestic, irrigated wheat production. Therefore, technical and scale efficiencies[3] of wheat production based on farm-level surveys conducted in two regions of Uzbekistan were analyzed. Technical and scale-efficiency estimations were obtained from a

---

1 About 30–35 % of the cotton areas were freed to grow irrigated wheat (Aminova and Abdullaev, 2012).
2 These refer to the "private" farmers who work on land leased from the state (Rudenko et al. 2012).
3 We use the term technical efficiency (*TE*) to measure farm capacity to utilize resources, including production technologies, in the most efficient way.

two-step output-oriented Data Envelopment Analysis (DEA) model[4] (Charnes et al. 1978, 1979, 1981), which allows estimating to what extent production can be increased given the level of utilized resources. In addition, the traditional *DEA* model has been extended by an application of the double bootstrapping method to improve the estimates (Simar and Wilson 2007).

4.3.2 Methodological Approach

The adopted *DEA* model has its theoretical foundations in the work of Farrell (1957), and relies on linear programming techniques initially developed by Charnes et al. *(1978). The method has been extensively used in efficiency studies, including those by Wadud and White (2000), Coelli et al. (2002), and Latruffe et al. (2005).* The use of the *DEA* method is justified for various reasons. First, the model enables division of the overall technical efficiency[5] into pure technical efficiency (TE) and scale efficiency (SE). This clarifies whether or not resource-use inefficiencies arise from diseconomies of scale due to SE or from factors associated with deficient farming techniques due to a low TE. Second, with the model there is no need for the specification of a functional form. Third, variables in the model (inputs and outputs) can be measured in different units and dimensions (e.g., continuous or categorical variables).

In a two-stage *DEA* model, the sampling variations, measurement errors and sample size can increase the bias in the efficiency estimators. Simar and Wilson (2007) proposed therefore a double bootstrapping technique to improve the consistency and reliability of the *DEA* estimators.[6] An advantage of this improvement is that it allows the use of bootstrapping measures of *TE* in the first stage, and via non-parametric statistics, to generate standard errors and confidence intervals for exogenous variables in the second step (Charnes 1978). Both single and double bootstrapping techniques were therefore used. The results were compared with those of the traditional *DEA* model, which relies on a *Tobit* regression in the second stage.[7] Single and double bootstrapping techni-

---

4 With this output-oriented approach, it can be estimated how much production is increased for a given level of utilized input resources.
5 The term overall technical efficiency is used for those farms that operate under constant returns to scale (*CRS*), while pure technical efficiency (TE) is utilized for the farms that operate under variable returns to scale (*VRS*). The difference between CRS and VRS technologies is that CRS also takes into account scale effects, and scale efficiency (*SE*) can thus indicate farm size optimality. Moreover, we can also determine farms operating under different returns to scale (constant, increasing and decreasing).
6 Bootstrapping allows estimating properties of an estimator by assigning measures of accuracy to sample estimates.
7 In the second stage of *DEA*, where we included explanatory factors which impact on *TE*, a

ques are similar in algorithms, except that in the case of double bootstrapping, efficiency scores obtained from the second stage are also bootstrapped.

4.3.2.1 Estimation of Technical Efficiency via a Double Bootstrapping Procedure

Initially, the standard *DEA* model is used to estimate *TE* scores for each individual farm (i) under variable returns to scale (*TEvrs*) as follows:

$$\hat{\tau}_i = max\left\{\tau > 0 | \tau y_i \leq \sum_{i=1}^{n} \lambda_i Y; x_i \geq \sum_{i=1}^{n} \lambda_i X; \sum_{i=1}^{n} \lambda_i = 1; \lambda_i \geq 0, i = 1, \ldots n\right\} \quad (1)$$

where, $\hat{\tau}_i$ is the *TEvrs* score for each (i); y_i and x_i as well as Y and X denote the input and output matrices of individual farms i's and their corresponding sample mean, respectively. The symbol λ captures a non-negative intensity variable that reflects returns to scale.

Secondly, the *TEvrs* scores estimated in the first stage are used as a dependent variable in a truncated maximum likelihood regression that takes the form:

$$\tau_i = \varepsilon_i \geq 1 - z_i \beta \quad (2)$$

where z_i is a vector of exogenous variables, β is a vector of parameters to be estimated, and ε_i, a continuous independent and identically distributed random variable.

Thirdly, several steps are needed for each individual farm ($i=1,\ldots, n$) S_n times to obtain the

$$\left\{\hat{\tau}_{i,s}^* = 1, \ldots S_n\right\} \text{bootstrap estimates:}$$

a. For each individual farm, $i=1,\ldots, n$, ε_i is obtained from the normal distribution function, and $\tau_i^* = \beta z_i + \varepsilon_i$ is computed.
b. A new pseudo data set (x_i^*, y_i^*), where $x_i^* = x_i$ and $y_i^* = y_i \hat{\tau}_i / \hat{\tau}_i^*$, is constructed and used to compute the *TEvrs* scores.

---

censored (Tobit) regression is usually employed. With single and double bootstrapping methods, truncated regression is used as recommended by Simar and Wilson (2007).

Fourthly, (bias corrected) for each individual farm is calculated as follows:

$$\hat{\hat{\tau}}_i = \hat{\tau}_i - \widehat{bias}_i \qquad (3)$$

$$\text{where } \widehat{bias}_i = \left(\frac{1}{S_n}\sum_{s=1}^{S_n}\hat{\tau}^*_{i,s}\right) - \hat{\delta}_i \qquad (4)$$

Fifthly, bias-corrected *TEvrs* scores are used as a dependent variable to estimate a truncated regression function employing a maximum likelihood framework. Then steps (a) and (b) below are repeated $S2$ times to obtain $\{(\beta^*_s, \sigma^*_s) s = 1,\ldots, S_2\}$:

a. For each individual farm, ε_i is obtained from $N(0, \sigma^2_\varepsilon)$ and $\tau^{**}_i = \beta z_i + \varepsilon_i$ is computed, and then
b. A second bootstrapping is employed on the truncated regression.

Lastly, bootstrap-based 95 % confidence intervals are computed for each parameter estimate. The $(1-\alpha)\%$ confidence interval of the *j-th* element of vector β is constructed as

$$\Pr\left(-b_{\frac{\alpha}{2}} \le \hat{\beta}^*_j - \hat{\beta}_j \le -a_{\frac{\alpha}{2}}\right) \approx 1 - \alpha \qquad (5)$$

such that the estimated confidence interval is $Higher_{a,j} = \hat{\beta}_j + \hat{b}_\alpha$ and $Lower_{a,j} = \hat{\beta}_j + \hat{a}_\alpha$.

4.3.2.2 Scale Efficiency Estimation

To estimate *SE* for each individual farm (*i*), the following expression is computed:

$$SE = TEcrs\ /\ TEvrs \qquad (6)$$

TEvrs is obtained from Eq. 1, whereas *TEcrs* (*TE* under constant returns to scale) is also calculated from the Eq.1 but without the convexity constraint ($\sum_{i=1}^{n} \lambda_i = 1$). Note that scale-inefficient farms may be operating under constant (*CRS*), increasing (*IRS*) or decreasing (DRS) returns to scale. Estimates of technical efficiency under non-increasing returns to scale (or *TEnirs*) need to be obtained by changing the convexity constraint to. If ($\sum_{i=1}^{n} \lambda_i \leq 1$) two scores are different from each other, and *SE* is less than one, then the farm is operating under *IRS*. In contrast, if the scores are equal and *SE* is lesser than one, the farm displays the *DRS* in its operation.

4.3.3 Data

This study was conducted in the Khorezm and Fergana regions of Uzbekistan with individual farms that grow wheat on land leased from the state (see Figure 4.3.1). The Khorezm region is located in the northern climatic zone according to the classification by FAO (2003) for agro-ecological zones. It is situated in the north-western part of the country, bordering the Karakalpakstan region and Turkmenistan. The main water source for irrigation is the Amudarya river. Soils belonging to this zone are hard and loamy, and have been partly influenced by human interventions since the introduction of a large-scale irrigated agriculture (Akramkhanov et al. 2012). The Fergana region is situated in the eastern part of Uzbekistan and, based on the FAO (2003) classification, in the central climatic zone. Fergana's main water providing river is the Syrdarya, which is formed from the joining of the Naryn and Kara-Darya rivers. The region comprises gley and meadow soils characterized by initial low salinity levels. It has better hydro-physical infrastructures in comparison to Khorezm.

Figure 4.3.1: Map of Uzbekistan in Central Asia indicating (squares) the two study regions Khorezm and Fergana. *Source: GIS laboratory of the ZEF/UNESCO Khorezm Project*

A farm-level survey was conducted among randomly selected individual farms, recollecting information about use of resource endowments in wheat production for the 2007 agricultural season. The input-output dataset was constructed such as to capture information on agricultural production activities from a sample of 180 farms representative of 8 districts in Khorezm and 164 farms from 9 districts in Fergana. The survey also collected qualitative information on farm activities as well as on socio-economic, demographic and location-specific characteristics to account for possible sources of efficiency differentials.[8] A statistical summary of input and output variables used in the *DEA* model is given in Table 4.3.1.

Table 4.3.1: Descriptive statistics of variables used in the analysis

Variable	Unit	Khorezm Region N=180		Fergana Region N=164	
		Mean	Standard Deviation	Mean	Standard Deviation
Output variable					
Yield	tons h^{-1}	4.2	1.04	4.8	1.01
Input variables					
Land	ha	11.7	12.1	11.7	12.0
Labor	man-days ha^{-1}	184	55	172	57
Seeds	kg ha^{-1}	224	19	222	19
Nitrogen fertilizer	kg ha^{-1}	175	30	170	32
Diesel fuel	kg ha^{-1}	172	34	170	31
Other expenses	1,000 UZS ha^{-1}	101.7	12.7	101.6	13.0
Farm Characteristics					
Bonitet score*	Index (1 – 100)	58	10	56	13
Farm size	ha	23	13	32	20
Water availability	Dummy (Enough Water = 1; Not Enough = 0)	0.54	0.50	0.54	0.50
Diversification index	Index	0.34	0.44	0.42	0.46
Dependency ratio	Ratio	1.13	0.84	1.26	1.42
Potential to work in larger land area	Dummy (Yes = 1; No = 0)	0.48	0.50	0.40	0.49
Obsolete canal	Dummy (Yes = 1; No = 0)	0.57	0.50	0.51	0.50
Distance to market	Km	9.7	2.5	9.5	2.2

*A *bonitet* score is on a scale 1 – 100 and is calculated based on several indicators for soil quality. Land with a higher *bonitet* score is considered more fertile and hence attractive.

In the first stage analysis of *DEA*, we included one output and six production factors. Output consisted of quantities of wheat both sold and kept for self-consumption. Inputs used in the analysis were land, labor, seeds, nitrogen fertilizer, diesel and 'other expenses'. Land input was defined as the total cultivated land utilized for growing wheat measured in hectares (ha). The labor input variable was specified in man-days where labor was divided into thirteen

---

[8] For a more detailed discussion on the sampling frame see Karimov (2012)

agronomic activities (e.g., land preparation, planting, fertilizer application, and weeding etc.). One man-day consisted of eight hours. Since information was based on recollection of farmers, it was not always possible to distinguish the labor used in wheat production by sex and age. The seeds variable was measured in kilograms (kg), and consisted of seeds purchased at the market or produced by the farmer. Nitrogen fertilizer is the most important nutrient necessary for wheat growth (Kienzler et al. 2011). This variable was calculated by estimating the share of nitrogen (in kg) in each fertilizer. Diesel fuel was included as a proxy for machinery services and was measured in kg. "Other expenses" is an aggregated variable that consists of the sum of expenses farmers paid to the Water User Associations (WUAs)[9] for the membership fee, for obtaining chemicals (other than chemical fertilizers), organic manure and machinery services, which all were measured in monetary values.

The second-stage analysis of the *DEA* model included eight independent variables and one dependent variable (*TE*), which was obtained from the first stage analysis. The first variable is a *bonitet* score, which captures land productivity. The second variable, farm size (in ha), is the whole-farm operating area. The third variable, water availability, is expressed as a dummy indicator. Since water usage was not directly observed, it was not possible to provide information on the intensity of water utilization during the survey. Instead, a water-related dummy variable was included under the assumption that farmers who reported adequate water sources were more efficient in the use of other inputs. The fourth variable is the Shannon diversity index (SHDI) that captures farmers' crop diversification (Eilu, et al. 2003, Shaxson and Tauer, 1992). It was calculated as:

$$SHDI = -\sum_{i=1}^{J}(P_i * lnP_i) \qquad (7)$$

where *J* stands for the number of grown crops, and the term P_i is the proportion of the area used for a particular crop. When there is only one crop, the index in Equation 7 equals zero and increases with the number of cultivated crops. The fifth variable is the dependency ratio, which is calculated as the ratio of family dependents aged 15 and younger and 60 and older to the number of adults of working age. The sixth indicator is a dummy variable that captures information on whether or not a farmer is interested in working on a larger cropping area. The seventh variable is a dummy, which is used as a proxy for the condition of the irrigation canals. Those farmers who reported a poor canal system were

---

9 Renamed to Water Consumers Associations in 2009

recorded as 1. Finally, the eighth variable indicates the mean distance between the farm and the nearest local market.

The output-oriented *DEA* model produces *TE* values starting from score 1, in other words they are bounded to 1 from below. However, for convenience, the inverse of the initial and bias-corrected average values of *TE* are calculated. As a result, *TE* values were between 0 and 1. With the model it was possible to estimate efficiency scores under different categories, although it should be noted that for each classification under each category, a frontier needed to be constructed separately. In this way, additional information for assessing policy implications was obtained. The *DEA* model is categorized by location (pooled sample, and regions), by cropped area, and by bonitet scores under two types of technologies: (1) assuming constant returns to scale (*CRS*) and (2) variable returns to scale (*VRS*). Efficiency results obtained from the model were estimated using the *FEAR* package developed by Wilson (2008) for the R, and Stata 12 statistical software.

4.3.4 Results and Discussion

Findings from the pooled results show that estimated values for the *TE* coefficient under *CRS* are 0.72, and 0.75 under *VRS* (Table 4.3.2). Since the highest efficiency in the use of resources is achieved at the score of 1.0, model results indicate room for technical efficiency gains with the current production technologies. A similar conclusion can be drawn when the *DEA* model is constructed for each region, but with varying efficiency scores. Individual farms in the Fergana region had a higher *TE* under *CRS* and *VRS* technologies in comparison to those in the Khorezm region. This suggests a regional divide in terms of efficient use of resources. Given minor differences between *CRS* and *VRS* technologies, we postulate that most of the productivity gaps monitored arose not from scale differences between farms but rather from farm-level factors (e.g., mismanagement, delay in agronomic activities, farm characteristics) and exogenous determinants such as location, and institutional and socio-economic attributes that are associated with farm activities.

TE scores were also highest for farms that had the largest wheat fields (e.g., group of farms with wheat areas of 30.1 ha and above); however, when the sample was grouped by *bonitet* levels, differences between the three farm size groups were not significant. It should be noted that the highest efficiency score was achieved in lands with *bonitet* scores below 50. This is an interesting finding, as it suggests that farms that operate in low-fertility lands were relatively more efficient in terms of resource endowment.

To verify the initial results from the *DEA* model, the single bootstrapping method was employed to obtain bias-corrected *TEvrs* estimates (column 4 of Table 4.3.2), illustrating that the uncorrected initial results were considerably biased. For instance, the initial for the Khorezm sample suggests farms could increase their production by 28.2 % (((1/0.78)-1)*100) if full efficiency were to be achieved, whereas the bias-corrected *TEvrs* estimates suggested a production increase in the order of even 41 %(((1/0.71 – 1)*100). Furthermore, in the case of Khorezm, the lower and upper bounds of the 95 % confidence interval for the bias-corrected *TE* obtained from the single bootstrapping suggests that an 'average' farm could increase its production by 30 % (((1/0.77 – 1)*100) to 47 % (((1/0.68 – 1)*100) by improving the use of existing resources.

Table 4.3.2: Estimates of Technical Efficiency

	Initial TE_{CRS}	Initial TE_{VRS}	% of farms with $TE_{VRS}=1$	Bias – Corrected TE_{VRS} Single	Lower -Bound 95 % CI Single	Higher -Bound 95 % CI Single
Categorized by Location						
Pooled sample	0.72	0.75	0.12	0.69	0.66	0.74
Khorezm region (north-western)	0.75	0.78	0.12	0.71	0.68	0.77
Fergana region (eastern)	0.78	0.82	0.24	0.74	0.71	0.81
Categorized by Wheat Area						
up to 10.0 ha	0.73	0.76	0.15	0.68	0.66	0.75
10.1 – 30.0 ha	0.75	0.81	0.25	0.72	0.68	0.81
30.1 ha and above	0.83	0.94	0.41	0.88	0.81	0.93
Categorized by Bonitet Score						
Up to 50.0 ha	0.76	0.81	0.26	0.71	0.67	0.80
50.1 – 60.0 ha	0.74	0.77	0.15	0.69	0.67	0.76
60.1 ha and above	0.75	0.77	0.20	0.70	0.67	0.78

Note: In the first and second column, TE under CRS and VRS technologies, respectively, are presented. Even though the analysis concentrates on the VRS technology, calculation of CRS helps to visualize scale efficiency immediately when efficiency scores from two technologies are compared. The third column lists the percent of those farms that constitute frontier under VRS. The bias-corrected efficiency scores in the fourth column are calculated based on the method suggested by Simar and Wilson (1998). Confidence intervals in columns 5 and 6 are related not for the initial efficiency scores but for those, which are bias-corrected. In all cases, biased-corrected efficiency scores are less than the original ones. This shows that initial average TE coefficients are overestimated, e.g., upwardly biased.

The width of the 95 % confidence intervals obtained by the single bootstrap approach for *TE* estimates is 0.09, which reflects a relatively higher statistical variability for the *TE* efficiency scores. Similar results can be found in other socio-economic environments (e.g., Brümmer 2001; Latruffe et al. 2005; Olson and Vu 2009). When farms were ranked by the bias-corrected *TEvrs*, from the lowest to highest scores, the quantitative differences were much more obvious. Figure 4.3.2 shows that the initial *TEvrs* does not provide such a smooth line as the bias-corrected *TEvrs*. A visible variability is also noticeable in the lower and upper bounds of the bias corrected *TEvrs*, even with similar efficiency scores.

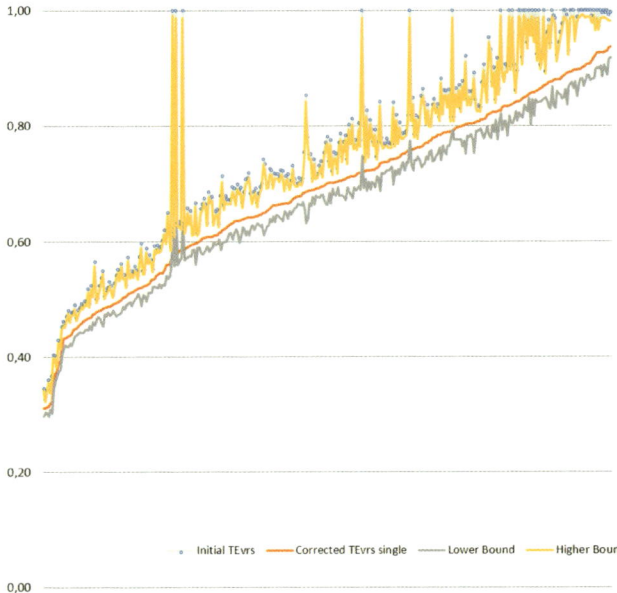

Figure 4.3.2: Distribution of TE with confidence intervals (single bootstrap)

As mentioned earlier, an increase in output appears to be possible by a more efficient use of resources. This can be further analyzed by calculating slack values[10] for all farms monitored (Table 4.3.3). Based on the slack values, the necessary input adjustment values to achieve 100 % efficiency can be estimated. The findings indicate that an average increase in production of 28 % (((1/0.78)-1)*100) in the Khorezm sample would be achieved if the technically inefficient farms optimally adjusted their input use by around 12, 15, 13, 10, 19 and 14 % of land, seeds, fertilizer N, diesel, labor and other expenses, respectively (Table 4.3.3). There would be an average increase in production of 22 % (((1/0.82)-1)*100) in the Fergana sample if the technically inefficient farms optimally

---

10 Zhu's *DEA* Excel Solver algorithm is used for the computation of 'slacks'. A 'slack' provides information on the inputs that are in excess supply and those that are effectively constraining production. Fully efficient farms would have no slacks.

adjusted their input use by 9, 11, 10, 16, 21 and 14 % of land, seeds, fertilizer N, diesel, labor and other expenses, respectively (Table 4.3.3).

On average, labor was the most inefficiently used input in both regions. For example, in Khorezm, 43 % of the farms on average could reduce ca. 19 % of their labor to achieve full efficiency. The least inefficiently used input was diesel (10.1 %) in the case of Khorezm, while it was land (8.5 %) in the Fergana sample (Table 4.3.3). This example indicates that 38 % of the farms in Khorezm would achieve full efficiency when reducing only 10 % of the diesel use.

Table 4.3.3: Potential percent of input adjustments needed to reach 100 % technical efficiency in two study regions

Study region	Input	Farms with slacks		Percent of input adjustments to give 100 % efficiency			
		N	% of total	Mean	Standard deviation	Minimum	Maximum
Khorezm (N=180)	Land	136	76	12.2	9.7	0.01	44.3
	Seeds	126	70	15.0	10.9	0.2	45.9
	Fertilizer N	91	51	13.4	11.0	0.5	52.0
	Diesel	69	38	10.1	7.8	0.2	34.6
	Labor	77	43	18.9	11.7	0.7	55.9
	Other expenses	80	44	13.7	10.6	0.1	38.6
Fergana (N=164)	Land	96	59	8.5	6.2	0.2	25.7
	Seeds	84	51	10.8	7.3	0.1	28.5
	Fertilizer N	76	46	10.3	8.8	0.2	44.1
	Diesel	70	43	15.8	10.5	0.01	48.2
	Labor	56	34	20.8	12.9	0.9	46.6
	Other expenses	70	43	13.8	9.1	0.1	36.2

Efficient farms attained higher crop yields in all cases. For example, while inefficient farms achieved wheat yields of 4.0 tons ha^{-1} and 4.5 tons ha^{-1} in the Khorezm and Fergana samples, respectively (Figure 4.3.3), efficient farms achieved 5.3 tons ha^{-1} and 5.7 tons ha^{-1}, respectively. Figure 4.3.3 depicts *TE* scores for six farm size intervals. The technically most efficient farms were those with the largest farm size in both regions. The technically least efficient farms in Khorezm were those in the second interval (farm sizes 5.1 – 10.0 ha). In the Fergana region, these farms were in the third group (farm sizes 10.1 – 15.0 ha).

4.3.4.1. Scale Efficiency

The results from the scale efficiency assessment (Table 4.3.4) show the percentage of farms with increasing (*IRS*), decreasing (*DRS*), and constant (*CRS*)

Assessing Efficiency of input Utilization 243

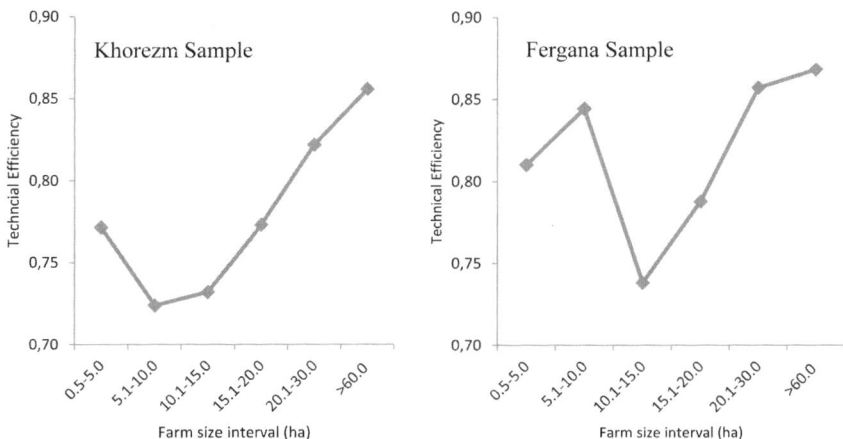

Figure 4.3.3: Technical efficiency in Khorezm and Fergana regions according to farm size

returns to scale. High scale efficiency *(SE)* was noted in both regions, thus leaving only little room for scale improvements. For example, in the case of Khorezm, *SE* was equal to 0.97, leaving room for only 3 % scale improvements in the production process. Moreover, the findings show that wheat producers operate mostly under *IRS*, thus indicating an underutilization of input resources. In contrast, farms that exhibited *DRS* would need to decrease the use of input resources to increase efficiency. In general, findings show no particular indication that wheat producing farms are too large or too small in their scale of operations. From the results, it can be confirmed that overall inefficiency mostly comes from the inefficient use of resources and not from ill-scaled production.

Table 4.3.4: Indicators of Scale Efficiencies

	Khorezm region (north-western)	Fergana region (eastern)
SE	0.97	0.95
SE=1	29.4	34.1
DRS	32.8	37.2
IRS	37.8	28.7

Farms were ranked by the bias-corrected efficiency levels using the double bootstrapping procedure (Figure 4.3.4). The double bootstrapped bias-corrected *TEvrs* ranged from 0.31 to 0.95, with average scores of 0.69. The scores for Khorezm and Fergana were 0.65 and 0.74, respectively. This suggests that wheat producers in both regions could significantly increase yield productivity even with the presently existing technologies.

Scale-inefficient farms are thus functioning at decreasing returns to scale, which reflects an overuse of input resources. This means that the initiated land consolidation process (cf. Djanibekov et al. 2014) should be carried out parallel to training programs for farmers while providing these with agricultural extension services to help them manage larger crop areas (Bekchanov 2009; Niyazmetov 2012).

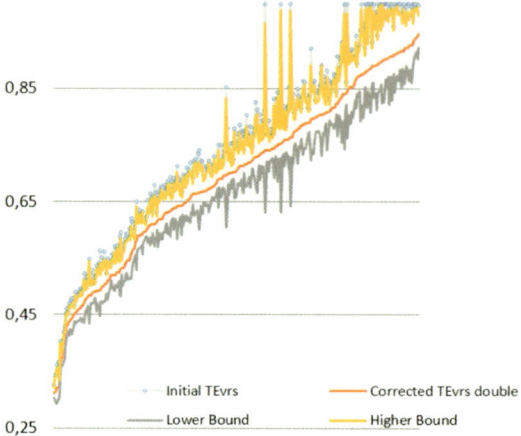

Figure 4.3.4: Distribution of TE with confidence intervals (double bootstrapping)

4.3.4.2 Factors explaining differences in technical efficiencies

The results from the second-stage analysis[11] show that comparisons of the *Tobit* and the truncated regression results are consistent with each other. The signs and statistical significance levels of the coefficients are similar with the exemption of *bonitet* scores, which turn out to be significant in the truncated regression. Our results support the recent findings by Latruffe et al. (2008) and Larsen (2010), who conclude that the *Tobit* and double bootstrap methods do not substantially differ with respect to sign and statistical significance.

---

11 The *Tobit* regression uses *TE* scores, while the truncated regression under the single and double bootstrap method utilizes technical inefficiency as a response variable. Because of this, their signs are opposite to each other, although they provide the same information. Table 4.3.5 also includes standard errors and confidence intervals that make non-parametric models as valuable as econometric models.

Assessing Efficiency of input Utilization

Table 4.3.5: Results from the truncated maximum likelihood regression in single and double bootstrap methods for the case of wheat production

Variable	Tobit regression Parameter estimate (S.E.) (95 % C.I.)		Bootstrap procedure I Parameter estimate (S.E.) (95 % C.I.)		Bootstrap procedure II Parameter estimate (S.E.) (95 % C.I.)	
Constant	0.672	***	0.881	**	1.072	***
	0.054		0.380		0.338	
	0.566	0.776	0.021	1.468	0.525	5.279
Region	-0.088	***	0.464	***	0.523	***
	0.015		0.124		0.113	
	-0.117	-0.058	0.224	0.691	-1.629	0.777
Bonitet score	-0.001		0.009	*	0.008	*
	0.001		0.005		0.004	
	-0.002	0.0003	0.0002	0.018	-0.014	0.019
Farm size	0.0005		-0.002		-0.002	
	0.0004		0.003		0.003	
	-0.003	0.001	-0.007	0.005	-0.009	0.011
Water availability	0.044	***	-0.404	***	-0.358	***
	0.015		0.112		0.099	
	0.014	0.073	-0.642	-0.186	-0.579	0.920
Diversification index	0.012		-0.070		-0.086	
	0.017		0.116		0.105	
	-0.021	0.045	-0.298	0.140	-0.328	0.774
Dependency ratio	-0.013	*	0.069	*	0.070	**
	0.007		0.041		0.038	
	-0.027	0.001	0.022	0.191	-0.167	0.168
Potential to work in larger land area	0.063	***	-0.295	***	-0.340	***
	0.014		0.110		0.100	
	0.035	0.089	-0.254	0.122	-0.555	0.679
Obsolete canal	-0.043	***	0.290	***	0.265	*
	0.014		0.108		0.096	
	-0.071	-0.015	0.116	0.507	-0.721	0.498
Distance to market	0.009	***	-0.046	**	-0.050	***
	0.003		0.023		0.021	
	0.003	0.016	-0.104	0.004	-0.084	0.080

Note 1: ***, **, * indicate significance at 1, 5 and 10 % respectively.
Note 2: Negative sign in bootstrap procedures must be read as factors positively influencing efficiency levels

Most of the variables used in the second stage regression are significant (Table 4.3.5), proving the following remarkable findings.

First, region had a positive influence on technical inefficiency (Table 4.3.5). Farmers located in Fergana were more efficient in the use of existing resources compared to those in the Khorezm region. This difference was statistically significant. Since lands in Fergana are more fertile than lands in Khorezm, this might be an influencing factor affecting efficiency scores. Differences in efficiency may also reflect the fact that input markets seem to be more developed in Fergana, perhaps because this region is more closely located to the capital city of Tashkent with the highest population in the country. This plays an important role in the delivery of farm resources, which in return affects efficient use of inputs and consequently crop productivity. Furthermore, virtually all cropland in the Khorezm region suffers from salinity (Tischbein et al. 2012), which negatively impacts crop yields. Indeed, Conliffe (this book) argues that in the Khorezm region, the spatial location is an important and often overlooked determinant influencing livelihood opportunities.

Empirical results also reveal a significant effect of *bonitet* score on *TE* (Table 4.3.5); however, the sign of the *technical inefficiency* is positive, which signals a higher inefficiency in more fertile lands. It should be noted here that under the state quota system, farmers receive soft credit and subsidized inputs for wheat production in exchange for supplying a minimum volume (or quota) of grains (Rudenko et al. 2012). Farmers with good-quality land might not be forced to use subsidized inputs very efficiently to achieve reasonably good wheat yields. Interestingly, while the crop diversification index is statistically insignificant, it is negatively correlated with technical inefficiency. This seems to reflect that farmers with lower land quality diversify crop production and use resources more efficiently. These farmers pay careful attention to the way agricultural inputs are used in order to meet the imposed quota and to avoid sanctions by the government.[12]

The correlation matrix between the *bonitet* scores and the diversification index is negative, which supports the argument that farmers operating in less fertile lands are more involved in crop diversification strategies than those that have fertile lands. Higher technical inefficiency in the more fertile lands can be explained through ratchet effects as introduced by Berliner (1952). Although there are incentives for farmers who manage to ratchet up the volume of wheat production beyond the established quota (e. g., remunerations, social bonuses, and the outstanding farmer award), they tend to avoid over-fulfilling the quotas, simply due to the fact that if they over-fulfill expectations, they are likely to have to increase the volume of production even more in subsequent crop seasons. Moreover, since farmers do not have property rights, and rewards are too small in comparison to the very high costs of future increased production plans, farmers operating in higher *bonitet* lands will aim to produce around the corresponding quota. Our explanation is in line with the evidence reported in Chertovitsky et al. (2007). In their study, the authors find that 80 % of the farmers operating in the Syrdarya region produced a volume of cotton close to

---

12 For a discussion on this particular issue, cf. Djanibekov et al. 2014

the state quota. They argued that the farmers were unwilling to increase the production of cotton above the quota threshold to avoid the expectation of having to meet even higher quotas in the future.

The effect of farm size is found to be negatively correlated with *TE* in the case of wheat production, which indicates that farmers with larger plots are more efficient in the use of resources than those farmers with smaller plots (Table 4.3.5). Not surprisingly, farmers who report a negative relationship with technical inefficiency also show a negatively significant result when the willingness to work in larger farmland areas is measured. Results suggest that these farmers are interested in working on larger farms to develop the necessary skills to improve production, which may open options for institutionalized farm-to-farm collaboration (cf. Djanibekov et al. 2014). They also seem to have sufficient financial resources and technologies at hand, although this recurrently has been denied (e.g., Djanibekov et al. 2012), which help them to use resources efficiently in larger plots.

Another variable is 'Obsolete Canal', which is significant and indicates that those farmers who reported that their water-canal systems were obsolete attained lower efficiency scores than those who reported access to good systems (Table 4.3.5). This spatial differentiation has been postulated before (e.g., Bekchanov et al. 2012, Conrad et al. 2007) and reflects the importance of irrigation infrastructure in influencing the better use of input resources. In general, the Water User Associations (WUAs) are responsible for maintaining the canal systems. However, due to financial constraints, the farmers use their own financial sources to clean and maintain adjacent canal systems.[13] It is therefore worth pointing out that there is a need for further investments in the maintenance and repair of canals and drainage systems to sustain crop productivity in the future.

In the case of the Shannon diversification index, which captures farm crop portfolio, the results show a negative correlation of the index with *technical inefficiency*, although the strength of the association was statistically insignificant (Table 4.3.5). Farmers who reported a higher crop diversification index were more efficient in the use of resource endowments. As mentioned earlier, additional cash from other crops could be spent on the production of strategic crops to meet the state quota. Crop diversification seems thus to reduce the risk of not meeting the planned target for wheat production.

Farmers with higher dependency ratios exhibited a more inefficient use of resources in wheat production, suggesting that family composition plays a critical role in time allocation (Table 4.3.5). In particular, farmers with young families seem to generate earnings from non-farming activities to meet house-

---

13 For a discussion, see cf. Saravanan et al. 2014.

hold needs, which is confirmed by Conliffe (cf. Conliffe 2014). The fact that under the quota system wheat is sold by farmers at prices below the market rate (Rudenko et al. 2012) also reduces the incentives for farmers to allocate resources evenly.

The findings also show that water accessibility is a highly significant factor in determining crop productivity as was concluded by Tischbein et al. (2012). Farmers who reported that they had sufficient access to water for crop production were more efficient in the use of resources. This is not surprising, as water plays a crucial role in decision making at all levels of agricultural production (Bekchanov et al. 2012). Prompt delivery of water must therefore be appropriately executed by the WUAs and monitored at regional levels. Since there is a private interest in the use of water, charges based on the excessive use of irrigation water could be introduced, although the effect is often disputed (Djanibekov et al. 2012). Appropriate attention is thus required to improve water use efficiency at canal and field levels, which is currently seen as a crucial step (Tischbein et al. 2012). Concessional lending schemes available to farmers could be extended to cover WUA fees and other expenses necessary for maintaining and repairing canals and drainage systems. All these aspects positively influence technical efficiency in crop production, which in return increases crop productivity at provincial and national levels.

Finally, the exogenous variable that captures the distance to key commodity markets indicates that as the market distance decreases, farmers display a more inefficient use of resources in the production of wheat (Table 4.3.5). Inefficiency came from at least two sources. First, farmers located in close proximity to local markets experience greater incentives to diversify their crop portfolio to meet the demand for agricultural goods that are not constrained by price ceilings. Second, closer locations to markets create job opportunities, which divert time for wheat production to other activities. The correlation matrix between dependency ratio and market distance is positive, which suggests that farms with high dependency ratios are also involved in off-farm activities.

4.3.5 Concluding Remarks

By extending the traditional *DEA* method, this study is a pioneer in the use of the double bootstrapping method in the field of agricultural economics and, in particular, in the context of the Central Asian region. It reveals that *TEvrs* estimates obtained from bootstrapping are lower than those in the traditional *DEA* method of estimating *TEvrs*, which reflects the bias of the *DEA* method. The extended methodology also shows that the use of *DEA* (with bootstrapping) as a benchmark to set up frontier farmers for a given sample is a useful approach to improve resource-use efficiency and increase competitiveness in crop production.

The results demonstrate that the level of *TE* among farmers differs across crop growing areas, location and *bonitet* scores, and that consequently crop producers could increase their technical efficiency considerably. Hence, there is room for potential increases in *TEvrs* even with the current state of technology in the production of wheat, although in terms of scale of operations, most of the farmers have achieved scale efficiency. Higher efficiency in arable land with lower *bonitet* scores implies that farmers with better land use resources *less* efficiently. Crop diversification seems to improve farm *TEvrs*, although this finding requires further research. Regional differences also show a geographical divide in terms of resource-use efficiency, with farmers in Khorezm being less efficient in wheat production, which could also call for an improvement via farmer-to-farmer exchange of experience. Access to adequate amounts of irrigated water is critical, as it substantially increases *TEvrs* in the production of wheat. To the extent that market-based reforms could take place, inefficient farmers could learn from best-farming practices and adopt explicit agronomic and innovative approaches.

References

Abdullaev I, De Fraiture C, Giordano M, Yakubov M, Rasulov A (2009) Agricultural water use and trade in Uzbekistan: Situation and potential impacts of market liberalization. Water Resources Development 25(1):47–63

ADB (2009) Uzbekistan: Implementation and Monitoring of PolicyReforms in the Agriculture Sector. ADB.

Akramkhanov A, Kuziev R, Sommer R, Martius C, Forkutsa O, Massucati (2012) Soils and soil ecology in Khorezm. Martius C, Rudenko I, Lamers JPA, Vlek PLG. Springer: Dordrecht, pp 37–58

Aminova M, Abdullayev I (2009) Water management in a state-centered environment: Water governance analysis of Uzbekistan. Sustainability 1(4):1240–1265

Bekchanov M, Kan E, Lamers JPA (2009) Options of agricultural extension provision for rural development in Central Asian transition economies: the case of Uzbekistan. 5th Annual Conference Proceedings Research for Sustainable Development held by Westminster International University Tashkent, 14th May, 2009 Tashkent. Tashkent: WIUT, 72–83

Bekchanov M, Muller M, Lamers JPA (2012) A Computable General Equilibrium Analysis of Agricultural Development Reforms: National and Regional Perspective. In: Martius C, Rudenko I, Lamers JPA, Vlek PLG (Eds.) Cotton, Water, Salts and Soums – economic and ecological restructuring in Khorezm, Uzbekistan. Springer, Dordrecht, pp 347–370

Berliner JS (1952) The informal organization of the Soviet firm. The Quarterly Journal of Economics, 66, 342–365

Brümmer B (2001) Estimating confidence intervals for technical efficiency: the case of private farms in Slovenia. European review of agricultural economics 28(3):285

Charnes A, Cooper WW, Karwan KW, Wallace WA (1979) A chance-constrained goal programming model to evaluate response resources for marine pollution disasters. Journal of Environmental economics and management 6(3):244–274

Charnes A, Cooper WW, Rhodes E (1981) Evaluating program and managerial efficiency: an application of data envelopment analysis to program follow through. Management Science:668–697

Chertovitsky A, Akbarov O, Yahshilikov Y (2007) Relaxing Control Over The Cropping Structure: The Next Step For Land Reform In Uzbekistan. IAAE- 104th EAAE Seminar. Budapest, Hungary: European Association of Agricultural Economists

Coelli T, Rahman S, Thirtle C (2002) Technical, Allocative, Cost and Scale Efficiencies in Bangladesh Rice Cultivation: A Non-parametric Approach. Journal of Agricultural Economics 53(3):607–626

Conrad C, Dech SW, Hafeez M, Lamers JPA, Martius C, Strunz G (2007) Mapping and assessing water use in a Central Asian irrigation system by utilizing MODIS remote sensing products. Irrigation and Drainage Systems 21(3):197–218

Davidova S, Latruffe L (2007) Relationships between technical efficiency and financial management for Czech Republic farms. Journal of Agricultural Economics 58(2):269–288

Djanibekov N, Bobojonov I, Lamers JPA (2012) Farm reform in Uzbekistan. In: Martius, C, Rudenko I, Lamers JPA, Vlek PLG (Eds) Cotton, Water, Salts and Soums – economic and ecological restructuring in Khorezm, Uzbekistan. Springer, Dordrecht pp 95–112

Eilu G, Obua, J, Tumuhairwe JK, Nkwine C (2003) Traditional farming and plant species diversity in agricultural landscapes of south-western Uganda. Agriculture, ecosystems & environment 99(1–3):125–134

FAO (2003) Fertilizer use by crop in Uzbekistan. Rome: FAO

FAOSTAT (2011) FAO Statistics Division, available at http://faostat3.fao.org/home/index.html (accessed 16 November 2011)

Farrell MJ (1957) The measurement of productive efficiency. Journal of the Royal Statistical Society. Series A (General) 120(3):253–290

Getu H, Storck H, Belay K, Vischt K (1998) Technical efficiencies of smallholder annual crop production in moisture stress area of eastern Oromia of Ethiopia: A stochastic frontier analysis. Ethiopian Journal of Agricultural Economics 2(2):91–115

Hopper WD (1965) Allocation efficiency in a traditional Indian agriculture." Journal of Farm Economics: 611–624

Karimov (2012) An Economic Efficiency Analysis of Crop Producing Farms in Uzbekistan: Explanatory Factors and Estimation Techniques. Göttingen, Germany: Cuvillier Verlag

Kienzler KM, Djanibekov N, Lamers JPA (2011) An agronomic, economic and behavioral analysis of N application to cotton and wheat in post-Soviet Uzbekistan. Agricultural Systems, 104, 411–418

Larsen K (2010) Effects of machinery sharing arrangements on farm efficiency: evidence from Sweden. Agricultural economics, 41, 497–506

Latruffe L, Balcombe K, Davidova S, Zawalinska K (2005) Technical and scale efficiency of crop and livestock farms in Poland: does specialization matter? Agricultural economics 32(3):281–296

Latruffe L, Davidova S, Balcombe K (2008) Application of a double bootstrap to investigation of determinants of technical efficiency of farms in Central Europe. Journal of productivity analysis, 29, 183–191

Niyazmetov D, Rudenko I, Lamers JPA (2012) Mapping and analyzing service provision for supporting agricultural production in Khorezm, Uzbekistan. In Martius C, Rudenko I, Lamers JPA, Vlek PLG (Eds) Cotton, Water, Salts and Soums – economic and ecological restructuring in Khorezm, Uzbekistan. Springer, Dordrecht pp 113–126

Olson K, Vu L (2009) Economic efficiency in farm households: trends, explanatory factors, and estimation methods. Agricultural economics 40(5):587–599

Rudenko I, Nurmetov K, Lamers JPA (2012) State order and policy strategies in the cotton and wheat value chains. In Martius C, Rudenko I, Lamers JPA, Vlek PLG (Eds). Cotton, Water, Salts and Soums – economic and ecological restructuring in Khorezm, Uzbekistan. Springer, Dordrecht, pp 371–387

Shaxson L, Tauer LW (1992) Intercropping and diversity: An economic analysis of cropping patterns on smallholder farms in Malawi. Experimental agriculture 28(02): 211–228

Simar L, Wilson P (1998) Sensitivity analysis of efficiency scores: How to bootstrap in nonparametric frontier models. Management science, 49–61

Simar L, Wilson PW (2007) Estimation and inference in two-stage, semi-parametric models of production processes. Journal of econometrics 136(1):31–64

State Committee of the Republic of Uzbekistan on Statistics (2008) Statistical Yearbook of Uzbekistan. Tashkent 2008

Tischbein B, Awan UK, Abdullaev I, Bobojonov I, Conrad C, Jabborov H, Forkutsa I, Ibrakhimov M, Poluasheva G (2012) Water management in Khorezm: current situation and options for improvement (hydrological perspective). In Martius C, Rudenko I, Lamers JPA, Vlek PLG (Eds.) Cotton, Water, Salts and Soums – economic and ecological restructuring in Khorezm, Uzbekistan. Springer Dordrecht pp 69–92

Wadud A, White B (2000) Farm household efficiency in Bangladesh: a comparison of stochastic frontier and DEA methods. Applied Economics 32(13):1665–1673

Wilson PW (2008) FEAR: A software package for frontier efficiency analysis with R. Socio-economic planning sciences, 42, 247–254

V. S. Saravanan, Mehmood Ul-Hassan, Benjamin Schraven

4.4 Irrigation water management in Uzbekistan: analyzing the capacity of households to improve water use profitability

Abstract

This paper questions the common "production functions" type approaches, which are based merely on bio-physical and economic relationships, to estimate water productivity at field level. Taking an institutional perspective, the paper critics the existing 'water productivity' approach by offering an alternative 'water profitability' approach defined as net value of products per unit of consumed water. Analyzing water profitability helps to understand the farmers' rationale in applying the existing water management practices in their physical and bio-physical settings. Based on the analysis of survey data from the Khorezm province of Uzbekistan, it is demonstrated how an innovative methodological approach using Bayesian Network analyses identifies factors beyond bio-physical and economic domains that influence water profitability. It is argued that the production function alone does not determine water profitability rather the interplay between endowment and contextual factors influences a production function. Maximizing overall annual farm profit over a longer period is only a part of a farmer's business objective. Water profitability should therefore be studied taking these endowment and contextual factors into consideration. It is concluded that agricultural water profitability cannot be explained by the optimization of a single objective but rather by analyzing the compromise between multiple objectives, because combination of the endowment, contextual and production factors that determine the space within which a farmer operates.

Keywords: Agriculture water productivity, households' capacity, Bayesian network analysis, Khorezm, Uzbekistan

4.4.1 Irrigation management institutions in Uzbekistan

Irrigation water management in Uzbekistan is of great significance given the country's geo-climatic conditions and political situation. The two major rivers, the Syrdarya and the Amudarya, that drain into the Aral Sea cross the country and are the main sources for irrigation. The waters from these rivers have been tapped since the Soviet era (1924–1991) especially following the extension of irrigation systems to secure the production of cotton, Uzbekistan's "white gold". But since its independence in 1991, Uzbekistan has been struggling to maintain cotton export earnings and grain-based food security, and at the same time protect its vulnerable environment. Searching for a path towards economic development, environmental protection and national cohesion, the country has introduced several institutional reforms to improve overall irrigation water management. One of the foremost reforms was the "Law on Water and Water Use" in 1993 (Table 4.4.1) that mandated the state to coordinate, regulate and promote efficient water use while protecting the quality of water. This law defines the functions of the state water management organizations, responsibility of the state organization with respect to water use and protection, different uses of water and types of water users, and regulates the design and construction of irrigation water structures.

Table 4.4.1: Water-related legislation since independence (1993–2009) in Uzbekistan

Year of Legislation/ Decree	Type of Legislation	Name of Decree/ Resolution/ Legislation	Objective
March 6, 1993	Laws of the Republic of Uzbekistan	Laws of the Republic of Uzbekistan – Water and Water Use	Regulation of water relations, rationalise water use, protection of water, and improvement of water management structures.
May 4, 2000	Decree of Cabinet of Ministers	Measures to Deepen Denationalization and Privatization of Enterprises of Water Construction	Privatization of water management structures to increase economic sufficiency
November 26, 2001	Decree of Cabinet of Ministers	Resolution to Support Agricultural Enterprises through International Bank for Reconstruction and Development (IBRD) Credit	Increase efficiency of agricultural production and develop village infrastructure for agricultural marketing.
June 21, 2003	Decree of Cabinet of Ministers	Decree on Main Water Resources Authority / Basin Irrigation Systems Authority	Set up Water Resources Authority

(Continued)

July 21, 2003	Decree of Cabinet of Ministers	Establishment of Basin Level Authorities	Restructure of existing institutional structures to manage water considering hydrological boundaries instead of administrative levels
August 26, 2004	Decree of the President	Amendments and Supplements to the law of the Republic of Uzbekistan "on Farming"	Consolidation of farmland
October 29, 2007	Decree of the President	Fundamental Improvement of Land Amelioration System	Improvement of cropland to increase agricultural production
October 6, 2008	Decree of the President	Optimization of Farm Sizes	Reconsolidation of land
October 21, 2008	Decree of the President	Optimization of Agricultural Land to Increase Food Production	Expansion of production through diversification of crops to meet the food demands and increase rural income.
March 17, 2009	Ministerial Order	Greenhouse Cultivation in Uzbekistan	Promotion of cultivation of vegetables during winter and spring.
August 30, 2009	Decree of Cabinet of Ministers	Plan of Integrated Control of Water Resources and Water Reservoirs in the Zarafshan river basin	Integrated water management in the Zarafshan river basin
25 December 2009	Laws of the Republic of Uzbekistan	Increasing Economic Reforms in Agriculture and Water Resources	Article 18–2 – Water Consumers Association Article 30 – Limits to withdrawal of water Article 8 & 55 – Allocation of land plots and terms for land allocation to *dehqon* farms Article 17 – On responsibilities of private farms Various other articles on land allocation and water uses

A second set of reforms that followed the initial state reforms focused on privatising the agricultural sector as illustrated by the decrees of May 2000 and November 2001 (Table 4.4.1). This was attempted with financial support from the International Bank for Reconstruction and Development (IBRD). The move to privatize the agricultural sector in the beginning of the 21st century followed by the institutional restructuring in 2003 was also driven by international financial institutions and advocated integrated water resources management (IWRM)

(Veldwisch 2008; Mukhtarov 2009; Abdullaev and Mollinga 2010). This resulted in the establishment of the Main Water Resources Authority (basin level) in 2003 (Figure 4.4.1), and subsequently the creation of primary water users (called Water Consumers Association, formerly known as Water Users Association – WUA[1]) at the lowest level in 2009 (Abdullaev et al. 2008; Veldwisch 2010).

The institutional reforms attempted, among others, to re-align the predom-

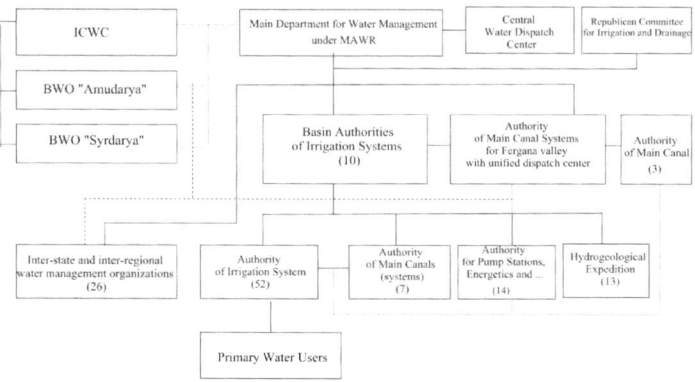

Figure 4.4.1: Institutional arrangement for water management in Uzbekistan

inating administrative units into hydrological units of water management. Through this albeit superficial realignment of water management, the Uzbek Republic gained international acceptability as it was considered as a step towards democratization of water management (Yalcin and Mollinga 2007). However, in reality, the state strategically exploited the internationally enforced "un-tested logic" of IWRM to legitimize its control over cotton production as well. While there may be a growing call to introduce IWRM from a decentralization perspective (Veldwich 2011), experience elsewhere gives enough evidence that the intended decentralization did not bring out efficient management of water (Saravanan et al. 2009; Mukherji et al. 2009).

The beginning of 2005 was marked by a Presidential Order aiming at increasing farming efficiency – 'more crops per drop'. Therefore, the large shareholder farms ('*shirkats*') were divided into smaller farming units ('*fermers*'). However, this reform was reversed during a land consolidation process, which started in October 2008 (Djanibekov et al. 2012). These changes were followed by a slight reduction in the area under cotton, thereby increasing the area under fruits and vegetables, and a revival of the Soviet form of greenhouse cultivation (in 2009) to ensure a year-round supply of agricultural produce

---

1 In this chapter, the term WUA is used as the study was carried out before it was renamed WCA in 2010.

to meet the growing domestic needs. The year 2009 also witnessed the attempt to re-evaluate some of the earlier reforms. In December 2009, while aiming at increasing economic reforms in agriculture and water resources management, the WUA's were renamed 'Water Consumers Association' (WCA) to reflect the actual role of the users as consumers and not as independent and autonomous entities as perceived in the global discourse. These reformation processes can also be understood as a legislative attempt to manoeuvre between local needs of food security, national needs for export-oriented cotton, and gaining international acceptance for promoting a democratically oriented water management regime. The current tendency can also be seen as a further move away from the exclusive state control and a step towards a state-regulated market-oriented water management regime.

Irrigation water management is influenced by a diverse set of factors, where water is one of many production factors and inputs. In the on-going national and international discourse, recommendations to improve irrigation water management include the 'more crop per drop' approach (Luquet et al. 2005), or technical water engineering approaches (Bouman and Tuong 2001). Farmers in Uzbekistan gain the water for irrigation through gravity irrigation by lifting water from canals and groundwater. A combination of crops is cultivated that depends on the farmers' resources, socio-cultural context and institutional environment. In other words, to understand a particular state of field-level water profitability[2], those factors beyond crop, soils and water use need to be understood as well. An increased understanding offers opportunities for improvement of the legislative and policy perspectives facing water management at the household level. Using data from the irrigated lowlands of the Khorezm province of Uzbekistan located in the lower Amudarya river basin, this paper takes further the water productivity studies by attempting to identify the explanatory factors that determine specific water productivity in a given context.

4.4.2 Water Profitability: A Network of Factors

A host of production and non-production factors influences profits in agriculture. Since these factors interact incrementally and cumulatively, profit can be seen as a complex process. The agricultural water management researchers in their attempt to define indicators for water and land productivity recognized the difficulties associated with the indicators when measured in biophysical terms, i.e. kilograms of biomass of a specific crop per unit of water used (Sakthivadivel

---

[2] This is defined as gross value of output per unit of water use (Molden et al. 2003; Boss et al. 2005).

et. al. 1999). They operationalized water productivity as the "*gross value of goods and services produced per unit of water consumed*" and developed various legislations supporting this paradigm. The water productivity as a paradigm assumes that water is the only limitation (or an input) to farm productivity and water productivity could simply be enhanced by either increasing the output from the use of same amount of water (popularly expressed as the "more crop per drop"), or reducing water input, implying water saving. Researchers have since long wondered "Is water productivity a useful concept for agricultural water management?" (Zoebl 2006). Since water productivity refers to gross values regardless of the cost, water profitability defined as *net value of products and services per unit of consumed water* thus helps to understand the farmer's choice in applying existing water management practices to suit their historic context, physical (land quality, and climatic factors) and bio-physical (socio-economic factors) conditions to benefit from their investment cost. This requires examining diverse set of factors (socio-political, historic, physical and bio-physical) that interact incrementally and cumulatively making it a complex system.

Unravelling the complexity remains central to identify opportunities and barriers to facilitate the agriculture water profitability at the household level. Researchers have attempted to classify these factors in different categories. Stern et al. (2002:453), for example, proposed four broad functional categories based on their possible theoretical relationship: (i) possible interventions, (ii) outcomes, (iii) contingencies, and (iv) mediators. Such a categorisation requires special attention in research and practical analysis by clarifying three issues: (i) determine how key intervening factors affect outcomes, (ii) identify contingencies under which mediators become critical, and (iii) identify the conditions under which particular interventions can successfully work. However, the categorization by Stern et al. (2002) demands that contextual factors that impact the interventions and outcomes are site-specific and thus need to be determined. The capacity of the farmers to derive adequate profit from the available water resources depends on how well they exploit the contextual factors given their endowments and production functions. In addition, many intermediate factors influence the endowments of farms and finally water profitability. Broadly, these factors are contextual, endowment, and production functions, intermediate factors, and outcomes (Table 4.4.2). Households draw on their available endowments to exploit the contextual factors, resulting in diverse production outcome. However, outcome depends also on those intermediate factors that influence the production functions, making the relationship between factors reflexive.

Table 4.4.2: Types of Indicators for Water Profitability

Factor	Description	Examples of indicators
Contextual	External factors in a system over which humans have less or even no control.	Rainfall, climate, quality of land and other geological phenomena, historical context, governmental policies and directives, etc.
Endowment	Physical and demographic variables that exploit the contextual factors and determine production functions.	Capital, size of landholding, number of cattle, number of male and female household members, ownership of machineries/pumps, etc.
Production	Variables that interact with the endowment variables to produce a certain outcome.	Education, knowledge, crop type, pricing of water, access to sources of water, cropping intensity, availability and affordability of other production inputs such as fertilizers, etc.
Intermediate	Variables that are not directly related to the production variables, but their aggregation or mere presence influence the outcome.	Cropping pattern, adequacy and timeliness of water availability, market behaviour.
Outcome	Final outcome from the management of water for agriculture.	Yields, water productivity, water profitability

4.4.3 Context and Setting

Water management in Khorezm falls within the jurisdiction of the Lower Amudarya Basin Department of Irrigation Systems (BUIS). It consists of three organizations that ensure water allocation and delivery in the province: 1) the Hydrologic Meliorative Expedition of Khorezm maintains the inter-farm collectors[3], 2) the Authority of Pump Stations and Electrical Power Communications maintains and operates large pumps, and 3) the Irrigation System Authorities (UIS) prepare plans for water allocation, distribution and maintenance of the inter-farm canals. These organizations have four irrigation system authorities and several WUAs at the secondary canal level (Figure 4.4.2). As of 2009, the Khorezm province had about 111 WUAs. The water distribution is not only based on the requirements of agriculture for the 'state quota' crops cotton and winter wheat and

---
3 "Inter-farm" refers to the canal/drain that passed through and hence served the areas of several Soviet-type collective farms. These terms are irrelevant, as in 2003 the agriculture land of these large collective farms was divided into several private farms. However, the terms 'inter-farm' collectors and 'inter-farm' canals used during the Soviet period are still valid today for the irrigation/drainage departments.

household needs, but also on water released by the Lower Amudarya Basin Authority[4] after meeting the hydroelectric and ecological requirements.

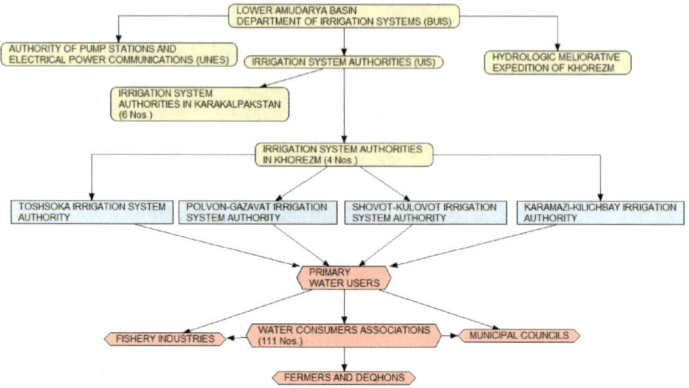

Figure 4.4.2: Organization of water management structure in Khorezm

Located in the northwest of Uzbekistan, Khorezm covers an area of about 6,800 km^2, of which about 270,000 ha can be irrigated (cf. Fritsch et al. this book). The region is characterised by a continental arid climate with hot and dry summers and cold winters with long-term annual average precipitation of about 100 mm (Conrad et al. 2012). Hence, agricultural production and rural livelihoods in Khorezm rely entirely on irrigation water supply, which stems from the Amudarya River. Over 70 % of the 1.7 million population of Khorezm live in rural areas and are mostly engaged in crop production, either as commercial farmers under a state plan[5], or as smallholder peasants. Unemployment rates are high, and about 28 % of the population lives below the poverty line of US$ 1 per day (Bekchanov et al. 2010).

Agriculture is the mainstay of Khorezm's economy and annually contributes

---

4 UPRADIK is a trans-boundary and inter-state organization that plans and regulates water use in the downstream of the Amudarya River.
5 Agricultural land in Uzbekistan is owned by the state, which assigns use rights to interested commercial farmers, who in turn are required to grow government-specified crops on a prescribed proportion of the land. The target production levels, input use, timing and frequency of agricultural operations are controlled. The farmers are required to sell the entire output of cotton, and a part of the winter wheat to the governmental procurement system at state-determined prices. They are entitled to receive part of the production costs as an advance that is settled after these two crops are harvested. The small household farmers have small parcels. Typical commercial farms in Khorezm are in the range of 10–70 ha, but sometimes more than 100 ha. The small household farms mainly own small plots as backyard kitchen gardens or a specified area within the main farmland of the farmers, and are free to choose their crops and sell at their own discretion. They manage these small farms as private entrepreneurs.

about 46 % to the gross domestic product, namely from raw cotton, wheat and horticultural products (Rudenko et al. 2009). The irrigated area increased from 200,000 ha in 1982 to 276,000 ha until the drought year 2000/2001 after which it decreased. Since then, the potential irrigated area has stabilised at around 230,000 ha each year (cf. Fritsch et al. this book). The cropping patterns reveal a drastic increase in the area under cotton, especially since 2008 when a farm-size optimization policy was pursued by the government.

4.4.4 Methodology

Structured interviews were conducted with 120 households (86 household farmers and 34 commercial farmers), and semi-structured interviews with 25 key informants and government officials equally spread over three selected WUAs[6]. The semi-structured interviews made it possible to identify issues facing the farmers' households and to assess and identify factors influencing water profitability. The WUAs are located in the upstream and mid-stream area of the main canals at different distances from the Amudarya river (Figure 4.4.3).

The structured interviews focused on details on the location of the households (e.g., location of the village), socio-economic and demographic factors (e.g., increasing population, sex ratio, etc.), agricultural production factors (e.g., agricultural inputs, agricultural knowledge), and the social networking of the households. The questionnaire was built on the rapport between the households and the research team through their decade-long collaboration. This helped to obtain reliable information, which was triangulated with key informants from the WUAs.

The Bayesian network analysis was applied to the information collected to explore the inter-linkages among the variables influencing water profitability. It offers advantages (Batchelor and Cain 1999; Robertson and Wang 2004), as it can integrate both qualitative and quantitative information, and quantify the probability of relationships amongst variables. The network variable indicates the actors or the contextual factors such as rainfall, climate, quality of land, etc. These variables are mapped in a network using three approaches. The first approach is based on information collected from field surveys through participatory methods, focus group discussions and semi-structured interviews conducted to investigate possible linkages between those factors influencing water resource management. The second approach adopts the standard quan-

---

6 The WUAs were selected based on their location in the lower Amudarya Basin. The Amudarya WUA was located in the head reach, the Shomohulum WUA in the middle reach, and the Ashirmat WUA in the tail end.

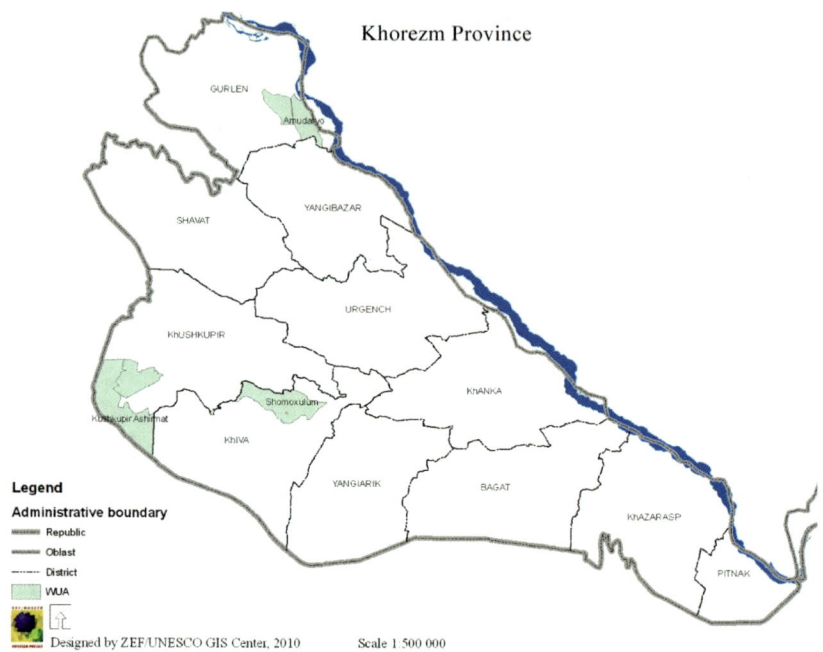

Figure 4.4.3: Location of Case Study WUAs in the Khorezm Province of Uzbekistan

titative survey, where quantitative information is tabulated and analysed using chi-square tests to determine the relationship among variables. In this study, chi-square tests were preferred over other tests, such as cluster analysis (Joshi et al. 2001), due to the relatively small number of respondents in the surveyed WUAs, the prevalence of semi-quantitative data (which required classifying the dataset into categorical and ordinal variable), and missing data. The third approach primarily uses the researcher's observations in the field and the researcher's logical reasoning to draw linkages among variables. The variables and their linkages were applied in probability models of the Bayesian Network using NETICA software (Norsys Software Corporation Canada).

With the help of the Bayesian network, complex inter-linkages among variables are described where a variable is both a cause and an effect of other variables. Further, it identifies the significant role of a combination of variables that, in this case, influence water profitability. The commercial and household farmers in Khorezm practice farming on different terms largely determined by the contractual arrangements between the state and the farmer. The commercial and household farms in the study region also differ with respect to the size of land parcels cultivated. The former cultivate at least 4 ha of farmland in contrast to the latter that cultivate less than 2 ha. Therefore, the data of the commercial farmers and household farmers were analysed separately.

4.4.5 Findings

The water profitability (net value of output per unit of water use) was divided in tertiles (Table 4.4.3): low (less than 90 UZS[7] m^{-3}), medium (between 91 and 330 UZS m^{-3}) and high (above 330 UZS m^{-3}). The distribution of these tertiles differed between commercial and household farmers. About 40 % (32/79) of the household farmers achieved higher water profitability compared to only 13 % (4/30) of the commercial farmers. Most of the commercial farmers reported lower water profits compared to the household farmers.

Table 4.4.3: Distribution of water profitability by type of farmer in three WUAs in Khorezm (n = 109)

Water profitability	Household	Commercial	Total
	Number of samples		
Low	19	17	36
Medium	28	9	37
High	32	4	36
Total	79	30	109

Note: Out of the total survey of 120 Households (household farmers and commercial farmers), 11 survey questionnaires were discarded due to poor recording of data.

Especially household farmers in the downstream WUAs (Ashirmat) achieved higher water profitability than those upstream (Amudarya) (Table 4.4.4). For commercial farmers, low water profitability was relatively equally distributed amongst the WUAs. This could be attributed in part to the fact that many of the household farmers supplement their usual surface irrigation water supplies through the use of groundwater pumps in critical times, an option which is not feasible for many of the commercial farmers due to the large size of their fields. The differences in water profitability indicate an influence of location of the WUA and farm type.

4.4.5.1 Household Farmers

The analysis of inter-linkages among different factors influencing farm water profitability was carried out separately for household and commercial farms. Therefore, the independent factors are categorized as contextual, endowment, production and intermediate, and the dependant or outcome factor as '*agri-*

---

[7] UZS is Uzbek Soum, the national currency of Uzbekistan. As of June 2011, 1.00 UZS = 0.000579710 USD (http://www.xe.com/ucc/convert/?Amount=1&From=UZS&To=USD)

Table 4.4.4: Distribution of water profitability levels (low, medium high) by type of farmer (household and commercial farmers) (n=109)

WUA	Location	Water profitability level							
		Household Farmers				Commercial Farmers			
		Low	Medium	High	Total	Low	Medium	High	Total
Amudarya	Upstream	5	6	7	18	6	5	3	14
Shomohulum	Mid-stream	8	8	13	29	6	2	1	9
Ashirmat	Downstream	6	14	12	32	5	2	0	7
Total		19	28	32	79	17	9	4	30

Water profitability level: low = less than 90 UZS m^{-3}, medium = between 91 and 330 UZS m^{-3}, and high = above 330 UZS m^{-3}.

cultural water profitability' (see Annex for list of variables). Various endowment factors such as *landholding* (land owned and cropped), *dependency ratio* (as a proxy for domestic labour availability), *source of irrigation*, *outside village income* (household remittance from migrant members), and *agricultural experience* of the household head play a prominent role in influencing the agricultural water profitability of household farmers (Figure 4.4.4). These endowment factors along with contextual factors such as the *household size* and cost of the access to water collectively influence the production function of the household farm type. The findings illustrate that households draw on their family labour to increase *cropping intensity* and, along with *non-village income* (remittance from outside the village), influence the production functions. Similarly, the *cost of water* and *source of irrigation* collectively influence the expenditures on water. Furthermore, the combined endowment context factors influence the production function of the households, i. e., *income from cotton/wheat, income from rice/maize, income from fruit/vegetables, income from livestock*, and *total expenditures on agricultural activities*. The production functions are integrated by the household to increase total *agricultural profitability* and the total *water use for agriculture*, which plays a key role in influencing the *agricultural water profitability* for household farmers in the region.

The findings demonstrate that given the current conditions in Khorezm, a 58 % probability exists for household farms for achieving 'good', a 28 % for 'medium', and a 15 % for 'less' *agricultural water profit*. Key factors influencing the *agricultural water profitability* are the contextual factors such as *source of irrigation* and *agricultural experience*, and the production functions such as *income from farm produce*. Of these, *income from farm produce* played a significant role in determining 'high' agricultural *income*. An increase in the *income from farm produce* subsequently increases the 'good' *agricultural water profitability* to 68 % from its current 58 %. This is mainly due to the higher contribution from *income from rice/maize* and *income from vegetables and fruits*. A second factor is

Irrigation water management in Uzbekistan

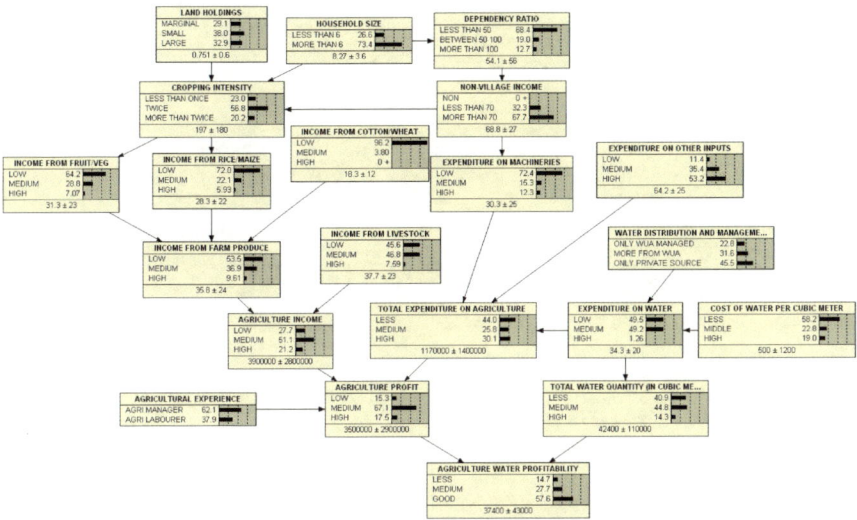

Figure 4.4.4: Factors influencing water profitability among household farmers in three WUAs in Khorezm

the *income from livestock*, as controlling 'high' increases the 'good' *agricultural water profitability* from its current 58 % to 63 %. The third influencing factor is the *source of irrigation*, as controlling *only from private source* increases the 'good' *agricultural water profitability* to 68 %, whilst controlling the *agriculture experience* of the households as 'managers' increases the 'good' *agricultural water profitability* by 1 %.

Controlling all these factors collectively significantly increases the 'good' *agricultural water profitability* to 92 % (Figure 4.4.5), with a noticeable decrease in 'less' and 'medium' *agricultural water profitability*.

4.4.5.2 Commercial Farmers

In contrast to the household farmers, the commercial farmers received less *agricultural water profit*. Only about 8 % of them were able to achieve a 'good' *agricultural water profit*, followed by 39 % who achieved a 'medium' profit, and 53 % achieved 'less' *agricultural water profit*.

The endowment factors again played a prominent role in influencing the agricultural water profitability of commercial farmers (Figure 4.4.6). These included *landholding* and *source of irrigation*. Interestingly, these two endowment factors have an impact through contextual factors such as *household size, percentage of females, experience in agriculture* and *cost of water*, which are beyond

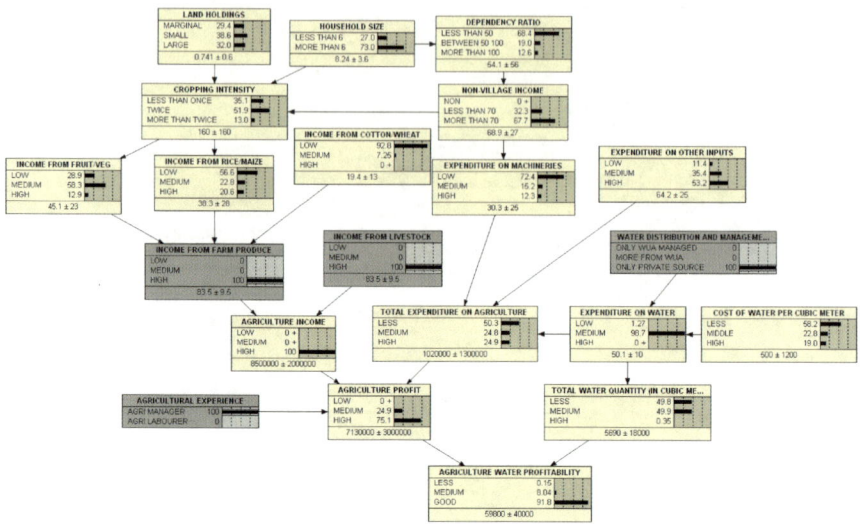

Figure 4.4.5: Key factors influencing water profitability among household farmers

the control of the households. The endowment factors *landholding* and *source of irrigation* along with the contextual factors collectively influence the production function. The households of commercial farms draw on their adult household members to increase *cropping intensity* and, along with higher *income from non-village sources*, significantly influence the production functions. The *cost of water* and *source of irrigation* also collectively influence the *expenditure on water*. The endowment factors and the context factor collectively influence the production function of the households, i. e., *income from cotton/wheat, income from rice/maize and fruit/vegetables, income from livestock*, and *gross expenditures incurred in agriculture*. The production functions are integrated by the household at various levels to increase the agricultural profitability and the total water volume used for agriculture, which plays a key role in influencing the *agricultural water profitability* for commercial farmers in the region.

Significant factors influencing the agricultural water profitability of commercial households are the *location of the WUA, cropping intensity*, and past *experience in agriculture*. However, the control of these factors only marginally increases *agricultural water profitability* from 8 % to 10 %.

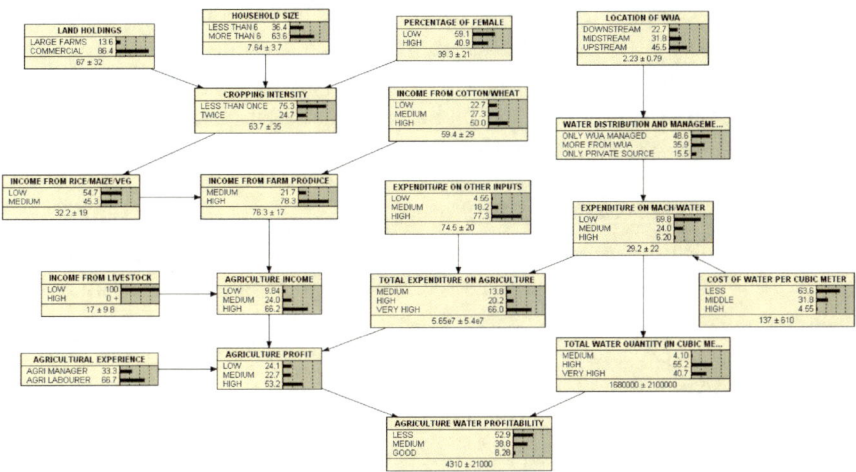

Figure 4.4.6: Factors influencing water profitability among commercial households

4.4.6 Discussion and Conclusions

Contemporary institutional arrangements primarily aim to increase 'more crop per drop' of water. In the light of farmers' decision-making process, the study finds that water profitability is influenced by diverse set of factors. Therefore institutions facilitating agricultural water management should facilitate the capacity of farmers to improve profitability from water use. This gives autonomy to the farmers to take decision given the current endowment and production functions. The study finds commercial farmers in Khorezm have lower water profitability compared to household farmers for two main reasons. Firstly, household farmers plan and produce predominantly for themselves, whereas commercial farmers produce cotton and wheat in the first place for the state. Therefore, household farmers have a much higher degree of autonomy and can make crop choices depending on their domestic need and marketability in contrast to the commercial farmers who produce the state crops at prices determined by the state. Furthermore, household farmers manage small- rather than large-scale irrigation schemes, which facilitates water management. Secondly, and perhaps more importantly, although household and commercial farmers draw water from the same irrigation infrastructure, previous research findings in the region and WUA showed that household farmers faced greater water scarcity compared to the commercial farmers, and therefore seemed to be more careful in using water specially in water-scarce years (Oberkircher et al. 2012). This is because the water is first allocated to commercial farmers, who pass it on to household farmers when their own irrigation needs are satisfied (Veldwisch 2010). A few households attempted to overcome this dependency by using their own pumps. This further helps to cultivate crops with a higher profitability and achieve in turn a higher water profitability, which is an additional reason explaining the difference in water profitability between commercial and smallholder farmers.

The location of the fields along the irrigation system also played an important role in determining water profitability, as evidenced by the fact that water profitability of households in downstream WUAs was higher compared to those located in upstream WUAs. As water is conveyed from upstream to downstream, the latter WUAs experience less reliability and reduced availability due to the abstraction upstream and additional conveyance losses as was previously suggested (Bekchanov et al. 2012). Farmers in downstream areas thus use water more cautiously, as the value of water becomes higher for them. The location within the same agro-climatic zone in Khorezm plays an overlooked role as was underlined recently (cf. Conliffe this book).

The analyses revealed a number of linkages of water profitability to non-agronomic contextual, endowment, and intermediate factors, such as *agricultural experience*, ability to reduce dependence on WUA as an irrigation service provider, and share of income from livestock (especially among household farmers), an agricultural activity that uses less water compared to crop production. The analyses furthermore showed that the interplay between various endowment and contextual factors determines the production function, which in turn determines agricultural water profitability. Therefore, it could be argued that lower water profitability, which is associated with higher water applications, is often based on a sound decision by the respective farmer, rather than on unawareness or unavailability of technological solutions to enhance agricultural water profitability. This hypothesis is supported by recent findings from an analysis of the spatial and temporal components of water productivity illustrating that in water-scarce years, water productivity increased regardless of the location (Bekchanov et al. 2012). Water productivity was also crop dependent, i.e., water productivity in cotton production was low almost everywhere. However, the lowest water productivity was usually found in tail-end regions when including the high conveyance losses in the accounting (Bekchanov et al. 2012). This factor was not considered in this research.

The present debate about enhancing agricultural water profitability at farm/field level recurrently blames the unawareness of farmers and the lack of access to innovations for the present low water profitability. Based on the present findings, this debate needs to be complemented in the light of farmers' decision-making processes. The actual behaviour of farmers can only in part be explained by the optimization of a single objective for maximizing water profits, as recurrently suggested and solved through technical interventions (on-farm developments) alone or through sourcing additional water. Most importantly, a compromise between multiple objectives (Sumpsi et al. 1997) and the endowment and contextual factors determine the space within which a farmer operates. Maximizing overall annual farm profit over a longer period of time is only one part of a farmer's business objective, and cannot be studied without considering endowment and contextual factors.

The policy implications of this research is that blaming farmers for achieving less by "overusing" water needs to be carefully analyzed through more studies to better understand the reasons for the overuse of water. Accusing farmers of failing to adopt technologies to enhance water management due to ignorance, neglect, or being environmentally unfriendly appears to be an excuse for the system and policy managers to shift the blame to the wrong shoulders. Attaining higher water profitability will thus need much more than campaigns to re-engineer user psychologies or to promote technological innovations. Any amount of teaching, convincing or demonstration and extension will not result in improved water profitability at field level as long as the contextual factors are not addressed.

The implications for research are that a mere estimation of water productivity through capturing water use and the output through production functions will not offer significant strategies to improve agricultural water management. The paper highlights the importance of examining water profitability and more in-depth analyses of diverse sets for improving agricultural water management.

References

Abdullaev I, Mollinga P (2010) The socio-technical aspects of water management. Emerging trends at grass roots level in Uzbekistan. Water 2(1), 85 – 100

Abdullaev I, Nurmetova F, Abdullaeva F, Lamers JPA (2008) Socio-technical aspects of water management in Uzbekistan: emerging water governance issues at the grass root level. In: Rahaman M, Varis O (Eds.) Central Asian Waters. Social, economic, environmental and governance puzzle, Helsinki: Water and Development Publications – Helsinki University of Technology

Batchelor C, Cain HO (1999) Application of belief networks to water management studies, Agri. Water Manage. 40, 51 – 57

Bekchanov M, Lamers JPA, Karimov A, Müller M (2012) Estimation of spatial and temporal variability of crop water productivity with incomplete data. In: Martius C, Rudenko I, Lamers JPA, Vlek PLG (Eds.) Cotton, water, salts and soums – economic and ecological restructuring in Khorezm, Uzbekistan. Springer, Dordrecht pp 329 – 344

Bekchanov M, Karimov A, Lamers JPA (2010) Impact of Water Availability on Land and

Water Productivity: A Temporal and Spatial Analysis of the Case Study Region Khorezm, Uzbekistan. Water 2010, 2:668–684

Bouman BAM, Tuong TP (2001) Field water management to save water and increase its productivity in irrigated lowland rice. Agricultural Water Management. 49(1): 11–30

CAWATERinfo (2011), Organizational structure of water management in the Central Asian countries. Accessed at: www.cawater-info.net/library/rus/spm/02.pdf)

Conrad C, Schorcht G, Tischbein B, Davletov S, Sultonov M, Lamers JPA (2012) Agrometeorological trends of recent climate development in Khorezm and implications for crop production. In: Martius C, Rudenko I, Lamers JPA, Vlek PLG (Eds.) Cotton, water, salts and soums – economic and ecological restructuring in Khorezm, Uzbekistan. Springer, Dordrecht, pp 25–36

Djanibekov N, Lamers JPA, Bobojonov I, (forthcoming): Land Consolidation for Increasing Cotton Production in Uzbekistan: Also Adequate for Triggering Rural Development? in: Labar K, Petrick M, Buchenrieder G. (Eds.) Challenges of Education and Innovation. IAMO Studies on the Agricultural and Food Sector in Central and Eastern Europe, Halle (Saale).

Djanibekov N, van Assche K, Bobojonov I, Lamers JPA (2012): Farm restructuring and land consolidation in Uzbekistan: New farms with old barriers. Europe-Asia Studies, 64 (6): 1101–1126

Luquet D, Vidal A, Smith M, Dauzat J (2005) More crop per drop: how to make it acceptable for farmers? Agricultural Water Management. 76(2): 108–119

Martius C, Froebrich J, Nuppenau EA (2009) Water Resource Management for Improving Environmental Security and Rural Livelihoods in the Irrigated Amu Darya Lowlands. In: Brauch HG (Ed.) Facing Global Environmental Change. Environmental, Human, Energy, Food, Health and Water Security Concepts. Springer, Berlin, Heidelberg, pp 749–761

Molden D, Murray-Rust H, Sakthivadivel R, Makin I (2003) A Water productivity framework for understanding and action. In Kijne JW, Barker R, Molden D (Eds) Water Productivity in Agriculture: Limits and Opportunities for improvement. CABI International

Mollinga PP (2003) On the waterfront: water distribution, technology and agrarian change in a South Indian canal irrigation system. Orient Longman Private Ltd., India: Hyderabad. 460p

Mukherji A, Facon T, Burke J, de Fraiture C, Faures JM, Füleki B, Giordano M, Molden D, Shah T (2009) Revitalizing Asia's Irrigation: to sustainably meet tomorrow's food needs. Colombo, Sri Lanka: International Water Management Institute (IWMI), Rome, Italy: Food and Agriculture Organization of the United Nations

Mukhtarov F (2009) The Hegemony of Integrated Water Resources Management: a Study of Policy Translation in England, Turkey, and Kazakhstan, PhD Thesis, Department of Environmental Sciences and Policy, Central European University, Budapest

Oberkircher L, Haubold A, Martius C, Buttschardt TK (2012) Water patterns in the landscape of Khorezm, Uzbekistan: A GIS approach to socio-physical research. In: Martius C, Rudenko I, Lamers JPA, Vlek PLG (Eds.) Cotton, water, salts and soums – economic and ecological restructuring in Khorezm, Uzbekistan. Spinger, Dordrecht, pp 285–307

Rudenko I, Lamers JPA, Grote U (2009) Can Uzbek farmers get more for their cotton? European Journal of Development Research, 21 (2) 283–296

Ul-Hassan MM, Hornidge AK (2010): "'Follow the Innovation' – The second year of a joint experimentation and learning approach to transdisciplinary research in Uzbekistan," ZEF Working Paper Series. Vol. 63. Bonn: Zentrum für Entwicklungsforschung.

Ul Hassan MM, Abdullaev I, Oberkircher L, Ismailova B, Abdullaeva F, Kudryavtseva A, Abdullaev D, (2010) Improving the Performance of the Ashirmat Water Users Association through Social Mobilization and Institutional Development. ZEF/UNESCO Khorezm project's internal documentation. Work package 720: Follow the Innovation. ZEF, Bonn.

Robertson D, Wang QJ (2004) Bayesian networks for decision analysis – an application to irrigation system selection. Australian Journal of Experimental Agriculture, 44, 145–150

Sakthivadivel R, De Fraiture C, Molden DJ Perry CJ, Kloezen W (1999) Indicators of Land and Water Productivity in Irrigated Agriculture. International Journal of Water Resources Development. 15(1&2): 161–179.

Saravanan VS, McDonald GT, Mollinga PP (2009) Critical Review of IWRM: Moving Beyond Polarized Discourse. Natural Resources Forum, 33, 76–86.

Saravanan VS, McDonald GT, van Horen B, Ip D (2010) Policies Are Never Implemented, But Negotiated: Analyzing Integration of Policies in Managing Water Resources in the Indian Himalayas Using a Bayesian Network. Journal of Natural Resources Policy Research, 2: 2, 117 — 136.

Stern PC, Dietz T, Dolsak N, Ostrom E, Stonich S (2002) Knowledge and Questions After 15 Years of Research, in NRC (Ed.) The Drama of the Commons. Washington DC, National Academic Press.

Veldwisch GJA (2008) Cotton, Rice & Water. The Transformation of Agrarian Relations, Irrigation Technology and Water Distribution in Khorezm, Uzbekistan, University of Bonn, Bonn. PhD. Dissertation.

Veldwisch GJA (2010) Adapting to Demands: Allocation, Scheduling and Delivery of Irrigation Water in Khorezm, Uzbekistan'. In Arsel M, Spoor M (Eds.), Water, Environmental Security and Sustainable Rural Development. London and New York: Routledge, pp 99–121

Yalcin R, Mollinga PP (2007) Institutional Transformation in Uzbekistan's Agricultural and Water Resources Administration: The Creation of a New Bureaucracy. ZEF Working Paper. 22, Department of Political and Cultural Change, Center for Development Research (ZEF), Bonn, University of Bonn.

Zoebl Dirk (2006) Is Water Productivity a useful concept in agricultural water management? Agricultural Water Management, 84, 265–273

Annex: Variables Used in the Bayesian Network

Indicator	Variable	Explanation of variables
Contextual	WUA_location_dummy	Whether the WUA is located in 'upstream', 'midstream' or 'downstream'. =1, if WUA location is upstream; =0 if not
Endowment	Female_ratio	Percentage of female household members in overall number of household members
	Agricultural Experience	If the head of the household is experienced, i.e., 'Agricultural manager' or not, i.e., 'agricultural labourer'.
	Land holdings	Average landholding of the household classified as 'marginal' (less than 0.20 ha), 'small' (between 0.20 and 1 ha), 'large' (1 to 2 ha) and 'commercial' (more than 2 ha)
	Dependency_ratio	Division of number of dependent household members (age<=15 | >=65) by number of productive household members (age=15–64 years old) multiplied by 100. Division of number of dependent household members (age<=15 | >=65) by number of productive household members (age=15–64 years old) multiplied by 100. Categorised as 'less than 50 %', '50–100 %' and 'more than 100 %'.
	Household size	Number of household members living under a roof. Categorized as 'less than 6' and 'more than 6'.
	Cost of water per cubic meter	Actual cost of water paid/ per cubic meter. Categorized as 'less' (less than 10 Uzbeck Soums), 'medium' (between 10 and 100) and 'high' (more than 100)
	Number_of_machineries	Number of machines owned by the household
Production	Share_non_village_earning	Share of the household income earned outside the village in overall household income. Categorised as 'non', 'less than 70 %' (when share is less than 70 %), and 'more than 70 %' (when share is more than 70 %).
	Water source	Categorized as 'only WUA managed' (when water distribution is only by WUA), 'more from WUA' (when water received in the year 2009 is more from WUA-managed irrigation source than private) and 'only private source (if water is obtained only through private pumps).
	Share_wheat_to_all	Percentage of area under wheat cultivation in overall cultivated area
	Share_others_to_all	Percentage of area cultivated by crops other than wheat or cotton in overall cultivated area

Annex *(Continued)*

	Share_of_govworkers	Percentage of productive household members (age=15–64 years old) working for the government in the total number of productive household members
	Agric_Manager_exp	Number of years a head of household has worked in an agricultural management position
	Agric_Labourer_exp	Number of years a head of household has worked in an agricultural labourer position
	Share_of_cropped_area/ Cropping intensity	Total cultivated area divided by total land holding multiplied by 100. Categorised as 'less than once' (when the value is less than 100), 'twice' (100–200) and more than twice (more than 200).
	WUACOMP_in_total	Share of non-private water volume (e.g., WUA) in total volume of water that has been used for agricultural production
	Income from fruit/ vegetables	Percentage of income from fruit/veg. of total farm income. Categorised as 'low' (less than 34 %), medium (between 34 and 67 %) and 'high' (more than 67 %).
	Income from rice/maize	Percentage of income from rice/maize of total farm income. Categorised as 'low' (less than 34 %), medium (between 34 and 67 %) and 'high' (more than 67 %).
	Income from cotton/wheat	Percentage of income from cotton/wheat of total farm income. Categorised as 'low' (less than 34 %), medium (between 34 and 67 %) and 'high' (more than 67 %).
	Income from livestock	Percentage of income from livestock of agricultural income. Categorised as 'low' (less than 34 %), medium (between 34 and 67 %) and 'high' (more than 67 %).
	Income from farm produce	Percentage of income from farm produce of agricultural income. Categorised as 'low' (less than 34 %), medium (between 34 and 67 %) and 'high' (more than 67 %).
	Agriculture income	Total agriculture income in Uzbek Soums. Categorised as 'low' (income between 200,000 and 2,000,000), 'medium' (between 2,000,000 and 5,000,000) and 'high' (more than 5,000,000).
	Expenditure on machineries	Percentage of expenditures for machines of total agriculture expenditure. Categorized as 'low' (less than 34 %), medium (between 34 and 67 %) and 'high' (more than 67 %).

Annex *(Continued)*

	Expenditure on other inputs	Percentage of expenditure on other inputs (such as labour, fertilizers, etc.) of total agriculture expenditure. Categorized as 'low' (less than 34 %), medium (between 34 and 67 %) and 'high' (more than 67 %).
	Expenditure on water	Total water expenditures per cubic meter (total water expenditures divided by total amount that has been used for agricultural production in m^3). Percentage of expenditure on water of total agriculture expenditure. Categorized as 'low' (less than 34 %), medium (between 34 and 67 %) and 'high' (more than 67 %).
	Total expenditure on agriculture	Total expenditure (in Uzbek Soums) incurred in the year 2009 for agriculture. Categorized as 'low' (less than 400,000), 'medium' (between 400,000 and 1,000,000) and 'high' (more than 1,000,000).
Intermediate	Agric_income_per_ha	Agricultural income per hectare (total agricultural income divided by total cultivated area)
	Total_water_quantity_cubic	Total amount of water that has been used for agricultural production (in m^3). Categorized as 'low' (less than 2000 m^3 ha^{-1}), 'medium' (2001 to 15000 m^3 ha^{-1}) and 'high' (15000 m^3 ha^{-1} and above).
	Agricultural profit	Gross profit gained by household to total income and expenditure on agriculture (in Uzbek Soums). Categorized as 'low' (-1,100,000 to 1000000), 'medium' (1,000,001 to 5000000) and 'high' (5,000,001 to 12,000,000).
Outcome	Agricultural water profitability	Total agricultural profit in (Uzbek Soums) divided by total amount of water that has been used for agricultural production. Total agricultural profit divided by total amount of water that has been used for agricultural production (m^3). Categorized as 'less' (-2,500 – (+)92), 'medium' (92 – 330) and 'good' (330 – 130,000)

Yadira Mori Clement, Anik Bhaduri, Nodir Djanibekov

4.5 Food price fluctuations in Uzbekistan: Evidences from local markets in 2002–2010

Abstract

This study identifies the main determinants of food price fluctuations in Uzbekistan with a focus on the Khorezm province. We investigate the price behavior of ten agricultural commodities using the weekly data collected from local markets in the Khorezm province during 2002–2010. For the analysis, we used Autoregressive integrated moving average (ARIMA) models with exogenous variables such as water inflow, oil prices and international prices of selected commodities imported to Uzbekistan such as rice and wheat, and market exchange rate. The results show two general patterns of agricultural price fluctuations. The fluctuations of rice, beef and wheat prices are more sensitive to external factors such as their respective international prices, market exchange rate and oil prices. On the other hand, the price movements of apples, onions, potatoes, and tomatoes are more locally determined, and are particularly affected by seasonal patterns in which the minimum price occurs during the harvest season and the maximum in the off-season. The results indicate that for reducing of food price fluctuations, the creation and development of storage capacities and processing facilities, which have deteriorated following independence, require more emphasis in national policies.

Keywords: price analysis, ARIMA models with exogenous variables, price volatility, world-price transmission

4.5.1 Introduction

Uzbekistan is one of the most rapidly expanding economies among of the countries of the Commonwealth of Independent States, and was expected to grow by more than 8 % in 2012 (World Bank 2011). However, one of the major economic challenges facing the country is the mitigation of the effects of rising food prices. Rural households, which are predominantly poor with limited access to natural resources and assets (Robinson 2008), are particularly vulnerable to the common seasonal and inter-annual price fluctuations. A similar situation applies for Uzbekistan (Musaev et al. 2010). The rural households in Uzbekistan are net buyers of food products, particularly of wheat, which has the largest contribution to their energy intake. The annual wheat production of an average household in Khorezm covers only around 30 % of its annual consumption requirements (Djanibekov 2008). As household income depends to a large extent on agricultural production and also as the largest share of the budget is spent on food consumption (WFP 2008; Musaev et al. 2010), price fluctuations will have a strong effect on the level of both production and consumption, and thus on the households' overall welfare. The rise and volatility in food prices may result in a negative net-income effect for these households, particularly for those without access to land and those who depend on off-farm activities and employment (von Braun and Tadesse 2012).

The general underlying causes of food price fluctuations are structural, environmental and global in nature (von Braun and Tadesse 2012). The transmission of the effects of world commodity prices to the country level depends on several policy-related factors such as taxes, price controls and subsidies, and on the openness of the economy of the country. In transition countries practicing food self-sufficiency policies, which are indicated by the proportion of food imports to total consumption, such as in Uzbekistan, the effects of local factors on price formation are high. Up to now, there have been only few studies in Uzbekistan on food price variability and its determinants at the local level. Such information is, however, required for understanding the household welfare impacts. A few studies are worth mentioning in this context. Bobojonov and Lamers (2008), for instance, present a statistical analysis of commodity prices in the Khorezm province using market survey data of 23 months. A World Food Programme (WFP) study presents a comparative analysis of poverty rates and factors determining food insecurity in different provinces of Uzbekistan (Robinson 2008). The UNDP analysis of food security in Uzbekistan (Musaev et al. 2010) presents an assessment of the food supply and demand. This study draws on official statistics and comprehensive household survey data from the World Bank's Uzbekistan Regional Panel Survey (URPS) of 2006 for three provinces in Uzbekistan. Although quite informative with respect to food security concerns, these studies do not address price movements of food commodities, which is a major issue for livelihoods of rural people that accounts for the largest share of the poor. This study attempts to fill this gap by looking into nine years of price movements of agricultural goods at the local market level of Uzbekistan using the example of the Khorezm province. Khorezm is home to about 1.7 million people, of whom 78 % are rural (OblStat 2007). The province has around 275,000 ha of land suitable for irrigated agricultural production using the water inflow from the Amudarya River (Conrad et al. 2012). This river is the most important water source for agriculture in the region, and usually supplies a sufficient amount of water to satisfy the regional demand (cf. Bekchanov et al. 2014). Yet, during the last thirty years, the province has experienced frequent water shortages (cf. Bekchanov et al. 2014) during the vegetation period (March-October), which have affected agricultural production and rural livelihoods.

The multi-year data allowed examination of seasonal and inter-annual price movements and determination of the influencing factors based on data from an agricultural commodity market in Khorezm. The selection of the determinants of commodity price fluctuations is based on the classical supply-demand theory, results from empirical studies, and data availability. Essentially, price determination depends on the interaction between supply and demand functions on markets. The basic model typically explains market equilibrium as an adjustment process between demand, supply, inventory and price variables (Labys 1973, 1999). The most comprehensive analytical methods for commodity markets stem from structural models, which trace the interaction between endogenous market variables and exogenous variables and explain market behavior and performance (Labys 1973, 2006). One advantage of these models is that with the incorporation of more variables the market model can be extended, thus providing a consistent framework for forecasting price movements and studying the effects of policies.

Supply-demand shifters can include both market-specific and broader macroeconomic factors. Supply shifters can be categorized into two main groups (Tomek and Robinson 2003): short-run, medium-run and long-run shifters. Water constraints are treated as a short-run change in output. As water is one of the most critical inputs in farming in Uzbekistan, shortages may decrease crop yields and/or lead to a decrease in sown area. Medium-run supply shifters can be attributable to changes in input prices (crude oil, fertilizer, pesticides), which are directly connected to production costs and to changes in prices of commodities competing for the same resource. In addition, as identified by several empirical studies, macroeconomic factors might also play an important role in crop-price determination, such as exchange rate, inflation and interest rates (Roache 2010) over different time horizons. Demand shifters can be grouped according to demographic factors such as population size, population distribution by age, ethnicity, among others; economic factors such as income and its distribution, prices and availability of substitutes, and consumer preferences. The latter can be influenced by educational level, life experiences, information and social context (Tomek and Robinson 2003). In this study, the explanatory variables were divided in two main categories according to the type of shifter (market or macroeconomic factors) and spatial influence (regional or international level). We chose water inflow, oil prices and international prices of selected commodities imported to Uzbekistan such as rice and wheat as market shifters, while the exchange rate was the only macroeconomic factor included as a regressor.

4.5.2 Material and methods

The main data used for the analysis originate from the 2002 – 2010 market survey conducted by the ZEF/UNESCO Khorezm project on a weekly basis for the key food products at Khorezm's most important agricultural markets in Urgench. The Urgench and Khiva universal/dekhqan markets are the central agricultural markets and among the biggest in Khorezm (Bobojonov and Lamers 2008). The data included product prices, number of sellers, and amount of food products brought and sold at both markets. The market survey covered 12 products, i.e., grains (wheat, rice and maize), fruits (apples, melons), vegetables (potatoes, carrots, onions and tomatoes), and animal products (eggs, sour cream, and beef). A weekly quantitative survey was conducted from May 2002 to May 2010 and constituted in total 97 uninterrupted monthly time-series observations for each commodity. The survey data thus included a drought year that occurred in 2008 (Figure 4.5.1), and also price movements after the 2000 and 2001 drought years. This allowed analysis of the relationship between variability of water supply in Khorezm and commodity prices. Prices were recorded in Uzbek soums (UZS) per kilogram for all products and in UZS per 1 hen egg.

Since the price of apples also represents the price movements of melons and watermelons, the latter two crops were not included. Maize grain was also excluded from the analysis, as it is not crucial in population diets in Khorezm. The price of sour cream (dairy product) was used as a proxy for milk prices. The evolution of the real prices of the selected food commodities over time was analyzed based on the market survey. The domestic prices for wheat and rice were analyzed based on their world market counterparts (Figures 4.5.2 and 4.5.3). Estimated were also their annual volatility indicated by the coefficient of

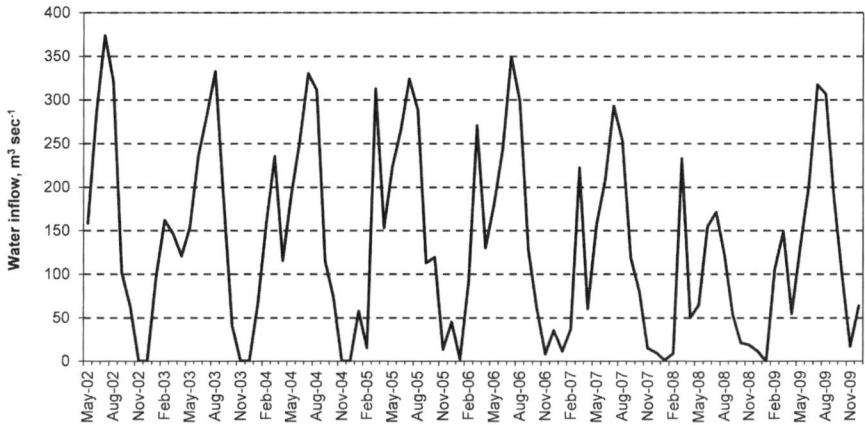

Figure 4.5.1: Monthly water inflow in Urgench (data source: CAWATER info)

variation[1] as well as the monthly price increases[2] over time. Complementary quantitative questionnaires were completed using face-to-face interviews with market sellers on the same day as that of the price survey (every Sunday).

Irrigation water intake for Khorezm at the regional level was used as a proxy for water availability. The data were obtained from the Portal of Knowledge for Water and Environmental Issues in Central Asia (CAWATER info) and Interstate Commission for Water Coordination (ICWC). The world oil prices were obtained from the International Monetary Fund (IMF). The international oil price is a simple average of three international spot prices (Dated Brent, West Texas Intermediate, and the Dubai Fateh), and is expressed in US dollars (USD) per barrel of crude oil (petroleum). World prices of rice and wheat (both in USD) were obtained from the database of the Food and Agriculture Organization of the United Nations (FAO). The world price of rice represents the monthly average of weekly prices for White Broken Rice, Thai A1 Super and FOB Bangkok, as reported by Jackson Son & Co. (London) Ltd on Wednesday of each week, while the world price of wheat is that of wheat No.1 Hard Red Winter, ordinary protein, FOB Gulf of Mexico, reported by the International Grain Council on Thursday of each week. Since indexes on monthly food prices were not available for Uzbekistan, all nominal prices were deflated using annual price indexes from the Asian Development Bank (ADB) and the IMF.

---

1 The coefficient of variation (CV) is a measure of dispersion defined as the ratio of the standard deviation and the mean. We also estimated CVs for two periods: 2002–2006 and 2007–2009. The latter period might include effects of food crisis and water variability.
2 The monthly price increase is defined as the monthly percentage change in prices.

Given the data restrictions on the exogenous variables used to run the regression, only four variables could be used. These variables mainly represent supply shifters. The mechanisms by which the selected exogenous variables might affect agricultural domestic prices on the Urgench market are summarized in Table 4.5.1.

Table 4.5.1: Explanatory variables included in regression models for market analysis

Variable	Type	Mechanism
Water inflow	Supply factor	Proxy of water availability. In this context, this variable might affect prices via expectations and thus storage.
International real rice price	Supply factor	Transmission via imports
International real wheat price	Supply factor	Transmission via imports
International real crude oil price	Supply factor	Via production costs.
Exchange rate	Macroeconomic factor	Can affect prices through a number of channels, including purchasing power and the effect on margins of producers with non-USD costs

In the framework of Autoregressive integrated moving average (ARIMA) with exogenous variables, the effects of selected determinants on commodity price behavior were analyzed. The Box-Jenkins methodology refers to the set of procedures for identifying, fitting, and checking ARIMA models with time-series data. The models were run using STATA software.

Before running the regression models, it is necessary to test the existence of stationarity. If the variables in the regression model are not stationary, the usual t-ratios will not follow a t-distribution, so it will not be possible to validly undertake hypothesis tests about the regression parameters. If a non-stationary process is detected, the variable must be transformed to stabilize the variance of the time series and thus to make it stationary. Table 4.5.2 displays the results of the formal procedures such as the augmented Dickey-Fuller test that tests the presence of non-stationarity in the time series. The basic objective of this test is to verify the null hypothesis that the series contains a unit root (or it is non-stationary) against the alternative hypothesis that the series is stationary. After testing various transformations, we found that the first difference is a stationary process for almost all of the selected time series.

Table 4.5.2: Stationarity in real and transformed variables

Variable	Real	1st dif
Real beef price		x
Real rice price		x
Real wheat price		x
Water inflow		x
International real rice price		x
International real wheat price		x
International real crude oil price		x
Exchange rate		x

Since markets such as those for apples, carrots, potatoes and tomatoes are more subject to seasonal features and also to the perishable nature of the products, the determinants of commodity prices in markets were explored where fluctuations may have been a result of price transmissions from other markets, as is the case of commodities such as rice, wheat and beef.

4.5.3 Results and discussion

4.5.3.1 Evolution of food prices over time

The prices of rice, wheat, beef and sour cream followed an upward trend, with a significant hike during 2007–2008 (Figure 4.5.2, 4.5.3 and 4.5.4), while prices for apples, carrots and tomatoes followed mixed patterns. The 2008 hike in prices may be explained by the extreme drought conditions that affected Khorezm during that year and reduced regional production (Djanibekov et al. 2012). In other cases, price hikes could be reinforced by price transmission effects from international markets.

Figure 4.5.2a depicts the monthly evolution of apple prices over the observed period showing periods of high volatility. During 2007 and 2008, high peaks occurred that can be partly explained by the drought conditions in 2008, which had reduced yields. On the other hand, the yearly coefficient of variation decreased over time from 0.55 in 2002–2006 to 0.28 in 2007–2009, partly as a response to the restrictive export policy that exists in Uzbekistan.

The price of carrots also showed periods of volatility (Figure 4.5.2b). During the first half of 2006, prices increased more than four-fold. The coefficient of variation decreased over time (0.55 for the period 2003–2006, and 0.30 for the period 2007–2009). The price peak in 2008 could be associated with irrigation water scarcity as a result of drought conditions in the region (cf. Bekchanov et al. 2014).

Figure 4.5.2c displays the monthly evolution of tomato prices. These showed clear seasonality: The price was lowest immediately after harvesting and highest in the off-season (winter-spring). The yearly coefficient of variation decreased over time from 0.76 in 2002–2006 and 0.64 in 2007–2009, indicating also an increased supply.

Before 2006, the average monthly increase in the price of onions was 11.1 %. After that, this rate increased slightly to 11.5 % in 2006 (Figure 4.5.2d). In 2009, prices were the highest of the entire observation period, and the monthly rate increased (14.8 %), obviously as a result of the drought in 2008. In 2009, the increase rate began to fall reaching 2.9 % at the end of the same year. The coefficient of variation increased slightly over time from 0.64 in 2002–2006 to 0.49 in 2007–2009.

Potato prices (Figure 4.5.2e) were also highest in the period of high volatility. Before 2006, the average monthly increase was 7.5 %, while during 2006 it declined to 4.8 % and to as low as 1.4 % in 2007 but rose again to 3 % in 2008. After 2009, prices increased steadily (4.7 % per month). The coefficient of variation decreased from 0.66 in 2002–2006 to 0.29 in 2007–2009.

Prices of eggs (Figure 4.5.2 f) as of other animal products, e.g., sour cream (Figure 4.5.2 g), which are mainly supplied by rural households (Djanibekov 2008), have a seasonal fluctuation pattern as a response to a natural increase in production during spring and early summer when prices fall. Prices also tend to increase during periods of fasting (*Ramadan*) when more eggs are used for the production of traditional sweets (*Nisholda*). Egg prices doubled between the first quarter of 2006 and the first quarter of 2008. Before 2006, the average monthly price increase was 2.8 %, after which it rose to 4.5 % in 2006. Although prices continued to increase during 2008, the rate of increase was lower (1.5 % per month) compared to the previous years. A reason could be the rising prices of poultry feed. In 2009, prices started to fall at a rate of 1.02 % per month. The coefficient of variation decreased over time from 0.33 in 2002–2006 to 0.29 in 2007–2009.

The price of sour cream (Figure 4.5.2 g) followed an upward trend. Before 2006, the average monthly increase was 3.1 %, and rose to 8.9 % during 2006, after which it fell to 4.1 % in 2007, and further to 2.1 % during 2008. This was due to several factors such as an increased supply of milk per capita (FAO 2012) as well as cross-price and income effects that offset the effect of drought on fodder prices. After 2009, prices continued to increase by 3.4 % per month. The coefficient of variation decreased from 0.27 in the period 2002–2006 to 0.24 in the period 2007–2009.

Beef prices (Figure 4.5.2 h) also increased considerably during the observed period. From the first quarter of 2006 to the first quarter of 2008, prices increased more than three-fold. Before 2006, the average monthly increase was 1.5 %, while in the following years prices rose monthly by 3.2 % (in 2006), 2.3 % (in 2007), and 2.6 % (in 2008). After 2009, prices continued to increase, but at lower rates (1.4 % per month). The coefficient of variation decreased slightly from 0.26 in 2002–2006 to 0.24 in 2007–2009.

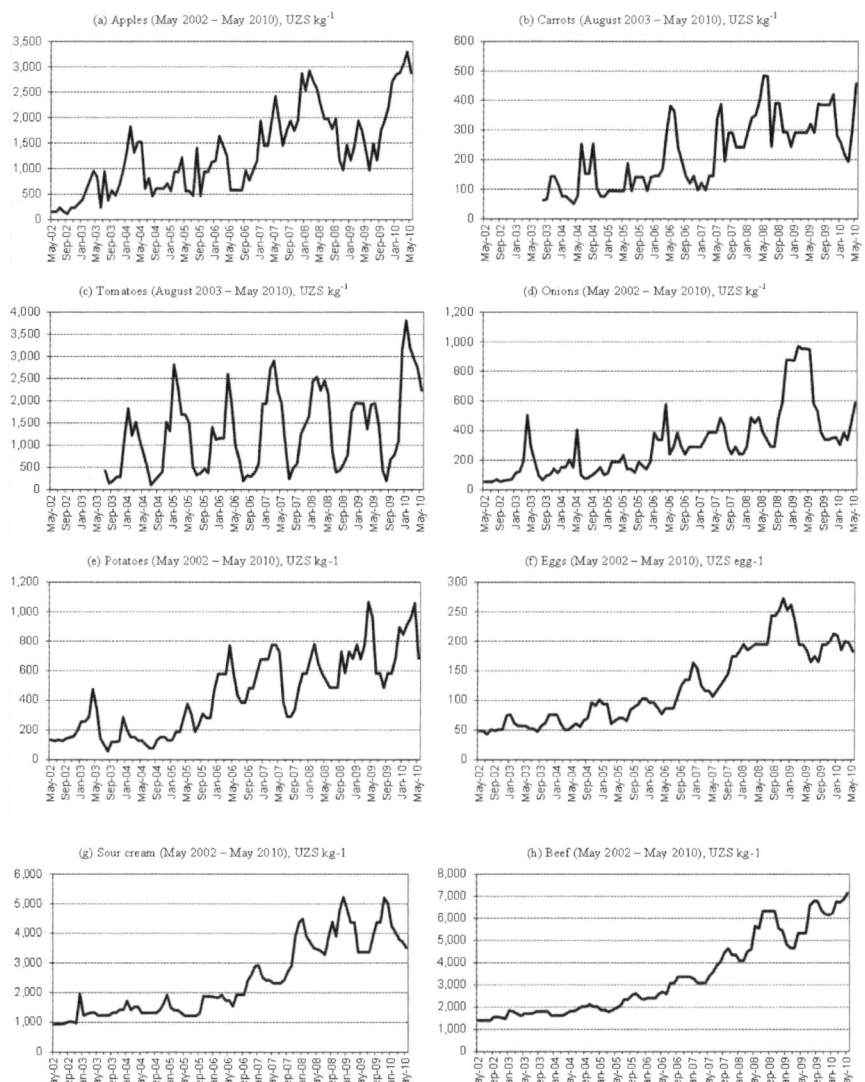

Figure 4.5.2 (a-h): Dynamics of real prices for eight market commodities during 2002 and 2010 at the Urgench market in Uzbekistan

Rice prices (Figure 4.5.3) showed not only an upward trend, but also a distinct increase by a factor of more than 2 from the first quarter of 2006 to the second quarter of 2008. Before 2006, the average monthly price increase was 4.4 %, while during 2006 this rate increased to 7.8 % and in 2008 to 10.4 %. During 2008, rice prices reached the highest peak during the observed period due to factors such as the 2008 drought, increased input prices, as well as price transmission from international rice markets. In fact, the imports of rice (milled) to Uzbekistan increased from 1,162 tons in 2007 to 8,249 tons in 2008 and further to 15,654 tons in 2009 (FAO 2012). After 2009, prices continued to increase, but at a slower rate (monthly by 4.3 %). The coefficient of variation decreased from 0.51 in 2002–2006 to 0.36 in 2007–2009.

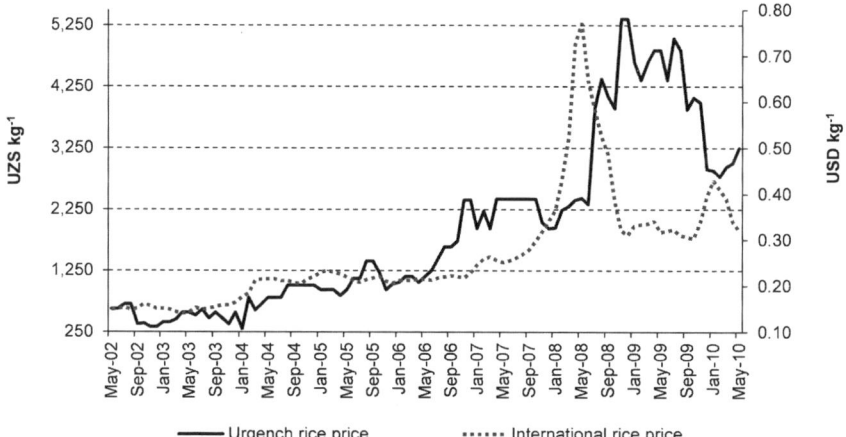

Figure 4.5.3: Urgench real rice price (UZS kg^{-1}) and international rice price (USD kg^{-1}) (May 2002 – May 2010)

Wheat prices (Figure 4.5.4) showed an upward trend during the entire period. The increase was quite large by a factor of more than 3 from the first quarter of 2006 to the second quarter of 2008. Before 2006, the average monthly increase was 2.6 %, while during 2006 prices increased monthly by 5 % and in 2007 by 8.3 %. During 2008, wheat prices were highest. This high price level could be associated with the transmission of the international wheat prices as well the drought in 2008 in Khorezm and in general in Central Asia. After 2009, the wheat prices decreased monthly by 5.1 %. The coefficient of variation of wheat price increased from 0.34 in 2002–2006 to 0.38 in 2007–2009.

In general, the coefficients of variation of all analyzed food commodities, except wheat, beef, potatoes and onions, declined over the observed period, indicating a decrease in price volatility in Khorezm (Table 4.5.3). Comparing the values of the two periods 2002–2006 and 2007–2010, the coefficient of variation

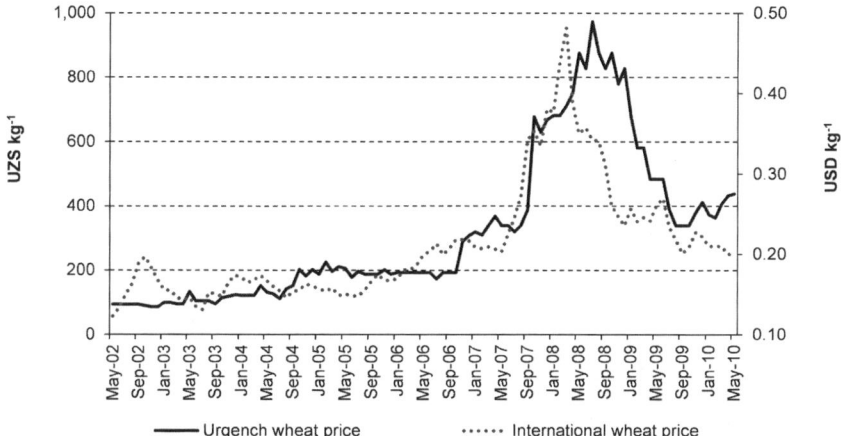

Figure 4.5.4: Urgench real wheat price (UZS kg^{-1}) and international wheat price (USD kg^{-1}) (May 2002 – May 2010)

increased only for wheat, while it decreased in the case of the other commodities. The decline in price volatility for vegetables and fruits can be explained by export restrictions that were introduced in this period in order to fulfill the local demand and support the growing local processing sector (Robinson 2008). As a result of increasing production and declining exports of fruits and vegetables (FAO 2012), prices at the local markets were stabilized.

Increasing price volatility can be partly assigned to the price situation in the world market. This can be particularly true for wheat, which is still imported to Uzbekistan as wheat flour (Kienzler et al. 2011) because the baking quality of Uzbek wheat is relatively low when compared to the quality of wheat on the world market. It should, however, be taken into account that, prior to baking, the locally produced (poor quality) wheat flour is mixed with imported flour (WFP 2008). Hence, even if the quality differs between world market and local wheat, the substitution effect between them is strong. Consequently, the international wheat prices can be included in the analysis in this study. Another reason for the increase in price variability over time, particularly of rice, is the drought that Khorezm experienced in 2008, which reduced yields and the area under rice production, while the period 2002–2005 was more or less stable in terms of water supply (Bekchanov et al. 2010). The increase in the inter-annual volatility of beef and sour cream (milk alternative) seems to be associated with the development of the processing sector in Uzbekistan, which increased the demand for these commodities. Another possible reason is that local production of these products remained below the demand. For instance, while a healthy diet as recommended in Uzbekistan requires an annual consumption of 295 eggs per capita (Musaev et al. 2010), in 2006 per capita production was only 80 eggs (FAO 2012).

Table 4.5.3: Coefficient of price variation for ten commodities in the Urgench market during 2002–2009

Commodity	Year								Period	
	2002	2003	2004	2005	2006	2007	2008	2009	2002–2006	2007–2009
Apples	0.32	0.39	0.48	0.37	0.38	0.22	0.30	0.30	0.55	0.28
Carrots		0.37	0.61	0.27	0.47	0.40	0.23	0.16	0.55	0.30
Eggs	0.18	0.16	0.26	0.19	0.25	0.19	0.15	0.15	0.33	0.25
Beef	0.05	0.05	0.10	0.14	0.14	0.16	0.16	0.15	0.26	0.24
Onions	0.12	0.72	0.59	0.22	0.28	0.24	0.42	0.45	0.64	0.49
Potatoes	0.16	0.58	0.27	0.34	0.21	0.34	0.19	0.26	0.66	0.29
Rice	0.32	0.16	0.26	0.18	0.32	0.10	0.36	0.14	0.51	0.36
Sour cream	0.32	0.05	0.13	0.19	0.20	0.24	0.15	0.16	0.27	0.24
Tomatoes		0.90	0.65	0.68	0.76	0.56	0.62	0.72	0.76	0.64
Wheat	0.04	0.12	0.22	0.07	0.20	0.35	0.11	0.25	0.34	0.38

Table 4.5.4 presents the seasonal price patterns of (a) potatoes and (b) tomatoes. The bold figures represent the months with the lowest prices, and the framed figures the months with the highest prices. A regular seasonal price pattern can be observed, where the minimum price occurs around harvesting and the maximum in the off-season, which reflects storage costs and capacities. In the case of potatoes, the lowest price was usually between July and August after the harvest period, while the highest price was in April. Similarly for tomatoes, the lowest price was after the harvest period between July and August, and the highest in February and March.

The seasonal fluctuations, often referred to as a sawtooth pattern, of fruit and vegetable prices showed a strong seasonal pattern in the regional production with supplies having to meet a relatively stable demand over the course of the year. This indicates that it is necessary to develop local storage (Bobojonov and Lamers 2008) and processing facilities and off-season greenhouse production, as well as to promote trade with other regions through relaxing trade barriers during the off-season and improving roads.

4.5.3.2 Interdependencies and relations

Results from the regression models for rice markets are displayed in Table 4.5.5. Four models with different explanatory variables[3] were run. In each

---

3 Model 1 only estimates the effect of water inflow on rice prices; model 2 estimates the effect of this variable plus the effect of international rice prices, while model 3 includes the effect of the exchange rate. Finally, model 4 estimates the effects of all previous variables (water inflow, international rice prices and exchange rate) plus the impact of crude oil prices on rice prices.

Table 4.5.4: Seasonal pattern of monthly prices (UZS kg^{-1}) of (a) potatoes and (b) tomatoes during 2002–2010 at Urgench market
(a)

Year	Month												Month low	Month high	Year average	Standard deviation
	May	Jun	Jul	Aug	Sep	Oct	Nov	Dec	Jan	Feb	Mar	Apr				
2002/2003	133	125	133	125	141	148	156	195	256	256	285	474	125	474	202	103
2003/2004	351	142	100	57	119	119	123	285	202	152	152	127	57	351	161	82
2004/2005	127	101	76	76	132	152	152	127	131	187	187	281	76	281	144	56
2005/2006	375	309	187	234	309	281	281	469	577	577	577	770	187	770	412	179
2006/2007	577	433	385	385	481	481	577	674	678	678	774	774	385	774	575	142
2007/2008	726	387	290	290	339	484	581	581	682	779	650	581	290	779	531	170
2008/2009	536	487	483	480	730	584	730	682	775	678	775	1,066	480	1,066	667	170
2009/2010	969	581	581	484	581	581	688	896	846	913	962	1,058	484	1,058	762	198
Average	474	321	279	267	354	354	411	488	518	528	545	641	267	641	432	118

(b)

Year	Month												Month low	Month high	Year average	Standard deviation
	May	Jun	Jul	Aug	Sep	Oct	Nov	Dec	Jan	Feb	Mar	Apr				
2002/2003																
2003/2004			427	142	190	285	285	1,139	1,822	1,215	1,518	1,113				
2004/2005	810	506	101	202	304	405	1,518	1,316	2,812	2,343	1,687	1,687	101	2,812	1,141	890
2005/2006	1,500	515	328	375	469	375	1,406	1,125	1,155	1,155	2,599	1,925	328	2,599	1,077	711
2006/2007	962	674	192	318	289	385	577	1,925	1,936	2,711	2,904	2,227	192	2,904	1,258	1012
2007/2008	1,936	968	242	484	581	1,258	1,452	1,646	2,434	2,532	2,230	2,450	242	2,532	1,518	819
2008/2009	2,142	876	389	438	584	779	1,753	1,947	1,938	1,938	4,360	2,907	389	4,360	1,671	1167
2009/2010	1,938	1,453	436	194	678	775	1,080	3,149	3,808	4,207	3,942	3,726	194	4,207	2,116	1542
Average	1,548	832	302	308	442	609	1,153	1,749	2,272	2,300	2,749	2,291	302	2,749	1,380	887

model (from 1 to 4), an explanatory variable was added to observe how much it explained price fluctuations in the rice markets. In all four models, rice prices were influenced by an autoregressive process [AR(2)]. However, since rice imports and exports in Uzbekistan are almost negligible (FAO 2012), it is surprising that the model results indicate an impact of international market prices of rice on domestic markets at the first lag in all models where this variable was included. Yet, Robinson (2008) postulated that up to 23 % of the national rice requirements could have been imported to Uzbekistan during the drought year of 2008, partly through unofficial channels. These imports during the drought years could establish a positive transmission across rice prices.

Oil prices also showed a significantly positive effect on rice prices at the first lag (Model 4). There is a strong connection between input costs (such as fertilizers, machinery and freight costs) and output prices in agriculture. Through fertilizers and fuel, oil prices comprise a large percentage of agricultural production costs, particularly in rice production. Consequently, any change in oil prices will be channeled into the agricultural commodity prices through the respective changes in farmers' variable costs.

Exchange rate (USD to UZS) is another variable that shows an immediate, direct and significant impact on rice prices in Khorezm (Models 3 and 4). This variable may impact agricultural commodity prices through a number of channels, including international purchasing power and the effect on margins for producers with non-USD costs (IMF 2008). Finally, our regression analysis reveals a negative and significant impact of water inflow on rice prices at its third lag in all models. This effect of water shortage might be transmitted to rice markets through farmer's expectations.

Table 4.5.5: Regression models for rice

Variable	Model 1	Model 2	Model 3	Model 4
L1_crude oil				0.485 ***
L3_water inflow	-0.019 ***	-0.019 ***	-0.017 ***	-0.019 ***
L1_international rice price		0.203 ***	0.208 ***	0.214 **
Exchange rate			0.108 **	0.096 **
AR (2)	0.261 ***	0.310 ***	0.387 ***	0.324 ***
R^2	0.269	0.335	0.435	0.668

* 10 % of significance, ** 5 % of significance, *** 1 % of significance

Results from the regression models for wheat markets are influenced by an autoregressive process [AR(3)] as well as by exogenous variables (Table 4.5.6). In this case, three models were run, and in each model (from 1 to 3) an explanatory variable was added to observe how much it can explain price fluctuations in wheat markets[4]. International wheat prices had a significantly positive effect on domestic market prices at its first lag (Model 3). As wheat is also imported from other countries as flour, a positive transmission across prices is expected. Similar to the case of rice, oil prices also show a significantly positive effect on wheat price at its first lag (Model 2 and 3). Finally, water inflow has a low but nevertheless significant negative impact on wheat prices at the third lag in all models. This indicates that wheat, which in Uzbekistan is mainly winter wheat, depends less on water variability that is the highest in summer season, but is rather influenced by its world price. Surprisingly, the exchange rate did not show any impact on the wheat prices. This can be explained by the fact that the governmental control over the prices of domestically produced wheat and flour as part of the national food security policy was successful in reducing the impact of exchange rates on wheat prices (Al-Eyd et al. 2012). For instance, in 2008, the grain distributed through the state organization at subsidized prices (USD 500 or less) made up 39 % of the total flour sold in Uzbekistan, while imports accounted for 20 % (Robinson 2008).

Table 4.5.6: Regression models for wheat prices

Variable	Model 1	Model 2	Model 3
L1_crude oil		1.537 **	1.523 **
L3_water inflow	-0.041 ***	-0.041 **	-0.033 **
L1_international wheat price			0.593 *
AR (3)	0.195 ***	0.207 **	0.195 **
R^2	0.110	0.379	0.443

* 10 % of significance, ** 5 % of significance, *** 1 % of significance

The results from the regression models for the beef markets reveal the influence by an autoregressive–moving-average process [ARMA (2,2)] (Table 4.5.7). In this case, two models were run. In each model, an explanatory variable was

---

[4] Model 1 only estimates the effect of water inflow on domestic wheat prices; Model 2 estimates the effect of this last variable plus the effect of crude oil prices and Model 3 includes the effect of the international wheat on domestic wheat prices.

added to observe how much it can contribute to explain price fluctuations in meat markets[5]. Again, oil price showed a significantly positive effect on beef prices at its first lag, (Model 1 and 2). Rice price also had a positive effect on beef price at its first lag. This positive and significant impact might be explained by the complementary relationship between these two agricultural commodities. For instance, they are both an essential part of a national Uzbek dish (plov). Finally, in none of the models do water inflow or wheat prices show a significant impact on beef prices. As livestock is mainly fed on forage such as maize stem and crop by-products such as cotton-seed cake and husk (Djanibekov 2008), it is expected that wheat prices do not play a significant role as a determinant of beef prices in Khorezm.

Table 4.5.7: Regression models for beef prices

Variable	Model 1	Model 2
L1_crude oil	0.163 ***	0.163 ***
L3_water inflow		0.0007
Rice price	0.357 **	0.356 **
Wheat price		0.982
AR (2)	0.445 **	0.440 **
MA (2)	0.356 **	0.353 **
R2	0.345	0.450

* 10 % of significance, ** 5 % of significance, *** 1 % of significance

4.5.4 Summary and conclusions

The results of the analysis of market prices serve as critical information for policymakers. This study is based on region-specific price information on ten major food commodities and assessed the temporal trends and variability of commodity prices. The empirical analysis confirms three general patterns of agricultural prices. The prices of apples, onions, potatoes, tomatoes are de-

---

5 Model 1 estimates the effect of crude oil prices and rice prices on beef prices, while Model 2 estimates the effect of all these variables plus the effect of water inflow and wheat prices on beef markets.

termined locally and follow seasonal fluctuations reaching a minimum during the harvest season and a maximum in the off-season.

Furthermore, the fluctuations of rice, beef and wheat prices are more sensitive to external factors such as their respective international prices, currency exchange rate and oil prices. A detailed analysis for these three commodities in Khorezm presents a case of the price transmission between international and local markets in Uzbekistan. The international price transmission of rice and wheat demonstrates that the markets of these products are integrated into world market, and the rising food prices on world markets will have a direct effect on the prices of these products. While rice prices were strongly connected to world prices, they also depended on the oil price as well as on the currency exchange rate. In the case of wheat, the state program on food security could offset the direct link between exchange rate and wheat price.

However, the observed trend of rice prices is also subject to the impact of water shortages as occurred in 2008. That climatic event could have added to the significance of the impact of the international price transmission in 2008 during which rice prices in Uzbekistan boomed. According to the model results, oil prices also had a significant impact on the beef price due to transportation costs. Surprisingly, the prices of rice and meat are positively correlated, which shows their complementarity in the food diets in Khorezm. The policy implication of this inter-commodity price transmittal is that the government of Uzbekistan could focus on, for instance, the price of rice to achieve its price objective also for the price of beef.

The model results show that, despite increasing regional and national production levels, an increasing price variability is evident for most crops. For some crucial food products such as wheat, rice, milk and beef, prices became more volatile, thus reflecting the unstable conditions in domestic supply and demand. These commodities account for the largest share in the regional agricultural revenues and in the expenditure of consumers, and these price volatilities will thus directly affect both consumers and producers in Khorezm. In addition to international and cross-regional trade within Uzbekistan, regional production patterns certainly influenced the domestic prices during the observed period. From the demand side, the income-driven changes in consumption diets also contribute to price trends. Hence, future work should take into account effects of local food production and consumption on the domestic prices.

The results of this study are relevant for policy- and decision-making discussions about prices, infrastructure such as storage and roads, as well as about food and trade policies in Uzbekistan. The data analysis reveals that the price increases were transmitted across commodities, in our case rice and beef. This cross-commodity integration implies that development of either trade or storage facilities of one food commodity (e. g., rice) will be beneficial to stabilize the price of other key food commodities (e. g., beef).

There is strong statistical evidence that market prices of all commodities have increased since 2002, but also that they have experienced a larger variation within a particular year. For example, December and January were the most common peak months for vegetables and fruits. This seasonal price movement reveals the necessity to develop storage and processing facilities and greenhouse production as well as to develop outside trade through relaxing the trade barriers during the off-season and improving roads.

References

Al-Eyd A, Amaglobeli D, Shukurov B, Sumlinski M (2012) Global food price inflation and policy responses in Central Asia. Middle East and Central Asia Department, IMF Working Paper WP/12/86, 25 pp

Bekchanov M, Karimov A, Lamers JPA (2010) Impact of Water Availability on Land and Water Productivity: A Temporal and Spatial Analysis of the Case Study Region Khorezm, Uzbekistan. Water, 2 (3), 668–684

Bekchanov M, Lamers JPA, Martius C (2014) Coping with water scarcity in the irrigated lowlands of the lower Amudarya basin, Central Asia. In Lamers JPA, Khamzina A, Rudenko I, Vlek PLG (Eds.) Restructuring land allocation, water use and agricultural value chains: Technologies, policies and practices for the lower Amudarya region. V&R unipress, Bonn University Press pp 199–216

Bobojonov I, Lamers JPA (2008) Analysis of agricultural markets in Khorezm, Uzbekistan. In: Wehrheim P, Schoeller-Schletter A, Martius C (Eds.), Continuity and Change: Land and Water Use Reforms in Rural Uzbekistan: Socio-economic and Legal Analyses for the Region Khorezm. IAMO Studies on Agricultural and Food Sector in Central and Eastern Europe, 43, Halle (Saale), pp 165–182

Conrad C, Schorcht G, Tischbein B, Davletov S, Sultonov M, Lamers JPA (2012) Agrometeorological trends of recent climate development in Khorezm and implications for crop production. In: Martius C, Rudenko I, Lamers JPA, Vlek PLG (Eds.) Cotton, water, salts and soums – economic and ecological restructuring in Khorezm, Uzbekistan. Springer, Dordrecht, pp 25–36

Djanibekov N (2008) A micro-economic analysis of farm restructuring in the Khorezm region, Uzbekistan. Dissertation, ZEF, Bonn University, Germany, 180 pp

Djanibekov N, Hornidge AK, Ul-Hassan M (2012) From joint experimentation to laissez-faire: transdisciplinary innovation research for the institutional strengthening of a water users association in Khorezm, Uzbekistan. Journal of Agricultural Education and Extension, 18 (4), 409–423

FAO (2012) FAO Statistical Database: http://faostat.fao.org/default.aspx, last access: 6 May, 2012

Labys W (1973) Dynamic Commodity Markets: Specification, Estimation and Simulation. Lexington Books: Lexington

Labys W (1999) Commodity markets and models: the range of experience. In Greenaway D, Morgan W (Eds.) The Economics of Commodity Markets. London. Edward Elgar. The International Library of Critical Writings in Economics

Labys W (2006) Modelling and Forecasting Primary Commodity Prices. Ashgate: Aldershot

Musaev D, Yakhshilikov Y, Yusupov K (2010) Food security in Uzbekistan. UNDP Mega Basin, Tashkent Uzbekistan, 64 pp

OblStat (2007) Socio-Economic Indicators for Khorezm, 1998–2006, Urgench

Roache SK (2010) What explains the rise in food price volatility? IMF Working Papers, 10:129

Robinson I (2008) Food markets and food insecurity in Tajikistan, Uzbekistan, Kyrgyzstan, Kazakhstan. World Food Programme, Regional Market Survey for the Central Asian Region, Report of Mission, 193 pp

Tomek W, Robinson K (2003) Agricultural Product Prices. Fourth Edition. Cornell University Press: Ithaca

von Braun J, Tadesse G (2012) Global food price volatility and spikes: an overview of costs, causes, and solutions. Center for Development Research (ZEF), Bonn University, ZEF-Discussion Papers on Development Policy No. 161, 42 pp

WFP (World Food Programme) (2008) Poverty and food insecurity in Uzbekistan, 57 pp

World Bank (2011) World Development Indicators and Global Development Finance. Available at http://data.worldbank.org/data-catalog/world-development-indicators, last access: 6 May, 2012.

Data sources

ADB (2011). Statistical Database System Online. Country Tables. Available at: https://sdbs.adb.org/sdbs/index.jsp

CAWATERinfo (2011). Amudarya River basin. On-line data of the BWO Amudarya. Available at: http://www.cawater-info.net/amudarya/index_e.htm#

FAO (2011). Trade and Markets. Statistical data (2002–2011). Available at: http://www.fao.org/economic/est/statistical-data/en/

IMF (2011). Primary Commodity Prices. Monthly data for 8 price indices and 49 actual price series (1980 – current). Available at: http://www.imf.org/external/np/res/commod/index.aspx

Section 5: Society, Policy and Institutions

Bekchanov Maksud, John P. A. Lamers, Kudrat Nurmetov

5.1 Economic incentives for adopting irrigation innovations in arid environments

Abstract

Water is getting scarce in many parts of the world, consequently challenging researchers, policy makers and practitioners to design options for a more efficient use of these resources, especially in irrigated agriculture. Although technical-economic efficiency of potential water-wise options and institutional restrictions for their implementation in the developing and less-developed countries are well documented, little evidence exists about the incentives for farmers and regional development agencies to adopt the efficient irrigation innovations. A linear programming model for optimizing regional agricultural income was developed to analyze the impact of water availability, water pricing, and investment accessibility on water-wise innovation adoption and conveyance efficiency improvement. The model was applied to the case of Khorezm, a region in north-western Uzbekistan that is part of the downstream Amu Darya River in the Aral Sea Basin. Model results indicate that improving conveyance efficiency is economically less attractive than improving field-level water use efficiency due to enormous investment costs for lining the canals. Water-wise options such as manuring cotton and potatoes, implementing hydrogel in wheat and cotton, and drip irrigation of melons and vegetables are among the most promising field-level improvement options to gain optimal regional incomes under decreased water availability and increased water prices. It is illustrated that despite the huge investments needed for a wide-scale implementation of modern irrigation technologies such as drip irrigation and laser-guided land leveling, their adoption will substantially improve water use efficiency, while their implementation costs can be compensated for by the additional revenues due to increased yields and reduced costs.

Keywords: water use efficiency, conveyance efficiency, water-wise options, water pricing, irrigation investment, linear programming, Khorezm region, Aral Sea basin

5.1.1 Introduction

Water scarcity has become a major challenge in many parts of the world, and more than 20 % of the global population is already living under water-stressed conditions. It is predicted that by 2025 this share may exceed 33 % (UN WATER 2007). While water demand is expected to increase due to population growth, irrigation expansion, and industrial development, the fresh water supply will decrease due to irreversible contamination and glacial surface area reduction. As a consequence, water scarcity will become unavoidable under a "business-as-usual" option, potentially leading to conflicts, wide-spread droughts, food insecurity, and poverty at national and regional levels. The World Economic Forum 2009 (WEF 2009) warns, for instance, that the consequences of the expected water crisis may be more severe than the current financial crisis unless instant and timely prevention and mitigation measures are taken. This is particularly crucial in regions with an arid and semi-arid climate located downstream of river basins.

Agriculture is a major culprit of the growing uncertainty: More than 70 % of the accessible fresh-water resources worldwide are consumed in the agricultural sector (WRI 2005), and average water use efficiency in irrigation is less than 40 % (Pimental et al. 1997). Therefore, adaptation and mitigation measures can enhance substantial water use reduction particularly in irrigated agriculture. Introducing irrigation innovations is essential to sustain livelihoods particularly in countries where economy and livelihoods heavily rely on irrigated agriculture. Efficient and sustainable irrigation water usage is the basis for sustainable food supply, stable food prices, healthy diets, and poverty reduction (von Braun et al. 2003; von Braun et al. 2009).

Many measures to increase water use efficiency have been developed so far, varying from measures for improving conveyance and distribution efficiency to measures for increasing irrigation efficiency at field level. Despite the existence of numerous technologies for increasing water use efficiency in irrigation settings, their implementation has been recurrently impeded due to inefficient institutions and governance, restricted financial means and insufficient knowledge on economic gains from technological change, especially in less developed and developing countries (Saleth and Dinar 2004).

Based on multi-year field experiments under different soil-climate conditions and with different crops, various studies have addressed the technical consistency of water management measures in Central Asia (e.g., Nerozin 2005, Horst et al. 2007, Ibragimov et al. 2007). The economic efficiencies of water conservation options have been analyzed taking crop type, yield and prices into consideration (Abdullaev et al. 2008; Mamatov 2009; Bekchanov et al. 2010). Institutional factors enhancing efficient water use have been analyzed through econometric approaches using cross-country data (Saleth and Dinar 2004). Parametrical (logit, probit) models have been used to assess the determinants of adopting selected technologies such as drip irrigation in the western USA and Spain (e.g., Alcon et al. 2011; Skaggs 2001). However, studies are rare that address incentives for a wide-scale adoption of efficient water-wise options by farmers while considering the complex water use and distribution and crop production relationships within the irrigation system.

Holistic approaches to water management and irrigation technology adoption have been implemented using hydro-economical models to analyze the impact of water availability on technology adoption in Chili and Vietnam (Ringler et al. 2004; Cai and Rosegrant 2004; Rosegrant et al. 2005). However, these models rely on an assumption of a continuous functional relationship between investments and irrigation efficiency. The models thus ignore that technology adoption is a discrete choice. Incentives to reduce water use have been also analyzed by examining the impact of increased water fees for water delivery on water demand, crop pattern changes, and regional and farm income in the case of Uzbekistan yet without considering the adoptability of potential irrigation innovations (Djanibekov 2008; and Bobojonov et al. 2008; Djanibekov et al. 2012). Only a limited set of water conservation measures demanding high initial investment costs such as drip irrigation and laser-guided leveling has been analyzed using the chance-constrained linear programming (LP) approach to assess the impact of water fees on water use efficiency at farm level (Bobojonov et al. 2010). Consideration of a broad set of water-saving options would further bear more efficient solutions to deal with water scarcity issues not only at farm level but also at regional level. Furthermore, improving conveyance efficiencies in addition to field-level water application improvements also may contribute to water use reduction.

This study aims to analyze the incentives for farmers and water management agencies to enhance adoption of a multiple set of water-wise innovations simultaneously considering the discrete nature of the innovation choice and employing a holistic (integrated) approach to model the irrigation system. The impact of economic incentives such as water pricing and irrigation investments on the magnitude of adopting different irrigation innovations to improve not only water application efficiency but also conveyance efficiency are examined under a limited availability of water resources. For this purpose, a specially developed LP model was applied to the case of the Khorezm region in western Uzbekistan. The region represents the arid and semi-arid irrigation zones of the world where the need for efficient water use is of utmost importance due to aggravating water scarcity conditions. Despite all the evidence on the technical and economic efficiency of water-wise options, these have not yet been adopted on a wide scale by the farmers in arid areas of Uzbekistan. This is often explained by a lack of incentives for farmers (Djanibekov et al. 2012) without indicating what incentives would be supportive. The LP approach was chosen since it is a simple but effective tool to address the complex relationships in the agricultural system (Hazel and Norton 1986). Since the regional water management organization in Uzbekistan decides on water distribution, cropland planning, and technology changes for the entire region (Bobojonov 2009), it is justifiable to implement the optimization model for the case of the entire Khorezm region rather than for an individual farm.

5.1.2 Agriculture and water use and supply in Khorezm

The Khorezm region in Uzbekistan is part of the lower reaches of the Amu Darya River and is located about 225 km from the remains of the desiccating Aral Sea (Conrad et al. 2007). The Aral Sea desiccation due to irrigation expansion in the past few decades is one of the worst human-induced ecological disasters known (UN 2010). It is the result of inefficient water management policies, which sacrificed ecological sustainability for attaining cotton self-sufficiency in the former Soviet Union in the last century. Despite many efforts by local and international communities during the past years to improve the ecological situation in the Aral Sea and in the territories adjacent to the sea, the affected area is gradually expanding, thus also influencing downstream regions as reflected in more frequent water shortages (Müller 2006) and increasing land and water salinity (Micklin 2000).

Khorezm is home to over 1.7 million people (as of 2011). Roughly 230,000 ha of the region are used for irrigation purposes (OblSelVodKhoz 2011). Irrigated agriculture in Khorezm contributes to more than 50 % of the regional income, provides more than 98 % of hard cash revenues, and employs more than 60 % of the economically active population (Rudenko et al. 2009). Cotton and wheat production dominate with a share of about 50 % and 20 %, respectively. Khorezm is also one of the main rice growing regions in Central Asia. Since more than 90 % of all water withdrawals to the region from the Amu Darya river is used in irrigated agriculture, which is the backbone of the regional economy (Rudenko et al. 2009; Bobojonov 2009), water availability is a key factor for the income level and welfare of the population in the region. However, water availability is gradually becoming uncertain due not only to high variations in natural river runoff but also to the neglected water needs of the downstream users by the upstream regions. The probability of adequate water availability has therefore been decreasing over the past years (Müller 2006). Seasonal variations in river runoff also decrease water availability during the vegetation period (Müller 2006).

Despite aggravating water scarcity problems, low conveyance and irrigation efficiencies are the main causes of irrigation water losses in Khorezm. Water is mainly conveyed by unlined irrigation canals (Tischbein et al. 2012). Furthermore, furrow and basin irrigation are the common water application practices at field level (Martius et al. 2009). The enormous water losses in the irrigation system call for an urgent action for improving water use efficiency to achieve sustainable agricultural production.

5.1.3 Methodology

5.1.3.1 Data Sources

Based on previous findings indicating the economic and water use efficiency enhancement potential of several water-wise innovations (Bekchanov et al. 2010), a series of options for water use reduction was examined as alternatives to conventional irrigation of cotton, cereals, rice, maize, potatoes, vegetables, melons, fodder crops, fruits, and grapes: (i) replacing paddy rice by aerobic rice, hydrogel application for cotton and cereals, (ii) manuring of cotton, cereals, and potatoes, (iii) laser-guided land leveling for cotton and cereals, (iv) drip irrigation of cotton, potatoes, vegetables, melons, fruits, and grapes, and (v) surge-flow, double-furrow, alternate dry-furrow, and short-furrow irrigation of cotton. Data on crop yields, initial (present) area, prices and production costs were compiled from various sources including OblStat (2011), Bobojonov (2009), and Djalalov (2005; see Appendix for more details). Information on crop water use requirements, conveyance efficiency, and overall regional water use was obtained from OblSelVodKhoz (2011). Data on technical and economic parameters of irrigation methods are based on Bekchanov et al. (2010).

5.1.3.2 Hydro-economic model

A static short-term optimization model allowed analyzing water use, water price, investments, and innovation adoption relationships. The objective function of the model optimizes regional crop production profit (TP) considering the revenues and variable costs under different irrigation options (t) and crops (c), costs of water delivery, and costs of conveyance efficiency improvement:

$$TP = \sum_{(t,c) \in TECH\_CP} (\overline{Pr_c}\,\overline{Y_{t,c}}A_{t,c} - \overline{VC_{t,c}}A_{t,c} - \overline{Pw_{t,c}}\overline{w_{t,c}}A_{t,c}) - \overline{CC}(CE - \overline{CE0})\overline{TW} \quad (1)$$

where $\overline{Pr_c}$ is crop price, $\overline{Y_{t,c}}$ is crop yield, $\overline{VC_{t,c}}$ is variable costs, $A_{t,c}$ crop area, $\overline{Pw_{t,c}}$ is water delivery costs, $\overline{w_{t,c}}$ is water use per hectare, $\overline{CC}$ is costs of increasing conveyance efficiency by 1 %, $\overline{CE0}$ is initial conveyance efficiency, and $\overline{TW}$ is total available regional water supply. The variables with bars are exogenous variables while the variables without bars are endogenous variables. TECH_CP is a set indicating which crops can be grown under the implementation of a particular water-wise innovation.

Several formulated model constraints address limits and conditions with respect to conveyance improvement, production of state-order crops, available water, investment, manure, and land resources. Improved conveyance efficiency is defined here as being higher than the initial (present) conveyance efficiency and lower than 0.95 (or 95 %), which is the potential maximum:

$$\overline{CE0} \leq CE \leq 0.95 \quad (2)$$

Considering the state control over cotton and wheat production (Rudenko et al. 2012), the production volume of these crops should be higher than the initially observed (present) production level $\overline{Q_c}$:

$$\sum_{(t,c) \in TECH\_CP} \overline{Y_{t,c}}A_{t,c} \geq \overline{Q_c} \quad (3)$$

Crop water use adds up to total water use while considering conveyance efficiency:

$$\frac{1}{CE} \sum_{(t,c) \in TECH\_CP} \overline{w_{t,c}}A_{t,c} = \overline{TW} \quad (4)$$

Since advanced technologies such as laser-guided land leveling and drip irrigation require substantial investments (Bekchanov et al. 2010), their implementation is restricted by total available financial assets $(\overline{TI})$ and depends on per hectare investment requirement $(I_{t,c})$:

$$\sum_{(t,c) \in TECH\_CP} I_{t,c}A_{t,c} \leq \overline{TI} \quad (5)$$

Manuring is restricted by the amount of total manure $\overline{TM}$ produced by the animal stock in the region and depends on per hectare manuring requirement $(M_{t,c})$:

$$\sum_{(t,c) \in TECH\_CP} M_{t,c} A_{t,c} \leq \overline{TM} \qquad (6)$$

Cropland areas are limited to the total available cropland area $\overline{(TA)}$:

$$\sum_{(t,c) \in TECH\_CP} A_{t,c} \leq \overline{TA} \qquad (7)$$

In addition, it is assumed that the area allocated to a particular crop can vary 20 % around the initial value. This assumption is needed to prevent the model from giving unrealistic solutions such as extreme increases in production of certain crops since a LP modeling framework does not consider responsive crop price reduction in parallel to increased output. The area under perennial crops is considered as fixed, since it takes several years to create orchards and vineyards that can be analyzed using a dynamic model rather than the static model used in this study. No land restriction is introduced when implementing an innovation.

The model is calibrated to the land and water use data of 2005, since a complete dataset was accessible and water resources was abundant to meet regional water demand in this year. Considering its strong grip on agricultural production, marketing and processing, the regional administration is perceived to be the first and most important decision maker when adopting innovations. The objective function optimizes overall regional agricultural profit while assuming an ideally functioning regional water managing agency.

5.1.3.3 Scenarios

Three scenarios were compared to the baseline. The baseline scenario reflects the initial water use and innovation adoption level observed in 2005. Other scenarios were run by maximizing the overall regional profit (Table 5.1.1).

Considering the crucial role of water availability on regional welfare (Bobojonov et al. 2008), Scenario 1 consists of seven sub-scenarios of water availability change with a sequential reduction in the initially observed water supply by 10 % in each sub-scenario (Table 5.1.1). Thus, 100 % of the water supply is available in the first sub-scenario, and only 40 % of the initial supply is available in sub-scenario 7. The lowest water supply level in the model scenarios is based on long-term observations of water availability in the region during the period 1980–2010, which showed that in the driest year 2001, the region was able to supply only 40 % of the required irrigation water (SIC-ICWC 2011; OblSelVodKhoz 2011).

Due to decreasing water availability over the years and increasing costs for reservoir management, the costs of water delivery to the region are likely to increase. Moreover, since water availability partly depends on water intake by upstream regions, decision makers in Khorezm can cooperate with the upstream regions and offer compensation for their reduced water consumption. In this case, compensation costs also would increase the cost of the water supply to the region. Thus, Scenario 2 addresses changes in water delivery prices. Eleven sub-scenarios address different water-fee levels ranging from no water fee in sub-scenario 1 to a fee of USD 0.02 m^{-3} of water in sub-scenario 11, i.e., an additional fee of USD 0.002 m^{-3} is added in each sequential sub-scenario (Table 5.1.1).

The role of irrigation investments is crucial for adopting advanced but at the same time expensive irrigation technologies such as laser-guided land leveling and drip irrigation (Bekchanov et al. 2010). Thus, Scenario 3 considers changes in total irrigation investments and consists of 11 sub-scenarios. Under sub-scenario 1, a total of USD 15 million was made available. This would allow producing 12,500 ha of cotton with drip irrigation if the complete investment was spent on irrigation efficiency improvement in cotton production. A further sum of USD 5 million was considered in each sequential sub-scenario (Table 5.1.1). The imposed investment amounts are arbitrary and do not show any planned amount by government or farmers.

5.1.4 Results

5.1.4.1 Scenario 1: Water availability impact

In response to a decreasing water availability (from 100 % to 40 %), optimal total net profit would decrease from the initial level of USD 71.3 million to USD 43.0 million (Figure 5.1.1). Parallel to water availability decrease, conveyance efficiency would need to increase from 58 % to 95 %.

Adoption of water-wise innovations and reduction in rice area enhance optimal profits under water-scarce conditions (Figure 5.1.2). For instance, in sub-scenario 1 of Scenario 1, optimal profits can be reached by applying double-furrow irrigation on 68,000 ha of cotton, manuring on 33,000 ha of cotton and potatoes, and implementing drip irrigation on 12,500 ha of melons and vege-

Table 5.1.1: Parameters used in three model scenarios considering different levels of water availability (Scenario 1), water supply pricing (Scenario 2), and investment levels (Scenario 3)

	Scenario 1			Scenario 2			Scenario 3		
	WPA	WP	IA	WA	WP	IA	WA	WP	IA
Baseline	100 %	0	0	100 %	0	0	100 %	0	0
Sub-scenario 1	100 %	0.002	15	70 %	0	15	70 %	0.002	15
Sub-scenario 2	90 %	0.002	15	70 %	0.002	15	70 %	0.002	20
Sub-scenario 3	80 %	0.002	15	70 %	0.004	15	70 %	0.002	25
Sub-scenario 4	70 %	0.002	15	70 %	0.006	15	70 %	0.002	30
Sub-scenario 5	60 %	0.002	15	70 %	0.008	15	70 %	0.002	35
Sub-scenario 6	50 %	0.002	15	70 %	0.01	15	70 %	0.002	40
Sub-scenario 7	40 %	0.002	15	70 %	0.012	15	70 %	0.002	45
Sub-scenario 8	-	-	-	70 %	0.014	15	70 %	0.002	50
Sub-scenario 9	-	-	-	70 %	0.016	15	70 %	0.002	55
Sub-scenario 10	-	-	-	70 %	0.018	15	70 %	0.002	60
Sub-scenario 11	-	-	-	70 %	0.02	15	70 %	0.002	65

WA – water availability (%); WP –water price (USD/m^3); IA – investment availability (million USD)

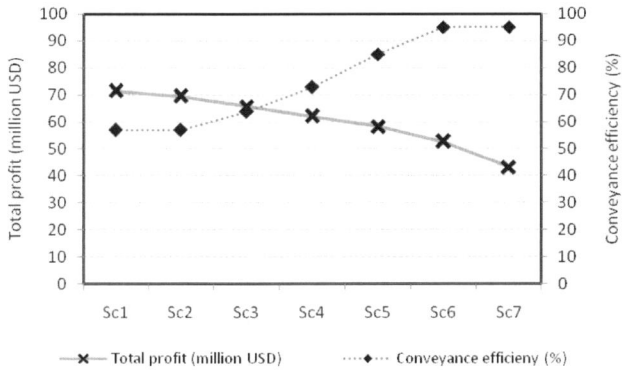

Figure 5.1.1: Impact of water availability on total profit and conveyance efficiency improvement. Note: Water availability decreased from 100 % (sub-scenario 1) to 40 % (sub-scenario 7); total available irrigation investment is fixed at USD 15 million and water delivery price at USD 0.002 m^{-3}.

tables. However, when assuming a 10 % decrease in water availability (sub-scenario 2), application of hydrogel for all 59,000 ha of cereals production is required for gaining optimal profits. Once the water availability dropped by 60 % (i. e., to 40 % of the baseline; sub-scenario 7), the use of hydrogel on 43,000 ha of cotton in addition to 59,000 ha of cereals would be advisable to reach the highest profits. Manuring on about 50,000 ha of cotton is needed when water availability falls below 80 %. Optimal levels of the share of double-furrow irrigation of cotton would decrease parallel with a decrease in water availability. Rice production would become less beneficial under decreased water availability or under increased marginal water profitability, which is in line with previous findings (Djanibekov et al. 2012). In general, it is advisable to decrease the total area of cropped lands from about 230,000 to 208,900 ha when water availability decreases to 40 % (sub-scenario 7).

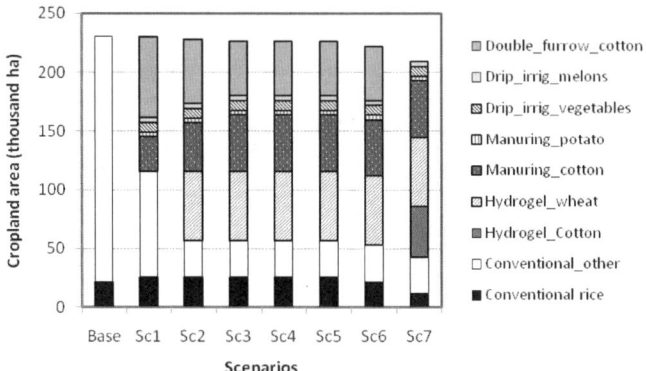

Figure 5.1.2: Impact of decreasing water availability on crop and technological structure. Note: Water availability decreased from 100 % (sub-scenario 1) to 40 % (sub-scenario 7); total available irrigation investment is USD 15 million and water delivery price is set at USD 0.002 m^{-3}.

5.1.4.2 Scenario 2: Water pricing impact

Following the increase in water price from USD 0 (no direct fees) to USD 0.02 per m^3, total optimal regional profits would decrease from USD 65.4 million to 32.4 million. Decreasing the area of croplands from 230,000 to 220,000 ha and concurrently reducing conveyance efficiency from 74 % to 60 % is recommended to gain optimal benefits (Figure 5.1.3).

Water fees above USD 0.016 m^{-3} most likely would increase the financial attractiveness of implementing field-level water-wise options rather than investments in conveyance improvement (Figure 5.1.3 and Figure 5.1.4). Similar crop pattern and innovation adoption changes occur under simulations with water fees as is shown in the different water availability scenarios.

5.1.4.3 Scenario 3: Irrigation investment impact

In contrast to Scenario 1 and 2, total profit is likely to increase from USD 62 million to 69.9 million when increasing irrigation investments in Scenario 3

Economic incentives for adopting irrigation innovations 307

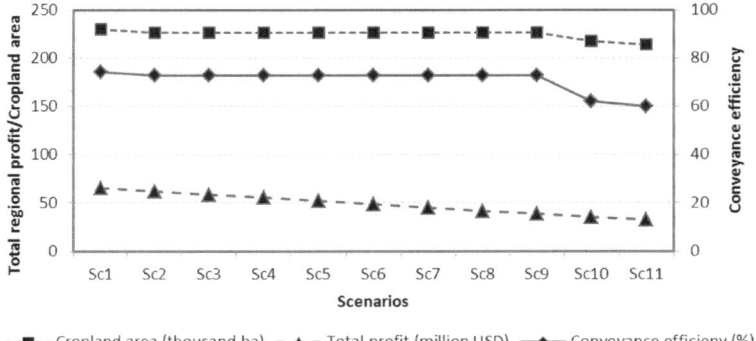

Figure 5.1.3: Impact of water pricing on total profit, total cropped area, and conveyance efficiency improvement. Note: Water delivery fees increased from USD 0 m^{-3} (sub-scenario 1) to USD 0.02 m^{-3} (sub-scenario 11); total available irrigation investment is USD 15 million and water availability is set at 70 % of the baseline water use level.

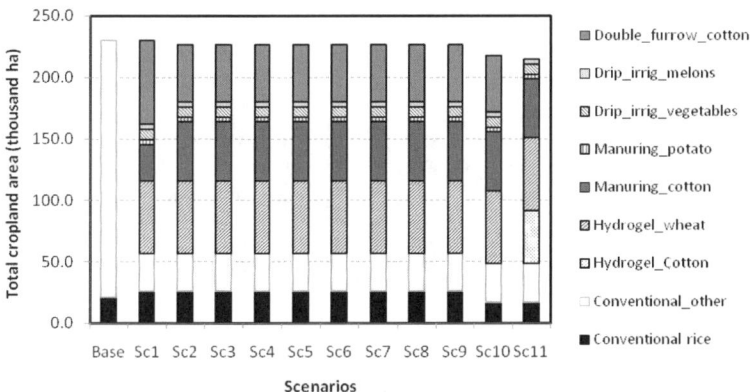

Figure 5.1.4: Impact of increased water fees on crop and technological patterns. Note: Water delivery fees increased from USD 0 m^{-3} (sub-scenario 1) to USD 0.02 m^{-3} (sub-scenario 11); total available irrigation investment is USD 15 million and water availability is set at 70 % of the baseline water use level.

(Figure 5.1.5). Concurrently, the cropland area is expected to decrease from 226,000 to 220,000 ha and conveyance efficiency from 73 % to 69 %.

Implementing drip irrigation while reducing conveyance improvement measures would enhance optimal regional crop production profits once the accessibility of financial assets increases (Figures 5.1.5 and 5.1.6). When irrigation investments are, for instance, increased from USD 15 to 65 million, increasing drip irrigation areas from 12,500 to 57,000 ha would lead to optimal profits.

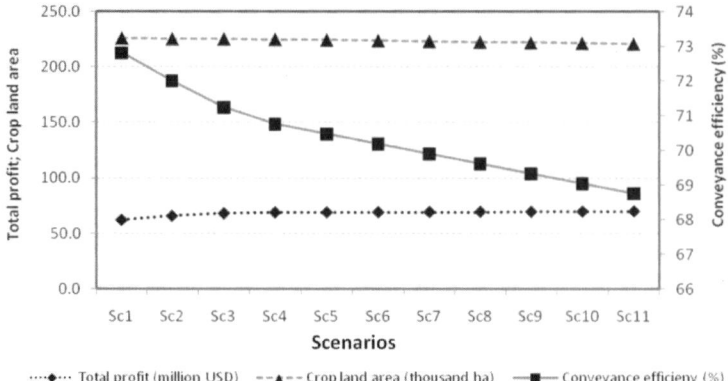

Figure 5.1.5: Impact of investment increase on total profit, total cropped area, and conveyance efficiency improvement. Note: Irrigation investments availability increased from USD 15 million (sub-scenario 1) to USD 65 million (sub-scenario 11); water delivery price is set at USD 0.002 m³ and water availability at 70 % of the baseline water use level.

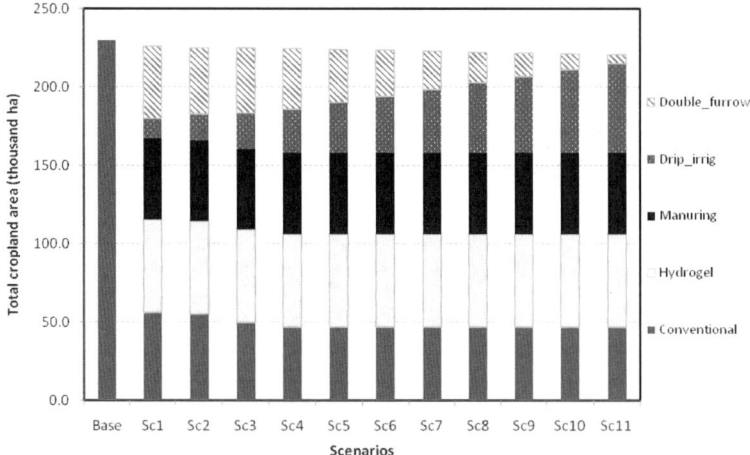

Figure 5.1.6: Impact of investment level on crop and technological structure. Note: Irrigation investments availability increased from USD 15 million (sub-scenario 1) to USD 65 million (sub-scenario 11); water delivery price is set at USD 0.002 m^{-3} and water availability at 70 % of the baseline water use level.

Summing up the findings of the three scenarios, profits from irrigated agriculture would decrease parallel to decreased water availability or increased water fees but would increase as a response to increased availability of investments in irrigation technologies, i. e., drip irrigation and laser-guided land leveling. This indicates that additional economic gains due to increased yields and reduced costs under these technologies can compensate their initial investment costs.

Therefore, increasing investments in modern irrigation technologies would boost sustainable and efficient irrigated agriculture.

5.1.5 Discussion

The present findings illustrate that various economic incentives such as water pricing and increased accessibility of irrigation investments under limited availability of water resources have the potential to enhance wide-scale implementation of irrigation innovations in Uzbekistan. The country is slow on implementing water-wise options despite improvements in water management is of utmost importance due to frequent occurrences of water scarcity. The adoption of water-wise innovations in a country where the national administration has a strong grip on crop production is not only a matter of convincing farmers and land users as often indicated (Hornidge et al. 2011), but in the first place an issue of convincing the regional and national administrations.

The availability of water resources during the growing season in downstream regions like Khorezm is bound to decrease due to the combined impact of increased seasonal fluctuations driven by climate change and increased demand for hydroelectricity generation and irrigation in the upstream countries (Micklin 2000). The findings show that, notwithstanding this perspective, improving conveyance efficiency (Tischbein et al. 2012) and adopting water-wise options such as manuring, drip irrigation, or the use of hydrogel (Bekchanov et al. 2010) have substantial potential to maintain sustainable agricultural production as previously postulated.

Yet incentives need to become available to farmers, and this is in the first place to be determined by the national and regional authorities. Water pricing, for example, is often mentioned as one of the incentives for wide-scale implementation of field-level water-saving measures due to the expected increase in the costs of water supply (e.g., Bobojonov 2008). Though water supply costs are currently covered by the government budget, and farmers are not paying directly for water aside from membership fees for the Water User Association (Bobojonov 2009), farmers may become obliged to pay the full costs of their water supply in the future. However, hurdles to the actual implementation of water fees should not be underestimated even when they are promoted by the national and regional administration. Determining water fees has been difficult worldwide, and even impossible without the support and agreement of the water users (Johansson et al. 2002). The reluctance of farmers to actually pay for the direct use of water can be a serious hurdle, particularly when the prices for agricultural commodities are low as is the case in Uzbekistan due to the state control over the production and trade of the strategic crops cotton and wheat.

Therefore, water pricing policies should be flanked by additional measures such as the liberalization of agricultural commodity prices. When water users (farmers) have a right to choose which crops to grow and when prices serve as a market signal, they may recognize the real economic value of irrigation water for different commodities as is expressed through the marginal water profitability.

Improved investment access can support the adoption of irrigation innovations and thus sustainable and profitable crop production. Yet at present, investment sources in Uzbekistan are limited to the national and regional administration. The government tends to implicitly tax the agricultural sector through paying low prices for state-ordered crops such as cotton and winter wheat rather than subsidizing it (Djanibekov et al. 2010). As a consequence, the low revenues stemming from farming make it difficult for farmers to build sufficient capital that can be used to invest in, for example, modern water-saving technologies. However, the need for sustainable incomes and food supply under decreased water availability and increased water supply costs should convince the Uzbek government of the necessity to enable investments in modern water-wise technologies and to modernize the agricultural sector. Initial investment costs of modern irrigation technologies such as drip irrigation and laser-guided land leveling are currently relatively high in Uzbekistan, since these technologies are imported to the country by individual retailers. However, the government can promote large-scale imports of these technologies at lower costs, which can be possible due to the scale effect, organized transportation, as well as reduced import taxes. Moreover, the potential to develop domestic production and marketing of the irrigation technologies at lower costs also exists and could enhance the wide-scale implementation of these water-saving technologies to combat water shortages in downstream regions like Khorezm.

Our results are based on the assumption of a static structure of the agricultural system. Since long recovery periods are perhaps needed when investing in, for example, drip irrigation systems and laser-guided land leveling, extending the present approach by considering a dynamic modeling component may improve the findings. Data limitations and assumptions may make the findings unbalanced and could mean restricted use for direct policy formulation. For instance, we assumed an ideally functioning regional water managing agency that optimizes overall regional agricultural profits. Since the administration in Uzbekistan has full control over all resources and their distribution, this assumption approaches reality but ignores that decisions are to a certain extent also made by farmers and land users. Nevertheless, since the regional administration is a major decision maker and decides which and to what extent incentives can be made available, the findings certainly are adequate for the situation in the case study region, and could be extrapolated to other parts of Central Asia and other arid zones of the world with similar conditions.

5.1.6 Conclusions

Adopting field-level water-wise options and improving conveyance efficiency have a high potential for increasing crop production revenues in Khorezm, a region greatly threatened by severe and frequent water shortages. Water pricing and irrigation investments could become relevant incentives supporting wider adoption of water-wise innovations. Under increased fees for irrigation water, investing in field-level water use efficiency improvement is more beneficial than investing in conveyance efficiency. Increased investment availability also contributes to a wider adoption of more advanced and efficient but presently expensive technologies such as drip irrigation and laser-guided land leveling. In spite of the high investment costs needed for their implementation, these costs can be compensated for by the additional revenues gained. Since water shortages will remain a major problem in Central Asia as well as in many other arid regions of the world due to high lifting and conveyance costs as well as increased competition for water resources use, the results in the case of the Khorezm region are also relevant to support the sustainability of the irrigated agriculture in these regions.

References

Abdullaev I, Ul Hassan M, Jumaboev K (2007) Water saving and economic impacts of land leveling: the case study of cotton production in Tajikistan. Irrigation and Drainage System 21: 251–263

Alcon, F., Miguel MDD, Burton M (2011) Duration analysis of adoption of drip irrigation technology in southeastern Spain. Technological Forecasting and Social Change 78 (6): 991–1001

Bekchanov M, Lamers JPA, Martius C (2010) Pros and cons of adopting water-wise approaches in the lower reaches of the Amu Darya: A socio-economic view. Water, 2(2): 200–216; Available online: http://www.mdpi.com/2073-4441/2/2/200/pdf

Bobojonov I, Franz J, Berg E, Lamers JPA, Maritus C (2010) Improved policy making for sustainable farming: A case study on irrigated dryland. Jounal of Sustainable Agriculture 34 (7): 800–817

Bobojonov I (2009) Modeling crop and water allocation under uncertainty in irrigated agriculture: a case study on the Khorezm Region, Uzbekistan. PhD dissertation, Bonn University

Bobojonov I, Rudenko I, Lamers JPA (2008) Optimal crop allocation and consequent ecological benefits in large-scale (shirkat) farms in Uzbekistan's transition process. In: Wehrheim P, Schoeller-Schletter A, Martius C (Eds.): Continuity and change land and water use reforms in rural Uzbekistan – Socio-economic and legal analyses for the region Khorezm. IAMO, Germany, 63–85

Cai X, Rosegrant MW (2004) Irrigation technology choices under hydrologic uncertainty:

A case study from Maipo River Basin, Chile, Water Resour. Res. 40, W04103, doi:10.1029/2003WR002810

Conrad C, Dech SW, Hafeez M, Lamers JPA, Martuis C, Strunz G (2007) Mapping and assessing water use in a Central Asian irrigation system by utilizing MODIS remote sensing products. Irrig. Drain. Syst. 21: 197–218

Djalalov, S (2005) The Competitive Advantages of Uzbek Agricultural Production on the International Market. Uzbekistan Economy: Statistical and Analytical Review for January-June, 95–100

Djanibekov N, Bobojonov I, Djanibekov U (2012) Prospects of agricultural water service fees in the irrigated drylands, downstream of the Amudarya. In: Martius C, Rudenko I, Lamers JPA, Vlek PLG (Eds.), Cotton, Water, Salts and Soums – Economic and Ecological Restructuring in Khorezm, Uzbekistan. Springer, Dordrecht, pp 389–411

Djanibekov N (2008) Economic Analysis of Introduction of Water Prices for Agricultural Producers in Khorezm, Uzbekistan. In: Qi J, Kyle E (Eds.): Environmental Problems of Central Asia and their Economic, Social and Security Impacts. NATO Science for Peace and Security Series. N-Y: Springer-Verlag, 217–241

Djanibekov N, Rudenko I, Lamers JPA, Bobojonov I (2010) Pros and Cons of Cotton Production in Uzbekistan. Case Study #7–9, In: Per Pinstrup-Andersen and Fuzhi Cheng (eds.), Food Policy for Developing Countries: Case Studies. 13 pp

Hazell PBR, Norton RD (1986) Mathematical Programming for Economic Analysis in Agriculture, MacMillan Co., New York

Hornidge A-K, Ul Hassan M, Mollinga PP (2011) Transdisciplinary Innovation Research in Uzbekistan – 1 year of "Following The Innovation", Development in Practice, 21 (6): 825–838

Horst MG, Shamutalov SS, Gonçalves JM, Pereira LS (2007) Assessing impacts of surge-flow irrigationon water saving and productivity of cotton.Agric Water Manag 87:115–127

Ibragimov N, Evet SR, Esanbekov Y, Kamilov B, Mirzaev L, Lamers JPA (2007) Water use efficiency of irrigated cotton in Uzbekistan under drip and furrow irrigation. Agricultural Water Management, 90: 112–120

Johansson RC, Tsur Y, Roe TL, Doukkali R, Dinar A (2002) Pricing irrigation water: a Review of theory and practice. Water Policy, 4(2): 173–199

Khorezm Department of Land and Water Resources of the Ministry of Agriculture and Water Resources (OblSel VodKhoz), (2011) Data on norms, limits and usage of irrigation water from 2000 to 2010.Urgench, Uzbekistan

Khorezm Region Department of Statistics (OblStat) (2011) Agricultural development indicators for Khorezm oblast, 2000–2010. Urgench, Uzbekistan

Mamatov SA (2009) Drip irrigation system. SANIIRI: Tashkent, Uzbekistan

Martius C, Froebrich J, Nuppenau EA (2009) Water Resource Management for Improving Environmental Security and Rural Livelihoods in the Irrigated Amu Darya Lowlands. In Facing Global Environmental Change: Environmental, Human, Energy, Food, Health and Water Security Concepts; Hexagon Series on Human and Environmental Security and Peace; Brauch HG, Oswald Spring U, Grin J, Mesjasz C, Kameri-Mbote P, ChadhaBehera N, Chourou B, Krummenacher H (Eds.); Springer-Verlag: Berlin, Germany, 4: 749–762

Micklin P (2000) Managing Water in Central Asia. London, Royal Institute of International Affairs

Müller M (2006) A General Equilibrium Approach to Modeling Water and Land Use Reforms in Uzbekistan, Inaugural-Dissertation zur Erlangung des Grades Doktor der Agrarwissenschaften der Hohen Landwirtschaftlichen Fakultät der Rheinischen Friedrich-Wilhelms-Universität Bonn

Nerozin SA (2005) Drip irrigation regime of agricultural crops. In: Proceedings of SANIIRI 80 years, 45–57. Tashkent, Uzbekistan

Pimental D, Houser J, Preiss E, White O, Fang H, Mesnick L, Barsky T, Tariche S, Schreck J, Alpert S (1997) Water resources: Agriculture, the environment, and society, BioScience 47(2): 97–106

Poluasheva G (2005) Dynamics of soil saline regime depending on irrigation technology in conditions of Khorezm oasis. In: Proceedings of the international scientific conference: science support as a factor for sustainable development of water management. Taraz, Kazakhstan

Ringler C, von Braun J, Rosegrant MW (2004) Water Policy Analysis for the Mekong River Basin. Water International, 29(1): 30–42

Rosegrant MW, Ringler C, McKinney DC, Cai X, Keller A, Donoso G (2000) Integrated economic-hydrologic water modeling at the basin scale: the Maipo river basin. Agr Econ 24(1), 33–46

Rudenko I, Lamers JPA, Grote U (2009) Can Uzbek farmers get more for their cotton?. European Journal of Development Research 21 (2): 283–296

Saleth RM, Dinar A (2004) The Institutional Economics of Water: A Cross-Country Analysis of Institutions and Performance, Edward Elgar, Cheltenham, UK. SIC-ICWC, 2011. CAREWIB – Central Asian Regional Water Information Base. Tashkent, Uzbekistan

Skaggs RK (2001) Predicting drip irrigation use and adoption in a desert region. Agricultural Water Management 51:125–142

Tischbein B, Awan UK, Abdullaev I, Bobojonov I, Conrad C, Jabborov H, Forkutsa I, Ibrakhimov M, Poluasheva G (2012) Water management in Khorezm: current situation and options for improvement (hydrological perspective). In: Martius C, Rudenko I, Lamers JPA, Vlek PLG (Eds.) Cotton, Water, Salts and Soums – Economic and Ecological Restructuring in Khorezm, Uzbekistan. Springer, Dodrecht, pp 69–92

UN WATER (2007) World Water Day 2007: Coping With Water Scarcity

UN (2010) Shrinking Aral Sea underscores need for urgent action on environment – Ban. News report on visit of UN Secretary-General Ban Ki-moon to the Aral Sea.Online at: http://www.un.org/apps/news/story.asp?NewsID=34276

Von Braun J, Swaminathan MS, Rosegrant MW (2003) Agriculture, Food Security, Nutrition and the Millennium Development Goals. New York: International Food Policy Research Institute (IFPRI)

Von Braun J, Hill RV, Pandya-Lorch R (2009) The poorest and hungry: a synthesis of analyses and actions. In von Braun J, Hill RV, Pandya Lorch R (Eds.) The Poorest and Hungry: Assessments, Analyses and Actions, Washington, D.C.: International Food Policy Research Institute, 1–61

World Economic Forum (2009) The Bubble Is Close to Bursting: A Forecast of the Main Economic and Geopolitical Water Issues Likely to Arise in the World during the Next

Two Decades. Draft for Discussion at the World Economic Forum Annual Meeting 2009. Available online at: http://www3.weforum.org/docs/WEF_ManagingFutureWater%20Needs_DiscussionDocument_2009.pdf (last accessed on 23.04.2012)

World Resources Institute (WRI) in collaboration with United Nations Development Programme, United Nations Environment Programme, and World Bank (2005) World resources 2005: the wealth of the poor-managing ecosystems to fight poverty. World Resources Institute, Washington

Appendix. Data used in the modeling analyses

	Cotton	Cereals	Rice	Maize	Potatoes	Vegetables	Melons	Fodder	Fruits	Grapes
Prices (USD ton^{-1})										
Conventional	224	141	452	175	145	82	137	86	129	191
Rice to maize			175							
Paddy to aerobic			452							
Hydrogel	224	141								
Manuring	224	141			145					
Laser leveling	224	141								
Drip irrigation	224				145	82	137		129	191
Surge flow	224									
Double furrow	224									
ADF	224									
Short furrow	224									
Yield (ton ha^{-1})										
Conventional	2.6	4.0	4.0	2.8	13.7	19.9	14.1	3.6	9.0	7.2
Rice to maize			2.8							
Paddy to aerobic			2.4							
Hydrogel	2.9	4.6								
Manuring	3.3	4.9			17.5					
Laser leveling	3.2	5.1								
Drip irrigation	3.1				22.6	32.8	23.2		11.7	9.4
Surge flow	2.3									
Double furrow	2.7									
ADF	2.3									
Short furrow	2.7									
Costs (USD ha^{-1})										
Conventional	625	347	736	444	718	689	572	241	698	685
Rice to maize			444							

Appendix *(Continued)*

Paddy to aerobic										
Hydrogel	725	447								
Manuring	801	523	736							
Laser leveling	771	493								
Drip irrigation	720			906	866	771			815	772
Surge flow	635									
Double furrow	631									
ADF	625									
Short furrow	638									
Water use ($m^3\ ha^{-1}$)										
Conventional	5800	4800	28500	5200	9600	9600	4500	8500	5100	5100
Rice to maize			5200							
Paddy to aerobic			17100							
Hydrogel	4060	3360								
Manuring	4350	3600			7200					
Laser leveling	4205	3480								
Drip irrigation	4060			4560	4560	4560	2138		2423	2423
Surge flow	4640									
Double furrow	4930									
ADF	4060									
Short furrow	5365									
Total land use (1000 ha)	110	49	21	1	3	8	4	26	7	2
Laser-leveling investment costs ($USD\ ha^{-1}$)	898	898	0	0	0	0	0	0	0	0
Drip irrigation investment costs ($USD\ ha^{-1}$)	1200	0	0	0	1200	1200	1200	0	800	667

Anastasiya Shtaltovna, Anna-Katharina Hornidge, Peter P. Mollinga

5.2 Caught in a Web – Travails of a Machine Tractor Park in Khorezm, Uzbekistan

Abstract

As part of the ongoing agricultural transformation in Uzbekistan, agricultural service organizations are in the process of change. This paper aims to analyze this process by offering empirical insight into a Machine Tractor Park (MTP) in the Khorezm region, western Uzbekistan. While finding its new way as a profit-making organization, the MTP is forced to redefine its role, tasks and relationships with the state, input providers and customers. Consequently, de jure it serves state-planned cotton production by rendering and repairing machinery and producing spare parts on a for-profit basis. Yet, de facto, it continues to serve old roles, which are still strongly embedded in society, by fulfilling a much wider range of tasks. These include various tasks requested by the state administration, such as providing machinery to farmers growing state-ordered crops, participating in numerous meetings dealing with the organization of state-ordered agricultural cotton and wheat campaigns, and organizational functions originating from Soviet times, such as acting as a social security net for its personnel and other societal actors. For the provision of these services, mandated by its former role, the MTP is not – or only to a limited degree – paid, which results in financial imbalance leading to the gradual demise of the organization. As such, the MTP exemplifies continued governmental control in agriculture 20 years after the breakup of the Soviet Union.

Keywords: agricultural transformation, agricultural service organizations, transition, Central Asia, Uzbekistan

5.2.1 Introduction

To ensure food security, rural employment as well as profits from cotton production, the government maintains strong control over agricultural production in Uzbekistan (Kandiyoti 2003; Spoor 2004; Khan 2005; Wehrheim 2008). Cotton – as in Soviet times – remains the 'white gold' of the country. In the Soviet Union, Uzbekistan produced two-thirds of all cotton (Rumer 1989). After the independence, cotton remains one of the foreign exchange earners of the national economy, reflecting that the contribution of agriculture to GDP reduced from 37 % in 1991 to 18 % in 2009 (Luong 2002; Spoor 2006; Veldwisch 2008; Spoor 2010; World Bank 2011).

In this paper we look at agrarian transformation from the perspective of agricultural service organizations (AGSOs). As part of the ongoing process of agricultural transformation, the AGSOs responsible for the provision of agricultural inputs, sales organizations and financial and insurance services underwent numerous changes (Niyazmetov 2012). They were established to serve state collective farms during Soviet times, and were centrally managed. Due to the agricultural reforms, and in particular the creation of individual farms from 2004 onwards, AGSOs have moved from being centrally managed service providers for a few state farms to providing services to a much larger contingent of individual farmers.

This paper sheds light on the changing roles of AGSOs during agricultural transformation in Central Asia. A Machine Tractor Park (MTP) was selected as a case study in the irrigated lowlands of the Khorezm province, western Uzbekistan. The study aimed to assess the factors affecting the changing role and ongoing functioning of the MTP by focusing on its internal and external organizational relationships, which determine the space within which it is forced to reinvent itself. Thus, we analyze (a) the MTP's internal relationships, (b) the MTP's relationships with service recipients (farmers), (c) the MTP's relationships with the state (including organizations supported by the state), and (d) the MTP's relationships with other organizations.

5.2.2 Methodology

Khorezm province, Uzbekistan, located in the irrigated lowlands of the Amu Darya River, serves as a case study for the transitional environment of Central Asia and for assessing the evolution of AGSOs. There are 1.3 million inhabitants in Khorezm with about 70 % living in rural areas (Wehrheim 2008). Agriculture is the main sector of employment. Cotton, winter wheat, rice, fodder maize, fruits and vegetables are the main agricultural crops. The main services available to agricultural producers in Khorezm are presented in Table 5.2.1. They are provided at the district level by water user organizations (WUAs), alternative machine tractor parks and machine tractor parks ((A)MTPs), fuel suppliers, fertilizer company, banks, veterinary stations, and bio-labs.

Table 5.2.1: Number of district branches of the agricultural service providers in Khorezm province

District	(A) MTP	WUA	Fuel suppliers	Fertilizer companies	Banks	Veterinary services	Bio-labs
Bagat	12	11	9	6	14	15	12
Gurlen	12	11	12	7	10	16	13
Kushkupir	15	17	14	14	11	19	13
Urgench	13	13	14	12	18	17	9
Khasarasp	13	12	10	8	11	21	11
Khonki	11	11	12	10	12	14	10
Khiva	11	10	10	8	11	16	8
Shavat	14	12	10	8	15	19	14
Yangiyarik	9	9	9	9	8	13	8
Yangibazar	6	6	9	10	9	14	10
Total	116	112	109	92	119	164	108

Source: Regional representative of Ministry of Agriculture and Water Resources of Uzbekistan (2010), based on data from 01.07.2009.

Taking a case study approach, a MTP in Urgench district, Khorezm, was studied in detail by the first author during an internship. She spent two months in the organization, from mid-August to mid-October, 2009. During this internship, mainly qualitative data were collected by means of daily participant observation of various processes in the organization and 35 semi-structured interviews with the staff. Furthermore, the author took part in 10 meetings of the MTP representative with state representatives and in 5 trips to farmers' fields. Additional quantitative and qualitative data on the service provision by AGSOs were collected in 2009 and 2010 through a farmer survey involving 50 farmers growing different types of crops in five districts of the province. As with the MTP, a private bio-laboratory and the regional fertilizer company were studied in depth. The three organizations were selected based on the results of the farmer survey and initial interviews conducted with different providers; 30 semi-structured interviews were conducted with decision makers and experts in agriculture on local, regional and national levels, with all agricultural service providers in Khorezm Province (Table 5.2.1) and with actors involved in extension services in other ex-Soviet countries. Interviews were performed in Russian and Uzbek, in most cases without a translator. For reasons of privacy, no names will be mentioned. A review of legal documents on agricultural service provision was carried out.

5.2.3 Agrarian change and rural transformation

Uzbekistan has gone through 60 years of Soviet collectivized and planned agriculture and 20 years of post-Soviet agriculture, which included three major reforms in which aspects of the Soviet and post-Soviet systems are intermixed. Soviet history thus continues to shape the rural social, political and economic landscapes. Soviet agriculture was organized in collective farms (*kolkhozes* and *sovkhozes*) that, for practical purposes, combined functions of a mega-farm and a local government (Humphrey 1998; Ioffe 2006; Trevisani 2007; Allina-Pisano 2008). Thus, the *kolkhoz* was a site of coordination for many sub-organizations and for policies and resources coming from outside (Shtaltovna, Van Assche et al. 2012).

Since 1991, Uzbekistan has experienced a chain of agricultural reforms beginning with the subdivision of the former *kolkhozes* and *sovkhozes* into joint stock companies (*shirkats*) between 1991 and 1998. Between 1998 and 2003, these were then 'privatized' and subdivided into small, individual farms (Veldwisch 2008; Lerman 2008; Trevisani 2008). This process of land de-collectivizing crucially modified inter-human relationships within the agricultural production system, as well as between the now diverse group of agricultural actors and the state. In November/December 2008 (within less than a month) farmland under the cotton and wheat state plan was re-consolidated again, merging several individual farm enterprises (10–25 ha each) into bigger farms (75–150 ha). The selection of farmers who continued to remain farmers or became landless depended on their performance with regard to the production of state-ordered crops in the previous years (Djanibekov, Lamers et al. 2010). Similar adjustments, but to a smaller degree, were made once more at the end of 2009.

Veldwisch (2008) distinguished farmers in Khorezm province into three categories according to the agricultural forms of production: (1) state-ordered production, which includes the production of cotton and wheat, (2) commercial production, which involves rice production, horticulture, poultry production and to a lesser extent the production of vegetables and fodder, and animal husbandry, and (3) household (subsistence) production farms). *Dehqons* grow fruits and vegetables in their gardens, and wheat and rice on their small plots of land. This kind of production primarily aims at home consumption and includes barter arrangements as well as petty trade at local markets.

Analysis of the reforms revealed that the Uzbek state is searching for an optimal kind of agriculture, which secures employment of the rural population and national food security as well as returns from cotton production (Kandiyoti 2003; Khan 2005; Trevisani 2008; Spoor 2010). Apart from keeping cotton and wheat production under control, the central government also sets low prices for the purchase of cotton and then sells it abroad at world market prices (Luong 2002).

The present system of governance, established during Soviet times, with highly centralized state power, strong vertical hierarchies and top-down rule, heavily relies on the use of state control[1], planning and intervention in many sectors of the economy, particularly in agriculture (World Bank 2011). Land remains the property of the state in Uzbekistan albeit under different tenure arrangements (Lerman 2008). The country has largely adopted the Soviet agricultural procurement system, where farmers have to fulfill the production goals assigned by the government for wheat and cotton. Yet, while in Soviet times the communist ideology supported the state's call to join in the cotton harvest, today this missing ideological embedding is compensated for only by the exercising of state control (Shtaltovna 2012). In doing so, the government relies on its regional and local state organizations, such as departments of local, district, and regional state administration, police, prosecution office, tax inspection and other state organizations, to ensure the fulfillment of the state goals. By maintaining control over the cotton sector, the Uzbek government has managed to sustain social and economic stability in the country (Ilkhamov 2000). The employees of the above-mentioned state organizations and the MTP used to work in the *kolkhozes* and *sovkhozes* during Soviet times, and thus have knowledge of agriculture and management during former days. In addition to their official functions, these state organizations consequently fulfill the unofficial function of exercising control in agriculture, which is, however, often more valid than their official functions. For the crops grown under the state procurement system, farmers and service providers alike are still dependent on government loans (3 % interest) granted to farmers to buy inputs (Ilkhamov 2000).

5.2.4 The machine tractor park – a case study

Through Soviet times until today, MTPs remain vital to the state-controlled agricultural system because their machinery is important for crop growing and their human capital for managerial roles. The history of the MTP dates back to

---

1 *State and government* are used interchangeably in this paper. We draw on definitions of 'state' and 'government', as defined by *Hyden, G., J. Court, et al. (2004). Making Sense of Governance. Empirical Evidence from 16 developing countires Lynne Rienner Publisher, Inc. State* refers to all institutions (government organizations) that comprise the public sector with responsibility for implementing policies. *Government* refers to elected or appointed officials serving in core institutions at the national, provincial, country, city, and local level. 'State' thus means any government organization acting in the name of the state strategic interest in cotton and wheat.

the 1930s when the Soviet agricultural system was initiated. At that time, the MTP was established and supplied *kolkhozes* with machinery. After the war, in the 1950s, when agricultural machinery became more common, the *kolkhozes* got their own parks and the earlier MTPs served as repair stations for machinery from all *kolkhozes* in the district. The 1960s and early 1970s were the boost period for the MTP, and became responsible for supplying all *kolkhozes* of Urgench district with machinery, spare parts, coal, fertilizers, and seeds. In addition, a land reclamation department was established as part of the MTP, which aimed at cultivation of virgin lands in the district. In the 1980s when these lands were taken under cultivation, this department was closed down and the MTP function of distributing most agricultural inputs was handed over to other organizations, leaving the MTP with the task of providing agricultural machinery, i.e. ploughs, sowing machines, and their spare parts. Particularly in the early years of independence (early and mid 1990s), the MTP focused on repairing agricultural machinery and the production of spare parts. In 1997, based on State Resolution #152, all MTPs were reorganized into joint stock companies (Cabinet of Ministers of Uzbekistan 1997). Additionally, 'alternative' MTPs (AMTPs) were founded in the late 1990s on the basis of former tractor parks that existed in every *kolkhoz* and *sovkhoz*. Both MTPs and AMTPs are regulated in accordance with the state law 'On Joint-stock Companies'. As MTPs are the main machinery service providers to farmers in Uzbekistan, the aim of these reforms when they were first introduced was to strengthen the material and technical agricultural base by increasing the technical service provision to agriculture. The supply of machinery and spare parts, previously received from the Regional MTP Union, stopped between 1991 and 1994. The few spare parts that were still received had to be paid for and thus were no longer supplied free of charge. The resulting poor financial situation of the MTPs has led to a steady decrease in their number in the Khorezm province (Regional Statistics Office 2010).

Nowadays, the MTP is a joint-stock company, with the government's share being less than 35 %. The main functions of the MTP are (a) rendering mechanical services, (b) repairing agricultural machinery, and (c) producing and supplying spare parts. The MTP's clients are the farmers, budget organizations, agricultural enterprises, *dehqon* husbandries, district AMTPs and other organizations. The MTP, a powerful organization in the Soviet period, has weakened under the transition period due to the high level of debt of the main clients, namely farmers and AMTPs, as well as its own debts to the tax inspection and input supply organizations. In contrast to Soviet times, when the MTP's debts were written off (for instance, dedicated to the 60th anniversary of the October revolution), today the MTP has to make a profit on its own and take independent decisions. Farmers growing state-ordered crops are the biggest group of MTP clients. Farmers often do not pay the MTP on time, as the allocated subsidized

Figure 5.2.1: The district Machine Tractor Park. Photo: Shtaltovna A.

loan is hardly enough to cover production costs, the (state owned) banks are slow in transferring it, or because farmers prefer not to pay and have reason to believe they can get away with it. This is further facilitated by the fact that many farmers are not used to bookkeeping. Moreover, the MTP is frequently pushed by the local state authorities to render services to these farmers without pay but at the same time is requested by tax inspectors to pay taxes. The resulting debts of the MTP were estimated in September 2009 as 253 million Uzbek Soums, of which 110 million were frozen[2] and 143 million had to be paid[3]. On April 24, 2010, the additionally incurred debt amounted to 19.5 million Soums. This is a vicious circle where, on the one hand, the MTP has to facilitate the cotton harvest by frequently providing services to farmers without being paid for them, and to "serve the state" in supervising farmers' fields and the cotton growing process, while being abused by many governmental bureaucrats (farmer survey, interviews, observations 2009 and 2010). Thus, the studied MTP, other MTPs and AMTPs cannot withstand the financial pressure and risk running into bankruptcy unless paid for their services.

Growing competition in providing machinery services in the Urgench district contributes to the financial problems of the MTPs. The machinery now available from a number of farmers and the regional fertilizer company is of better price and quality than the often worn-out machinery offered by the MTP. In addition

---

2 This sum is frozen by state tax inspection, meaning that it can be paid later and without additional fines or interest.
3 The official exchange rate on September 15, 2009 was 2,138 Soum for 1 Euro. Source: Central Bank of Uzbekistan (2010). www.cbu.uz.

to the above-mentioned problems, the MTP repair workshop is outdated, and a large part of the available tools and machinery (welding apparatus, a milling machine and a stand for fuel) originated from 1968. Furthermore, the production processes in the repair workshop are slow because of difficulties in obtaining the required inputs such as gas and metal. MTP managers and workers buy old or worn-out water and gas pipes from farmers. At the time of the internship, the supply of natural gas, which is centrally provided by the state, was terminated because of the MTP's debt. Hence, the MTP workers purchase the gas from neighboring organizations and transport it in combine-harvester tires. The prices of spare parts vary on daily basis (10 – 30 %), which again affects the price of services provided by the repair workshop.

The transition process and financial problems of the MTP triggered innovative decisions as a 'way out of the deadlock' (interview with the manager of the repair workshop, 2009). For instance, the MTP employees have designed and produced different caps for machinery from foreign manufacturers, and then adjusted their ploughs, cardans, clutches, etc., accordingly. Such modifications solved the problem of having to purchase expensive spare parts for the imported machinery. Moreover, the manufactured nozzles facilitated additional operations in the field, which in turn resulted in profits for the MTP. To cope with the current problems, the MTP director intends to strengthen two main business processes, diversify service provision, and pay off held back wages to the workers as well as debts on taxes and a bank loan by selling an administrative MTP building, subletting parts of the MTP's land, and selling old machinery (interviews with MTP officials, 2009). In this way, the MTP is mobilizing its long-term capital to cope on a short-term basis with the impacts of the ongoing processes of agrarian transformation and change.

5.2.5 Internal governance practices and relationships

In view of the above-described challenges, one of strengths of the MTP collective is the strong feeling of the employees of belonging to their organization. Most of them have been employed for more than 20 years, and the MTP's management and work experience has been passed on from the first to the following generations.

The former Soviet training is still reflected in a disciplined performance by the MTP staff, i.e., timely fulfillment of the state plan for cotton and wheat. The dedication of managers and workers along with the solidarity and discipline of the collective have made the MTP highly respected by the clients and collaborative organizations (Spoor 2010). Nevertheless, the continuing transition processes have negatively impacted the organization. A number of specialists left, and the remaining staff has had to combine multiple functions to keep the organization alive. Given the financial deadlock, workers are regularly asked to take unpaid 'vacations' in the winter period due to the lack of work; many employees have not received wages for 4–9 months. In order to gain additional income, the director allows the workers to take on extra jobs (earning not more than 2000 Soums per order).[4] The manager of the repair workshop legitimizes this practice by stating that "One cannot be strict with workers; otherwise they will run off" (interview, August 2009). Despite these attempts to 'outsource' the salaries of his staff, the MTP director had to release ten staff members in 2009. Similar effects were found by Ilkhamov in Ferghana Valley, Uzbekistan, where 57 % of the rural residents were not paid for an average of 14 months (2000).

Another major challenge for the present MTP management is the lack of experience and training in governing an organization in a market economy. Management nowadays differs a great deal from that in Soviet times where orders were issued in a top-down manner, i. e. from the republic-level MTP to the regional MTP union to the district MTP. These orders were fulfilled with little discussion. Today the MTP has to compile its own plan, taking into consideration its capabilities and indicators from the previous year (plan-actual comparison), and is responsible for generating its revenue independently from the state budgets. The MTP director performs ahead of other AMTPs managers and directors (interviews with the directors of the AMTPs, 2009 and 2010). He explains this by referring to his education and work during Soviet times, when the employees were encouraged to do things well: those who fulfilled or overfulfilled the plan were granted with certificates of appreciation, one's picture was put on the wall, one's name would be used as a good example to others, a trip to a resort or a financial remuneration was granted, etc. Nowadays, receiving fewer 'reprimands' also encourages better performance (interviews, August 2009).

Besides being responsible for all business processes in the entire organization, the director is the only decision maker in the MTP. This is in contrast to the past, when managers of different departments were responsible for their work.

The present director of the MTP has approximately 40 years of experience in agricultural management[5]. Due to this long-term experience and the MTP's formerly powerful role in distributing machinery, running the repair station and

---

4 In 2009, the average income of a worker in the MTP amounted to 120,000 per month, whilst a manager earned approximately 240,000 Uzbek Soums per month.
5 As a part of the state procurement system, the practice of assigning directors to lead AGSOs exists (Trevisani, 2008). During our research, such cases were noted in the AMTPs and other AGSOs. However, this was not the case in the Urgench MTP where the director was assigned by the main shareholder of the MTP. Therefore, in this respect this case is exceptional, and illustrates how difficult it is to reform the business during the process of agrarian change.

supplying spare parts to collective farms, he is a carrier of knowledge on farmers and farmlands. Even though the *kolkhoz* system of supply no longer exists, the *hokimiyat*[6] continues to use the MTPs' knowledge and localized expertise to serve the needs of the current agricultural system. For example, the MTP representatives together with other former *kolkhoz* co-workers is still involved in the state agricultural campaign and mobilized by state officials to fulfill state ordered tasks based on his former rather than present roles and knowledge. During such field visits, he also provides advice on agricultural practices and machinery to farmers (field diary, August 2009).

Securing financial stability of the MTP is a process of continuous negotiations for the MTP director. As pointed out by Roniger (2004), patron-client relationships are sensitive to local sentiments and may solve existential problems. Our study data illustrate that it continues to be common for managers in AGSOs to find patrons among local state authorities with whom they enter into a mutually beneficial clan-like relationship. This patron-client network pulls in other influential forces, such as heads of law enforcement agencies, *bazarkoms* (administrators of local bazaars), as well as representatives of the central government (Ilkhamov 2000) The official tries to establish contacts within the tax inspectorate and uses his best communicative qualities to convince the state officer not to withdraw money from the MTP's account.

> "The MTP official contacted Erkin (a representative from the tax inspectorate) and asked [begged] him to stop the collection of funds from the MTP's account[7]. So he asked whether it would be possible to take 10 million Soums only and leave 5 million for the MTP, as they also have to pay for gas, which otherwise will be disconnected – further impeding their ability to make a profit. He explained to Erkin that the workshop cannot work without gas, and assures that there will be money this month, so that the MTP will be able to pay out the remaining sum at the end of the month" (field diary, August-September 2009).

Oi (1985) showed that in developing countries, where formal channels for meaningful participation and interest articulation are weak, individuals regularly pursue their interests through the use of informal networks built upon personal ties. The same phenomenon can be found in the case study MTP. The director tries to establish friendly relationships with the 'bosses', which the MTP director usually refers to as district or regional *hokims*[8] or highly-ranked state representatives. While talking to these, he tries to show the advantages for both the 'boss' and his organization (as well as for him personally in the end). Thus, in order to run the MTP and maintain its capacities during the period of agrarian

---

6 *Hokimiyat* (in Uzbek) is the state administration organisation.
7 The money on the MTP's account will be transferred to the tax inspectorate to cover the debts.
8 A *hokim* (in Uzbek language) is the head of the district/regional state administration

change, the MTP director appeals to his sound Soviet managerial knowledge, his intelligence and communication skills as well as to a broad network of contacts, access to information and the ability to constantly improvise through 'learning by doing'.

By fulfilling the old roles in state-ordered agriculture, the director of the MTP secures his position and the potential benefits in patron-client-like relationships with the state authorities. The loyalty of the MTP director to the state regularly leads to increased freedom in developing business opportunities for the MTP (apart from servicing the state goals), and facilitates access to state support programs, for example, for purchasing new combines for the MTP at a subsidized rate. While both developing business opportunities for the MTP and accessing the state subsidized programs form the basic pillars of the MTP's work performance. Consequently, AGSOs and their managers find themselves "oscillating between the interests of their own enterprises, demands from the authorities, and demands from those who were excluded from land, *fermers* had complex relationships with their communities" (Trevisani 2011: 215).

5.2.6 The MTP and its service recipients

Farmers require machinery for both state-ordered and commercial crops. This dual purpose determines different types of relationships between the MTP and farmers. Growing state-ordered crops is associated with the Soviet, state-planned production system, hence market mechanisms are pushed out by long-in-use clientelist and informal arrangements largely practiced by state bodies (farmer survey, 2009). An exception to the contract between the MTP and a farmer is due to an unwritten rule that the MTP has to prioritize services to the state crop-producing farmers so that these farmers do not have to queue or pay for it. Next in line are farmers occupying 'privileged' positions. These include membership of boards of state organizations (e.g., schools, rural councils, public prosecutors' offices) or other AGSOs, close and well established contacts with representatives of the local government or history of leadership in *kolkhozes* (farmer survey, 2009; interviews with AGSOs, 2010). Although the current law in Uzbekistan does not allow occupying a position in the public organization as well as heading a farm at the same time, farmers might bypass the rule by registering the farm in the name of the wife or other relatives (Trevisani 2008). In most cases when the heads of households were employed as public officers, i.e. doctors, teachers, policemen, employees of state administration, they also controlled and led the farm (Trevisani 2008). Yet, as pointed out by Veldwisch (2006) with regard to the official salaries of a *hokim* or any other public position (teacher, nurse, doctor or lecturer), the salaries are low, therefore they require additional sources of income. In *kolkhoz* times, *kolkhoz* representatives formed links with the communist party hierarchy, and individuals made careers moving from *kolkhoz* to politics and service organizations and the other way around (Hann 2003; Verdery 2003; Ioffe 2006; Shtaltovna et al. 2012) Thus, they are the first to receive services from the MTP. Furthermore, farmers whose farmlands are located along the main road are similarly in the position to receive MTP services without queuing or paying, as their land will be the first to be observed during the state checks. Trevisani (2008) stated that farmers growing state-ordered crops counted as a privileged class. The order with regard to who is served first was also affirmed by a farmer in Gurlen district:

> "First, the MTP has to provide services to the farmers growing state crops. Before them, only the heads of other state organizations can get services (e. g., gas providers, *hokim*, director of the school, director of the water management organization, bank, state inspection). If the head of the MTP refuses to follow the above mentioned order, there is a high chance that he will meet obstacles while receiving services provided by those institutions" (farmer survey, Gurlen district, June-July 2009).

In order to avoid running into bankruptcy due to the little pay received for the delivered services, the MTP director tries different incentives to encourage farmers' payments. To this end, before going to a farmer's field, the tractor driver double-checks with the director if this farmer has any outstanding debts. By doing so, the MTP tries to avoid providing services without payment, or, if services cannot be held back due to the farmer's contacts, at least to remind the clients about the due payment. Yet the fact that, despite outstanding payments, the MTP still has to provide the services due to the farmer's position in the system of cotton production indicates that this informal mechanism of MTP functioning prevails over the formal procedure, i.e., providing services on a payment basis. Additionally, barter and informal arrangements are accepted due to the poor paying capacity on both sides. Farmers can pay in kind (e. g., grains, livestock or externally purchased spare parts for MTP machinery), and consequently either receive machinery service or are moved along in the queue. These in-kind payments are then passed on to the staff, but neither contribute to the reduction of the monetary debts of the MTP nor show up in the formal accounting system.

A different pattern of relationships emerged between the MTP and non-state crop-producing farmers, although financial relationships or the exchange of goods and services (due to the lack of cash) also come into play. Some local markets serve as an outlet possibility for *dehqons* to sell products from crops grown outside the state procurement system (Kandiyoti 2005). Ilkhamov (2000), further supported by Trevisani (2008), outlined that the free trading of rice on local bazaars in Khorezm was becoming increasingly important for the small producers (largely subsistence oriented) as well as for large farm holders (largely cash-crop oriented). The crops are marketed either outside the province (such as rice) or sold to the state organizations for processing (Wegerich 2006).

This increased market integration of agricultural small- and large-scale producers furthermore helps the MTP as a service provider. The MTP workers are eager to work during the night to secure the provision of machinery in particular for the cultivation of the commercially attractive crop rice (interviews, 2009 and 2010). These farmers always pay for any machinery service or input required for rice production in contrast to state-ordered cotton or wheat crops. These payments bring the highly needed cash income for the MTP. Consequently, the director strongly encourages his employees to provide the services reliably. The earned money is used to pay the MTP salaries as well as debts to the tax service and other state organizations (interview with the MTP officials, 2009 – 2010).

5.2.7 The MTP and the State

The AGSOs were part of the one-state agricultural mechanism, functioning as one and largely governed and administered from Moscow. Despite 20 years of Uzbek independence and its transformation into a legally independent entity, the MTP continues to serve state orders, which prevents the MTP from building an economically independent organization that works according to market mechanisms. Being strongly embedded in the informal system of reciprocal relationships, and due to the current state-planned crop production system, the MTP has established a multitude of relationships with the state. They consist of links between the MTP and the *hokimiyat*, the MTP regional union (a superior organization of the MTP), the state technical supervision, the tax inspection, the police and other state organizations.

These multiple relationships are not just business related. They are guided by differing rationales, i.e., help, control, supervision, scolding, abuse and friendship. In most transactions, the state (as the patron) gets its will fulfilled, whereas the MTP gets distracted, time- and money-wise, by having to interact with many state bodies. However, the MTP, as a client, gradually learns how to turn this unequal cooperation into an advantage for itself. The relationships between the local authorities and the AGSOs are very similar to the relationships between the farmers and district authorities described by Trevisani (2008), being both antagonistic and of mutual support.

The concept of patron-client relationships helps to illustrate the political behavior of low-status actors, particularly peasants, as they are incorporated, recruited, mobilized or inducted into the national political process (Powell

1970). The position of the MTP director and his functions in the current settings can be described by the functions of a *rais*[9], as "the *rais* emerged from Soviet modernization as the key representative of the local agricultural elites, and as an implementer of centrally directed policies, he influenced the way in which modernization shaped rural society" (Trevisani 2011: 66).

From all interactions between the state and the MTP, involvement of the MTP and other AGSOs in the process of controlling cotton and wheat (both state-ordered crops) production seems to be the most intensive. Each year, the MTP representative participates in approximately 200 meetings related to the preparation and actual harvesting of cotton and wheat (interview with the MTP officials, 2009). During the wheat and cotton harvests, the MTP official is mobilized by the *hokimiyat* to assist in controlling the harvests. He has to compile statistics of how much cotton was collected daily and delivered to the cotton factory. Apart from participating in the meetings, sometimes held at 6.00 a.m. and 11.00 p.m., he constantly has to be accessible by phone (daily observations, 2009). The work process of the MTP is negatively affected when the official is busy with state assignments.

The MTP cannot refuse to fulfill state orders as this may negatively influence the fulfillment of the future need for protection or favors to the MTP by the state. The *hokim* is responsible for ensuring agricultural production and fulfilling the plan on the district level, and in charge not only of managing the municipal economy and local infrastructure but also of employing all locally available means and resources to harvest and deliver cotton and grain (Ilkhamov 2000; Trevisani 2008). A MTP employee outlines the *hokim*'s position as follows:

> "The *hokim* is the landlord of the territory. Everything and everyone is subordinated to him. Medicine, markets, all sectors are subordinated to him. We need to adjust to all conditions dictated by the *hokim* in order to continue to exist as an organization" (field diary, September 2009).

In Khorezm, the firm authority of the *hokim* leaves little space for maneuvering. Wegerich (2004) refers to the *hokim* as a trouble shooter for a variety of problems. Furthermore, there is an attempt by the district authorities to strengthen their control on the agricultural production process as exerted in former *kolkhozes* and *shirkats* (Trevisani 2008). Whenever economic levers failed, coercive measures were applied to *kolkhozes* – pressure and persecution from the side of militia and prosecution authorized by the government to execute control over the sales of cotton and grain (Ilkhamov 2000). Amongst these enforcement methods, there are unpleasant instruments such as shouting, scolding, menacing and intimidation. These methods apply to the MTP, other AGSOs and

---

9 *Rais* (in the Uzbek language) is the head of the former collective farm, and now of the MTP.

farmers to ensure that all actors involved in the agricultural sector prioritize state orders above their other activities. As such, they demonstrate path dependency on clientelism formed in post-communist states (Trevisani 2008). This kind of meetings between the state representatives, AGSOs and farmers have little content purposes but are perceived by many participants as time disturbers only. As the previous citation indicates, on such occasions the *hokim* blames the directors of AGSOs who then blame their subordinates for their shortcomings, who in turn do not dare to counter the shouting even if asked to do the impossible. The audience (the directors of AGSOs and farmers) increasingly perceive these overwhelming control measures as crossing all limits, and do not remember instances of such a high degree of shouting during Soviet times, as indicated by the MTP employee.

> "The punishment during Soviet times was without offensive words, it was so strong and it had an effect. Now, it is different. At the beginning it touched us, now we do not care anymore. It doesn't have such an effect as during Soviet times" (field diary, September 2009).

Trevisani (2008) assessed that the pressure exerted by the district authorities, even if unspoken, has a strong capacity of intimidation. Active participation in meetings is not welcomed, passive listening is demanded, as illustrated by the following.

> "During the meetings it is all the same. Even if someone knows what to do, one has to sit quietly; it is not appreciated" (daily observations, September 2009).

The director of the MTP, having experienced Soviet times and now working intensively with the state, has decided to obey the state. Thus, he has redefined his position as well as that of the MTP according to the requirements of the current system. He is the first to carry out orders, and he makes sure that his subordinates do so, too. By trying to satisfy the state, he is used by the *hokim* as a reputable example to others. This kind of good relationship is useful for the director and the organization. For instance, when he has some private or business-related problems, he may appeal to the *hokim*.

In addition to these traditional patrons, who have been transformed into brokers, other local people with 'outside connections' also can assume brokerage functions, i.e., bourgeois landowners, schoolteachers, physicians, pharmacists, priests, tax collectors and other local officials (Powell 1970). These people are referred to as the 'small intellectuals' of society, whose status and role functions place them in the 'strategic middle' of the social structure (Paulson 1967). In the Uzbek context, organizations such as the tax office, public prosecutor's office, the police and state technical supervision departments play the role of 'state brokers'. Apart from their direct responsibilities, they have to make sure that

AGSOs and farmers follow state orders. Markowitz (2008) pointed to the '*prokuratura*' (public prosecution) as informally one of the most powerful offices within Uzbekistan's state apparatus.

Clientelism involves complex (often pyramidal) networks of patron brokerage selectively reaching different strata, sectors and groups, and selectively pervading political parties, factions and administrations (Roniger 2004). This translates into the constant pestering of the MTP by the state organizations. Occupying state positions and partially fulfilling state control function by so-called 'small intellectuals' has resulted in establishing non-business-related relationships with the MTP, which equates to an abusive use of the MTP's services. Officials exercise their power over the MTP to pursue private goals. The strong position of the bureaucracy is exemplified by the MTP manager as follows:

> "If we say 'no' to the *hokim* or someone from above, they will take revenge on us by means of tax inspection or by public prosecution, easily!" (Field diary, September 2009).

The MTP offers its services in order to avoid conflict. Due to the poor economic conditions of the organization and the excessive misuse of the MTP's services by state officials, the MTP is not enthusiastic about maintaining its machines in good shape or purchasing new machinery, which negatively impacts on the quality of provided services to other clients and results in poor MTP performance. Therefore, the MTP takes preventive measures such as selling old machinery rather than letting it serve the state bureaucrats for free. The MTP manager responds to the question about why he does not buy a defoliation machine for renting this to farmers:

> "It is not profitable. When people in the state organizations find out that the MTP has this kind of technique, they will immediately start asking for it themselves, for their relatives, for other officials and they won't pay for it!" (Field diary, September 2009).

The informal relations prevent investments in business opportunities that fit the formal for-profit designation of the MTP. The MTP is strongly embedded in the contemporary version of the rules as maintained and adapted since Soviet times. During Soviet times, when everything belonged to the state and everyone served the state, the MTP was expected to fulfill any request demanded by or in the name of the state without being paid for it. This is similar today, where old rules are still in use for whoever comes to the MTP in the name of the state, using it as a password to obtain any service for free. For instance, two men from the state administration came to the MTP and asked for a tractor to plough 0.5 ha. The official fulfilled the order. Even though the MTP manager always finds a solution to any problem, he tries to avoid additional misuse of his services. Yet, at the same time, reinventing the MTP remains a difficult task – caught between the structures, institutions and actors of the past and the so far unknown future. Trevisani (2008) described the change of institutions through the reform processes arguing that as the people remain the same, the managing principles also have not changed a lot since Soviet times:

> "Once there was the *shirkat rais* (before he was the *kolkhoz* manager), and his staff (the agronomist, the engineer, the land measurer, the deputy *rais* and others), after the end of the *shirkat* they were affiliated to the MTP and continue to exercise the tasks which they used to have in the past during the *shirkats*".

5.2.8 The MTP and public service providers

In contrast to the Soviet past when all inputs were supplied by the state and the mutual settlement of accounts (if such occurred) was easily solved with the help of the state, now the MTP has to achieve financial independence. The legal status of some input providers has changed, formally reducing state interventions into everyday business. Informally, nevertheless, and as seen above, these state interventions continue to be part of everyday work. Just like the MTP, many of the state input-providing organizations are in a poor economic state and face financial problems due to the transfer from the state supply system to the private management of their businesses (interviews with the MTP officials, daily observations, 2009). For instance, the state electricity company and a repair factory are in a bad economic condition, as the organizations they provide services to owe them money. Due to the status of the MTP and input-providing organizations, the MTP director mobilizes his personal agency, kinship and other networks to negotiate the extension of arrears terms. For example:

> "The representatives of the repair factory, while driving around and checking the factory's debtors in Urgench, arrived at the MTP and talked to the director about paying back the debt. The director asked them to wait for another 30–40 days when the MTP would have sold old machinery and would be able to pay. The MTP director: 'We also have a huge debt to the tax inspection (40 million). First we will sell machinery, after this the (administrative) building, and then we can pay out the bills" (Daily observation, September 2009).

The MTP director appeals to non-financial relationships to settle the financial problems of the process of redefining its MTP. For instance, he suggested to a Water Supply Company that the MTP repairs its equipment and writes off the outstanding water bill. Both sides are interested in this cooperation (field diary, September 2009).

Informal arrangements become less possible with the emergence of market principles. Representatives of Western machinery providers, i.e., New Holland

and Claas, vendors on the open-air markets or merchants from Turkmenistan (who drop by to the MTP) only accept financial relationships. While price negotiation and bargaining is possible, merely communication like in the case of state organizations (electricity, repair factory) is not feasible. Thus, relationships with organizations or individuals that are not state agents require financial emollients, otherwise they do not cooperate.

When working with banks, the MTP director uses all kinds of negotiations to reach his goals. Most of the banks that provide services to agricultural actors (e.g. AGSOs and farmers) are under state control. Establishing relationships with bank employees by giving them presents improves the situation and quickens transactions.

In cash-short situations, the director attempts to instigate informal arrangements or use his negotiation skills or some of his former links in order to find a solution for his organization and workers. For example, when there was no gas for more than one month in the repair workshop due to debts, the director agreed with a neighboring organization to obtain gas from them for a symbolic price, as outlined by the manager of the repair workshop:

> "One worker went to get some gas from the neighboring organization. He brought it in huge tires" (field diary, September 2009).

5.2.9 Concluding discussion

The studied MTP in this paper was presented as a case study of AGSOs in Uzbekistan during the ongoing processes of agrarian transformation. We focused in particular on assessing how the MTP redefines its position during the time of change with regard to its relationships with the state, its clients and input providers, as well as to its own staff. The agrarian transformation has brought many changes to the daily life of the organization and its workers. The broader structural working conditions of the MTP qualitatively illustrate post-socialism, the present political regime of Uzbekistan with strong state control over agriculture, and the MTP in particular. Consequently, the relationship between the MTP and the state crucially shapes the internal and external working environment and forms of governance of the MTP, and thus co-shapes the MTP's working relationships with its clients, its service providers and its staff.

We discuss three structuring characteristics of the MTP's internal and external relationships as well as its maneuvering through the times of change, i.e. the role of patronage and clientelism, the interplay of formal and informal spheres of governance, and the degree of personal agency in sustaining the organization. At the level of internal work organization, the data furthermore illustrates the difficult financial situation of the MTP, the cessation of state supply, the stalled working processes, the temporary employment contracts, further exacerbated by outstanding salary payments, and the constant pestering of the MTP manager by the state cotton-grain campaign exercised by state bureaucrats.

Despite the MTP's legal independence from the state and from organization according to market principles, the MTP official finds himself and his organization being continuously entangled in a larger web of patron-client relationships, which continue to influence the MTP's relationship with the state. Consequently, and as shown above, in Roniger's understanding (2004) the MTP director regularly takes on the role of a broker and adopts principles of clientelism as strategic tools for assuring the survival of his organization during these times of change. Yet, as further confirmed by (Markowitz 2008), the state's continued adoption of patronage principles to rule Uzbekistan's regions has promoted different patterns of rent seeking within the country, which in some localities further entrench the authority of local elites while in other localities they facilitate predation by state officials. Under these parameters, clientelism proves to be highly adaptive to changing market logics, individualistic strategies and capitalistic considerations, while at the same time it can be tuned to the agenda of politicians, brokers and citizens willing to make claims on grounds other than their only partially realized citizenship. As local patterns of rent seeking multiply and become embedded within the state's territorial infrastructure, their diversity may pose a significant challenge to the central state's command over resources in the future.

The director of the studied MTP accordingly finds himself in a situation pulled between pleasing the state and bureaucrats by continuing to fulfill tasks of the former MTP head and investing the time and resources for developing his organization into a fully privatized, market-oriented and market-based organization. Consequently, he appeals to a multitude of established relationships with state bodies, which indicates the importance of these relationships and the MTP's reliance upon them to counterbalance shortages during transition. These relationships can be largely characterized as either formal or informal in nature. While formal relationships exist on paper and are used by the state while promoting state interests, informal arrangements seem to be much more functional during the transition process, as they come to the fore when formal rules do not provide grounds for transactions. State organizations rely heavily on these informal rules due to the lack of finances, as well as due to the fact that the state continuously demands services according to the former, and by now informalized, roles of the MTP director. The MTP and similar organizations appeal to the barter system, the exchange of services and favors, and in-kind payments.

Longstanding, good relationships between the relevant actors assist in such transactions.

Despite the fact that the MTP is an agricultural service provider meant to serve individual farmers, the main client remains the state. The data indicate how the MTP continues to first and foremost fulfill state tasks and orders, the orders of bureaucrats and of farmers under state plan. Service provision to other clients is pushed to the second place and regularly has to suffice with lesser quality partly determined by MTP's poor financial situation. Additionally, the MTP increasingly faces competition in the machinery-providing sector and thus risks running into bankruptcy. Therefore, the main paradox of existence of MTPs and other state-ruled service providers is the continuation of control in agriculture. The presented case reveals that the organization, rather than focusing on its business, is busy with tasks related to the state procurement system. Due to their continued wide employment, these informalized institutions hamper change along the lines of formal restructuring and transformation. Evidence for this we could see in the case study of the MTP and its relationships with farmers, governmental organizations and input suppliers. We discussed the wide and often state-requested employment of former, but today no longer valid, roles and institutions. This suggests a discrepancy between the formal and informal, with the informal largely representing what was formal in the past and is still demanded by the state.

The above-mentioned spectrum of informal relationships with other social actors signifies the reason for the existence of seemingly unprofitable organizations like the MTP, which, as a former state organization, provides a platform for many people connected with agriculture.

As to the future of MTPs and AMTPs in Uzbekistan, our research suggests that the majority is not likely to survive the ongoing changes. Even today, many are on the brink of bankruptcy and merely seem to be kept alive to serve the state procurement system (interviews with AGSOs, 2009 and 2010). However, as the case of the MTP shows, there are organizations that try to redefine and reinvent themselves in face of the challenges. The MTP does so by diversifying its activities, trying to provide better quality services, and being more flexible in the provision of services, including payments and times of delivery. It is difficult for AGSOs to reinvent themselves, as, despite numerous formal changes to their legal set up, they are informally tied to the state procurement system and hence under the state control. Therefore, a little 'help from a friend' is needed for business development (Wegerich 2004), meaning continued clientelist practices and strengthening informal patronage tendencies.

According to Klyamkin (2002), the significance of informal arrangements declines with the emergence of market principles, although only fragile motivation in combating illegality and corruption might come from businessmen, especially in small businesses. At the moment, informal relations compensate institutional shortcomings of the Soviet past as well as the state-ordered system of crop production. Consequently, the space for development according to market principles remains limited. Thus, the development of private business will potentially contribute to the weakening of existing informal arrangements, which is highlighted when viewing the relationships between the MTP and non-state organizations, and examples of the trading of rice and of spare parts for agricultural machinery, which all work on a financial basis. With respect to relationships between the MTP and actors related to state-ordered crop production, given the dramatic split between formal and informal institutions, the linkages enabling survival will in many cases remain informal in nature, with survival depending on political and economic support, both usually materializing via informal channels.

References

Allina-Pisano J (2008) The post-soviet potemkin village: politics and property rights in the black earth, Cambridge University Press
Cabinet of Ministers of Uzbekistan (1997) On Measures of Strengthening the Material and Technical Basis and Increasing Efficiency of Machine-Tractor-Parks. Tashkent
Central Bank of Uzbekistan (2010) www.cbu.uz
Djanibekov N, Lamers JPA, Bobojonov I (2010) Land consolidation for increasing cotton production in Uzbekistan: Also adequate for triggering rural development. Challenges of education and innovation, for agricultural development: Proceedings of the Fourth Green Week Scientific Conference on the Agricultural and Food Sector in Central and Eastern Europe. Halle (Saale)
Hann C (2003) The postsocialist agrarian question. Property relations and the rural condition. Munster: LIT Verlag
Humphrey C (1998) "Marx went away. But Karl stayed behind. Ann Arbor, University of Michigan Press
Hyden GJ, Court, Mease K (2004) "Making Sense of Governance: Empirical Evidence from 16 Developing Countries. Lynne Rienner, Boulder, London
Ilkhamov A (2000) Divided Economy: *Kolkhoz* System vs. Peasant Subsistence Economy in Uzbekistan. Central Asia Monitor 4: 5–14
Ioffe G, Nefedova I, Zaslavsky J (2006) "The end of peasantry? The disintegration of rural Russia. Pittsburgh, University of Pittsburgh Press
Kandiyoti D (2003) The Cry of Land: Agrarian Reform, Gender and Land Rights in Uzbekistan. Journal of Agrarian Change Vol. 3 (Nos. 1 and 2): pp. 225–256
Kandiyoti D (2005) The Cotton Sector in Central Asia: Economic Policy and Development Challenges. SOAS University of London: pp. 1–238
Khan AR (2005) Land system, agriculture and poverty in Uzbekistan. Land, poverty and

livelihoods in an era of globalization: perspectives from developing and transition countries 1: 221

Klyamkin I (2002) Burokratiya i Bizness. Paper presented at the Centre d'études et de recherches internationales, Sciences-Po, Paris, 15 November 2002

Lerman Z (2008) Agricultural Development in Uzbekistan: The Effect of Ongoing Reforms. The Hebrew University of Jerusalem #7.08: pp. 1–29

Luong PJ (2002) Institutional Change and Political Continuity in Post-Soviet Central Asia: Power, Perceptions, and Pacts, Cambridge University Press

Markowitz L (2008) Local elites, prokurators and extraction in rural Uzbekistan. Central Asian Survey 27(1): 1–14

Niyazmetov D (2008) Efficiency of Market Infrastructure Development for Farmers of the Khorezm region. Urgench State University, Uzbekistan Master thesis: pp. 1–125

Niyazmetov D, Rudenko I, Lamers JPA (2012) Mapping and analyzing service provision for supporting agricultural production in Khorezm, Uzbekistan. Cotton, water, salts and soums – economic and ecological restructuring in Khorezm, Uzbekistan. Martius C, Lamers JPA, Rudenko I, Vlek PLG, Springer Dordrecht pp. 113–126

Oi JC (1985) Communism and Clientelism: Rural Politics in China. World Politics Vol. 37 (No. 2): pp. 238–266

Paulson B (1967) The Role of the Small Intellectual as an Agent of Political Change: Brazil, Italy, and Wisconsin. Paper delivered at the American Political Science annual meeting in Chicago

Powell JD (1970) Peasant Society and Clientelist Politics. The American Political Science Review Vol. 64, No 2 (June, 1970): pp. 411–425

Regional Statistics Office (2010) Information on Machine-Tractor-Parks in the Khorezm region. Statistics Department. Urgench

Roniger L (2004) Political Clientelism, Democracy and Market Economy. Comperative Politics Vol. 36, No 3 (April 2004): pp. 353–375

Rumer BZ (1989) Soviet Central Asia: "A Tragic Experiment", Unwin Hyman Boston

Shtaltovna A (2012) Servicing Transformation: Agricultural Service Organisations and Agrarian Change in Post-Soviet Uzbekistan. Department of Political & Cultural Change, ZEF. Bonn, University of Bonn. PhD: pp. 1–219

Shtaltovna A, Van Assche K, Hornidge A-K (2012) Where did this debt come from? Organizational change, interdependence and role ambiguity in rural Khorezm. Internationales Asienforum Vol. 43(Nr. 3–4): pp. 179–197

Spoor M (2004) Agricultural Restructuring and Trends in Rural Inequalities in Central Asia. A Socio-Statistical Survey. Programme Paper. Geneva, United Nations Research Institute for Social Development (UNRISD)

Spoor M (2006) Agriculture Reform Policies in Uzbekistan. In: Chandra Suresh Babu and Sandjar Djalalov (Eds.) Policy Reforms and Agriculture Development in Central Asia. Boston: Springer: pp. 181–203

Spoor M (2010) Agrarian Reform and Transition: What can we learn from 'the East'?

Trevisani T (2007) Communities in Transformation: Fermers, Dehqons, and the State in Khorezm. In: Patterns of Transformation In and Around Uzbekistan, DIABASIS: pp. 185–217

Trevisani T (2008) Land and Power in Khorezm. Farmers, Communities and the State in Uzbekistan's Decollectivization Process. Freie Universitaet zu Berlin

Trevisani T (2011) Land and Power in Khorezm. Farmers, Communities and the State in Uzbekistan's Decollectivisation. Berlin
Veldwisch G (2006) Organisation of Water Distribution. O'zbekiston qikhloq xo'jaligi Vol. 10
Veldwisch G (2008) Cotton, Rice, Water: The Transformation of Agrarian Relations, Irrigation Technology and Water Distribution in Khorezm, Uzbekistan. Bonn University, Centre for Development Research, ZEF, Bonn
Verdery K (2003) The vanishing hectare: property and value in postsocialist Transylvania, Cornell University Press
Wegerich K (2006) A little help from my friend? Analysis of network links on the meso level in Uzbekistan. Central Asian Survey March–June 2006: pp. 115–128
Wehrheim P, Schöller-Schletter A, Martius C (2008) Continuity and Change: Land and Water Use Reforms in Rural Uzbekistan: Socio-Economic and Legal Analysis for the Region Khorezm. Studies on Agricultural and Food Sector in Central and Eastern Europe. IAMO: pp. 1–203.
World Bank (2011) World Bank Tajikistan. www.worldbank.com/tj

Anna-Katharina Hornidge, Mehmood Ul-Hassan,
Laurens van Veldhuizen[1]

5.3 Follow the innovation: transdisciplinary innovation research in Khorezm, Uzbekistan

Abstract

Agricultural innovations produced by research projects without interaction with potential end users often fail to match the real-life complexities of local farmers. To overcome this lack of fit between scientifically generated innovations and local realities, the interdisciplinary research project 'Economic and Ecological Restructuring of Land and Water in the Region Khorezm, Uzbekistan' initiated a participatory approach to innovation development and diffusion together with local stakeholders. From early 2008 until early 2011, selected agricultural innovations, developed by the project and identified as 'plausible promises', were tested jointly by teams of local farmers, water managers and researchers under real-life settings. This chapter outlines the transdisciplinary innovation research experience, coined 'Follow the Innovation' by the team. It discusses the challenges faced in this process of joint experimentation and learning as well as the lessons learnt with regard to the innovations and the participatory innovation processes.

Keywords: transdisciplinary, participatory innovation research, institutional and technical innovations, Follow the Innovation, Follow the Technology, Uzbekistan, Central Asia

---

[1] The authors would like to thank the transdisciplinary FTI-teams (project members and local stakeholders) mentioned in this article for the continuous learning process achieved. We would like to thank particularly Iskandar Abdullaev, Farida Abdullaeva, Akmal Akramkhanov, Nodir Djanibekov, Oybek Egamberdiev, Bashorat Ismailova, Elena Kan, John P.A. Lamers, Ahmad M. Manschadi, Lisa Oberkircher and Inna Rudenko for their engagement in fostering the FTI process.

5.3.1 Introduction

The 'Follow the Innovation' (FTI) approach aimed at fostering participatory processes of testing, jointly with local stakeholders, institutional and technical innovations and adapting them to the local context of agricultural production in Khorezm, Uzbekistan. The innovations, or so-called 'plausible promises', had been scientifically assessed during the first two phases of a research project (Wehrheim et al. 2008; Martius et al. 2012), yet largely in isolation from the real-life situation of local stakeholders and potential end users. Additionally, this participatory process of innovation research was thought to empirically test and thus contribute to the development of a concept for innovation diffusion in Uzbekistan (Ul Hassan et al. 2011; Hornidge et al. 2011). In Uzbekistan particularly, farmers under the state plan continue to receive detailed instructions from the state on what and how to plant, when and how to irrigate and how to carry out agricultural operations to fulfil the plan. The farmers are thus not the sole decision-makers regarding land and water use. The bottom-up component of the FTI approach was thus complemented by an agricultural policy component in the project, which aimed to feed project innovations into the top-down channels of national decision-making and agricultural policy-shaping (Turaeva-Höhne/Hornidge 2012; Turaeva-Höhne 2012).

This chapter provides an overview over the processes of joint experimentation and learning between researchers and local stakeholders. The key features of the process design are presented in section 2. Section 3 elaborates the operationalization and implementation in practice. This is followed by a presentation of the outcomes, based on stakeholders' perspectives on the innovations tested. The paper ends with a discussion and conclusion.

5.3.2 The 'Follow the Innovation' Approach

Earlier research stresses the importance of innovations being deployed in a specific social, political and cultural context, to guarantee their local functioning and achieve a sufficient degree of acceptance amongst potential users for increasing the likelihood of future use (Bijker and Law 1997; Oudshoorn and Pinch 2003; Duncan and Barnett 2005; Rath and Barnett 2006; Hall 2007). Furthermore, previous research in the study region has shown that farmers in Khorezm, Uzbekistan actively experiment in order to improve their yields (Wall 2008). This experimenting takes place within the agricultural processes themselves as well as in the social and cultural realm in which agriculture is performed. The FTI component consequently focused on the integration of scientific knowledge on the one side and local and tacit knowledge and concerns of stakeholders on the other (Nonaka and Takeuchi 1995) to attune the developed innovations.

Table 5.3.1: FTI 'Steps'

Main phase	'Steps'
I. Initiation The project organises and prepares itself before starting to engage with other stakeholders	Choosing promising innovations Forming and building teams Team planning
II. Joint experimentation and learning Stakeholder engagement and mobilisation	Stakeholder analysis and initial selection Systematic stakeholder engagement towards agreement to collaborate
Planning, implementation and monitoring and evaluation of joint experimentation and learning activities	Participatory planning and design Implementing joint experimentation and learning M&E and impact assessment
III. Follow-up Sharing the results of FTI widely and strategically	Strategic documentation and communication of key findings on innovations *and the FTI process* Creating favourable conditions for continued use of the innovation *and FTI*

Source: Ul Hassan et al. 2011: 7.

The approach (Mollinga et al. 2006; Hornidge et al. 2009; Ul Hassan et al. 2011) therefore aimed at addressing the 'lack of fit' by integrating stakeholders' knowledge and the innovations at hand through a series of steps (indicated in Table 5.3.1) systematically bringing stakeholders into the innovation process. Optimally, each step builds on the results of the previous step and provides input for the design and implementation of the next step, eventually allowing the transdisciplinary teams to 'follow the innovation'. As such, experimentation may lead to innovation adaptations and the involvement of other stakeholders, which might feed into another round of experimentation, i.e. cycles of learning.

5.3.3 From Design to Implementation

5.3.3.1 The Design

In designing this transdisciplinary research component, the project borrowed and broadened the 'Follow the Technology (FTT)' framework of Douthwaite et al. (2001; Douthwaite 2002) to include not only technologies, but additionally institutional or social innovations. The 'Follow the Innovation' approach, as we termed it, was therefore elaborated in a series of steps (Hornidge et al. 2011; Ul Hassan et al. 2011), leaving conscious room for reiteration (the so-called 'validation loop') to ensure that the concerns of partners were included into the approach itself.

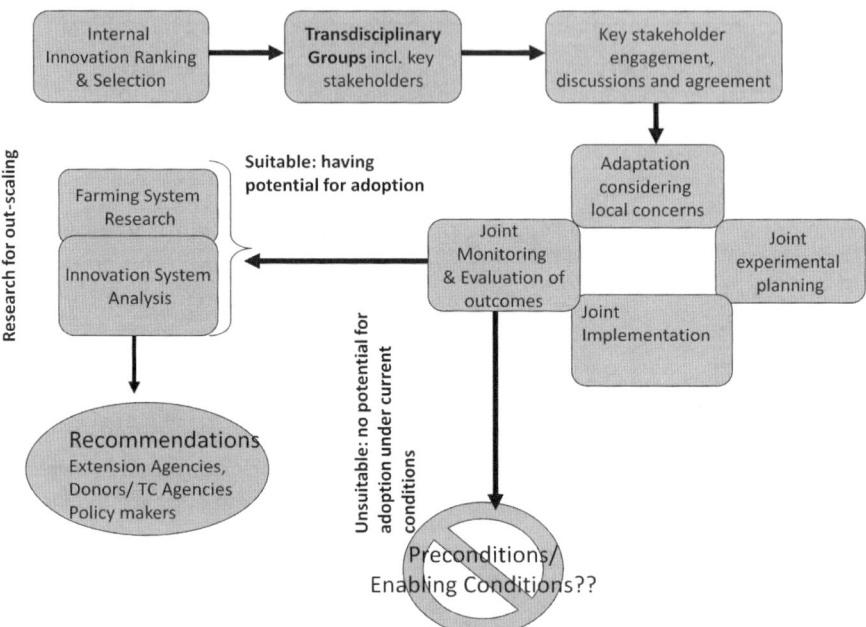

Figure 5.3.1: The FTI approach. Source: Ul Hassan 2008, Hornidge et al. 2011: 838.

Through participatory monitoring and evaluation (PME) of the implementation cycle it was to be verified whether or not the innovation in its revised form would continue to hold 'plausible promise', or would need to be further refined. Further research would be required to understand the entire innovation system and eventually form recommendations for the extension and technical assistance agencies for further out-scaling. In case the innovation failed to yield the anticipated benefits, enabling conditions to the successful implementation of the respective innovation could be identified.

5.3.3.2 Capacity-Development for Transdisciplinary Research

The FTI component was put into operation from early 2008 until early 2011 based on two pillars: (a) a series of externally facilitated trainings on participatory research methods, skills for stakeholder involvement, communication and facilitation, as well as participatory monitoring and evaluation tools, as displayed in Table 5.3.2; as well as (b) the step-wise and systematic joint experimentation with the local stakeholders around each of the four innovations. During the three years of transdisciplinary innovation research, the four teams were guided in their participatory interaction by a full-time FTI facilitator continuously in/at Urgench, Uzbekistan.

During FTI Workshop II, seventeen innovations were proposed for inclusion into the FTI process (Hornidge et al. 2009). Of these, the participants selected four based on an agreed set of criteria and formed interdisciplinary teams around these. When relevant stakeholders joined these they became transdisciplinary teams. The selected innovations comprised (a) strengthening Water Users Associations (WUAs)[2] through an adapted Social Mobilization and Institutional Development (SMID) approach; (b) conservation agriculture for irrigated low-lands (CA); (c) afforestation as an alternative land use for marginal farmlands (AF) (Khamzina et al. 2012); as well as (d) rapid salinity assessment using EM38, an electro magnetic induction device (SA) (cf. Akhramkhanov et al. 2014).

5.3.3.3 The selected Innovations

The focus of the WUA-FTI team was to test the transformation of existing weaker Water Users Associations (WUAs) through a social mobilization and institutional development approach (SMID) into well functioning organizations. In the implementation of the action research it concentrated on one selected WUA. SMID approaches have been successfully deployed to create bottom-up WUAs elsewhere in Central Asia as well as within Central Asia's Ferghana Valley. Since WUAs were established in Uzbekistan using top-down approaches, most WUAs in Uzbekistan remain paper organizations and the water users develop little ownership in the associations. Donor funded projects using SMID approaches within Uzbekistan were able to establish bottom-up WUAs that performed better compared to those which originated due to administrative actions in Uzbekistan. Hence, SMID appeared to offer the promise of improving WUA ownership by water users. The premise of the chosen approach was that an initial joint experimentation with the WUA staff and local elite to build their capacity in social mobilization would encourage continuous institutional development of the WUA under a 'laissez-faire' situation and the inclusion of a large share of water users and their concerns into the decision making processes of the WUA (Abdullaev et al. 2008), when the project team would gradually withdraw its

---

2 On December 25, 2009 renamed into water consumers association (Article 18 – 2).

Table 5.3.2: FTI Capacity-Development

Training title/ location	Timing	Focus	Participants
FTI Workshop I (Bonn)	February 2008 (4 days)	Concepts and approaches to innovation development and diffusion Multi-, inter- and transdisciplinary research hard- and soft-systems thinking Working in teams	20 staff from both Bonn and Urgench
Research discussions (Bonn, Urgench)	9 literature discussions 2008–09	Presentation and discussion of key conceptual papers on innovation, adaptation, adoption, policy development	Variable depending on availability
FTI Workshop II (Urgench)	May 2008 (4 days)	Stages and activities of the FTI approach Participatory research methods & tools Selecting innovations for FTI Formation of transdisciplinary teams around innovations	22 staff from both Bonn and Urgench
Communication and facilitation training (Urgench)	August 2008 (0.5 day)	Skills for effective communication and facilitation of teams	14 Urgench-based staff
Teambuilding (Urgench)	August 2008 (0.25 day)	Activity-based team-building exercises	21 Urgench-based staff
FTI Workshop III (Urgench)	November 2008 (4 days)	Review and reflect on initial FTI implementation, lessons learnt Additional participatory research methods and tools for use in FTI (participatory monitoring and evaluation) and skills in using them, also through field study Re-assess FTI team organisation and develop measures to improve/re-strategise	21 staff from both Bonn and Urgench including 3 stakeholder representatives
Interim Review- I (Urgench)	May 2009 (2 days)	Critically review the FTI progress and its constraints Plan further steps	22 staff, mostly Urgench-based, 3 stakeholder representatives

(Continued)

FTI Workshop IV (Urgench)	November 2009 (4 days)	Critically review FTI implementation Participatory impact assessment methods and tools; practice through field study Process documentation Review of FTI teams and their functioning Discussion of additional innovation areas for inclusion in FTI programme	15 staff from both Bonn and Urgench, 7 stakeholder representatives
Interim Review II (Urgench)	April 2010 (2 days)	Critical review of progress Plan further steps	11 Urgench-based staff
FTI Workshop 5: 'Writeshop' (Bonn)	January 2011 (4 days)	Present and review first draft of papers for each of the FTI processes Lead authors improve drafts based on comments to be ready by end of training Discussion of main content of FTI Guidelines	4 staff lead authors, one per group 3 FTI process facilitators

Source: Ul Hassan et al. 2011: 15.

intellectual inputs. The innovation process' limitation lay in the assumption that participation in decision-making processes leads to improved water management.

The CA-FTI team aimed at testing the basic tenets of CA – minimum tillage, retention of crop residue and appropriate crop rotations – with interested farmers to ascertain its fit within local farming realities and the extent to which it would maintain its promise of resource conservation with respect to water, cost of inputs and enhancement of soil organic matter (Egamberdiev 2008). The project's earlier research had indicated that it could be successfully applied in the irrigated areas of Uzbekistan and could potentially enable farmers to grow more food, feed and fiber crops in an environmentally sustainable way using less labor and fossil fuel and at a lower cost while at the same time gradually increasing fertility and water-holding capacity of the soil. The innovation process' limitation lay in the time required to build up organic matter (exceeding the given time frame) as well as the legal situation of farmers in Khorezm being bound by the state plan on cotton and wheat to state prescribed agricultural practices and timings.

The aim of the AF-FTI team was to test the project's afforestation strategy in the real-life setting of farmers and adjust it according to the farmers' reflections and varying concerns of farmers' (Lamers et al. 2008). After evaluating ten tree species on the marginal lands in Khorezm, the project identified that at least three species of trees, namely *Elaeagmus Angustifolia, Ulmus Pumila and Populus Euphratica*, could be grown profitably on marginal and agriculturally abandoned lands, offering fruit, feed, fodder and fuel benefits (Khamzina et al., 2006a; 2006b; 2008). These species start paying off in four to five years. A four year old plantation could yield up to 14 tons of oil equivalent per ha (poplar). These findings were considered significant for the Uzbek context as availability of fuel wood, and livestock fodder remain key challenges, especially for rural areas. On the other hand, thousands of hectares of lands become marginal or unfit for crop production annually in Uzbekistan. The innovation process' limitation lay in the given time frame, too short for yielding the benefits of grown trees and thus demonstrating their usefulness to land users to the fullest extent.

And finally, the aim of the SA-FTI team was to create awareness about the use of the electromagnetic induction meter (EM38) as an express salinity mapping tool amongst stakeholders, and analyze its comparative advantage (and disadvantage) compared to salinity mapping methods currently used in Uzbekistan (Akramkhanov et al. 2008). Considerable parts of irrigated lands in Uzbekistan are salt affected. To manage salts, the soils are annually leached using salinity estimations at a regional level resulting in excessive use of water for leaching locally. Traditionally, salinity levels are assessed through soil samples from saline lands. However, this approach is time consuming as the results of sample analysis can only be available after long periods of time. Thus the soil salinity maps produced always remain outdated due to the time lag between sampling, analysis and mapping. EM38, used in many parts of the world, can effectively measure soil salinity up to 1.5 meters of soil depth. Salinity in upper 1.5 meters of soil is the most relevant for both deep- and shallow-rooted crops. The project's earlier research demonstrated that this device could accurately map the spatial distribution of soil salinity in a much shorter time span. Additionally, salinity mapping through EM devices does not destroy the soil as samples do not need to be taken.

5.3.3.4 Identifying Stakeholders, Designing Joint Experiments

Three of the four teams – WUA, CA, and AF – identified farmers as their primary stakeholders, whereas the fourth team identified salinity mapping organizations as cooperation partners. Additionally, each of the four teams elaborated innovation specific stakeholder engagement strategies comprising a series of sequential steps (Ul Hassan and Hornidge 2010).

The WUA stakeholders (6 WUA representatives from one selected WUA) argued for a broader approach of WUA strengthening, namely a blend of "hard" and "soft" interventions, arguing that minor repairs of the WUA office were needed to give the WUA a 'face of recognition' while the provision of bicycles would facilitate the mobility of the water managers and thus a better and more timely interaction with water users. Additionally, a computer and printer were provided to facilitate the WUA's record keeping. Consequently, the WUA stakeholders came up with a 12-step WUA improvement plan (Abdullaev et al. 2008) covering further steps, timing, inputs, finances, and division of responsibilities. The ZEF project was responsible for: a) carrying out trainings with the WUA; b) a study tour to well functioning WUAs in the Ferghana Valley (East Uzbekistan); c) financing office refurbishment and bicycles; and d) collection of hydrological and other monitoring data. The WUA on the other hand was responsible for: a) identifying social mobilizers and undertaking social mobilisation efforts in the WUA; b) inventorising WUA assets; c) convincing farmers to build hydroposts; d) carrying out its routine operation and maintenance tasks; e) approaching higher level authorities for the provision of canal cleaning equipment; and f) sending monthly progress reports to the project.

Within the CA-team, the CA expert and team leader held several rounds of informal discussions with 3 farmers regarding joint experiments and the division of roles and responsibilities. This was then followed up through regular field visit phone calls and clarifying discussions. During these interactions, main attention was given to the technical aspects of the collaboration and less on the process of collaboration and (joint) experimentation. So while the objectives of the interaction were formulated as to: a) conduct joint experimentation; b) monitor and evaluate CA adoption; and c) develop further and disseminate CA on a regional scale, the roadmap to implement the process was written first by the scientists for later discussion with farmers only.

The AF-team from the onset closely collaborated with a forestry researcher from Tashkent, who – keen to learn about the process of implementation of scientifically based recommendations in farmers' real-life situations – became the team leader. Important determinant for the timing of cooperation was the availability of a sufficient number of saplings of the selected species in local nurseries. Between November 2008 and March 2009, the farmer selection process was finalized through field visits initially involving 100 farmers of various profiles during which farmers' interests were explored and characteristics of marginal cropping sites proposed by farmers for afforestation were examined. Three farmers expressed interest in practicing afforestation on their marginal lands. The major concern of all surveyed farmers was obtaining permission to grow trees on highly saline, unproductive cropping sites within the cropland area that was registered by the state as fertile. Initially the farmers agreed to approach the authorities themselves but the project's support was ultimately instrumental in soliciting a written permission from the local authorities to release these cropland parcels from the state crop production for afforestation.

The SA-team initially struggled with approaching the identified stakeholders by mid 2009. Yet after a respective review, the team undertook a series of systematic meetings with stakeholders where the project researchers presented the idea and the previous results from the use of EM38. Although several contacted stakeholders showed interest, only the Central Asian Irrigation Research Institute (Russian Acronym: SANIIRI) followed it up, resulting in a formalized collaboration.

5.3.3.5 Joint Innovation Testing and Adaptation

While the implementation of joint experiments had already begun, the formal agreements of collaboration were signed between January and July 2009 in English and Uzbek. Overall the stakeholders provided land, manpower and other on-farm/ institute facilities free of cost, while the project covered small investment costs, as well as training and monitoring costs, and counterbalanced the farmers' risks for example in the case of CA-FTI of weed infestation.

In January 2009, the cooperating WUA convened its first ever general assembly approving the 12-step WUA improvement plan as well as the following proposals from the WUA management: a) nominating hydro-technicians according to the hydrographic layout of infrastructure thus making them responsible for a group of farmers within their hydrological jurisdiction; b) approving the budget and the proposed water plan, c) appointing a conflict resolution committee and a water inspector; d) considering the problems in the recovery of electricity bills from farmers due to cash unavailability – farmers whose lands were irrigated by pumps were to be informally responsible for the operation and maintenance of pumps; e) providing labor by farmers to clean canals and drains; f) requesting the water management organizations (WMOs) to provide machinery for the canals budgeted by WMO but managed by the WUA; and g) rotating the water supply on a turn-by-turn basis between the two WUA canals in case of water scarcity. Since outstanding water payments continuously led to outstanding salary payments of WUA staff, the WUA additionally began to charge kitchen garden water users a flat cash rate per irrigation (UZS 2000 per user[3]). Following from here, the 'joint experimentation' took the form of using these agreed on strategies and jointly monitoring whether they indeed significantly improved the WUA's functioning.

---

3 According to the official conversion rate on September 2, 2010 this converts into Euro 0.97 (Source: http://coinmill.com/UZS_calculator.html).

In the case of the CA-team, three farmers agreed and allocated plots for testing CA practices, including land leveling using laser equipment. The trials were subdivided into two parts: an experimental plot and a conventional plot. While the farmers grew crops on the conventional plots using their usual agronomic practices, their experimental plots were cultivated according to the advice of the CA-expert from the project. Some elements of the experiments were adapted based on the farmers' suggestions. For example, the seed rate for winter wheat was increased in one case. In another case, herbicides were applied during the second crop growth to suppress the weed population. Furthermore the fertilizer doses in small amounts were adjusted in some cases by mutual agreement.

For the AF-team, the start of the field activities was somewhat delayed by the final selection of the three farmers and negotiations about species choice and planting methods. Although late planting (exacerbated by the early arrival of spring) was feared to cause poor tree survival, it was mutually decided to initiate the implementation process and planting was accomplished in three marginal, highly saline cropping sites on March 20 and March 26, 2009. One of the farmers insisted on including hybrid poplars, as a commercially attractive addition to FTI-team recommended species. Another adjustment of the original afforestation strategy was a partial use of cuttings instead of saplings as planting material and a single row, borderline planting instead of a dense stand at one of the sites. There was a concern about water availability for post-planting irrigation. Due to the specifics of water turns assigned by the WUA to the farmers, the irrigation was delayed by two weeks at one of the sites, which might have resulted in coupled salt and water stress for the plantings. By mobilizing social capital, one of the two affected farmers was able to get irrigation water on time, whilst the other farmer acquired the project support in approaching the responsible authorities (here the Water Users Association and the Hokimiyat). The plantings were followed up in regular field visits during which the trees' conditions and survival rates were observed and discussed with the farmers.

The SA-team, after signing the agreement with SANIIRI, assisted SANIIRI in testing the equipment at a research station of SANIIRI itself, from where SANIIRI had salinity assessment data through its soil sampling techniques. In the field, the FTI researcher explained the technical details of the device, measuring principle, depth of signal penetration depending on device orientation mode, and the method of calibration. Additionally a project's field assistant demonstrated the device calibration before taking measurements and assisted with further EM38 and GPS measurements where needed. The SANIIRI researcher was assisted by two assistants for soil sampling to compare the results. Altogether 20 locations were sampled and measured using the conventional methods as well as EM38. Soil samples were taken to SANIIRI for further analyses. By mid December 2009 the collected data were analyzed and a draft report shared with the project in mid January 2010. SANIIRI felt that the equipment needed to be tested in several locations, and proposed to undertake similar measurements at their own cost in the Syrdarya region, to which the project agreed. These measurements were underway at the time of writing.

All four teams – to varying degrees – struggled with high degrees of staff and expertise transitions that influenced their performance. Amongst the senior staff of the project, two senior economists, two senior water management specialists, and a tree specialist changed their jobs during the course of the initial two years. Some of these vacancies were re-filled, but the replacements had missed the earlier discussions and trainings, and thus lacked the interest and required exposure to the FTI process. Overall, and for various reasons, the number of project staff involved in or supporting FTI in 2008 sank drastically until the end of 2009 in all 4 cases (WUA from 10 to 5; CA from 14 to 4; SA from 9 to 3; and AF from 7 to 5).

5.3.3.6 Participatory Monitoring and Evaluation

Within the WUA-team the monitoring arrangements were two-fold: a) the lead social mobilizer was responsible for submitting a brief monthly progress report to the project, and b) project staff would collect data on water availability, distribution and use and provide these to the WUA. The WUA's own financial and water management records were also monitored. The progress reports by the social mobilization group were submitted on time, but merely reported success stories while neglecting the challenges the WUA continued to face. A WUA performance discussion meeting on July 24, 2009 with farmers and WUA staff and a water user perception survey were used as additional tools to assess the outcomes.

The CA-team largely regarded monitoring and evaluation as a scientific evaluation, whereby the CA researchers identified indicators and assessed those. Regular data collection activities in the field had been launched to monitor input and water use as well as crop growth. To supplement these scientific assessments of the innovation (not the transdisciplinary process), the research team arranged a field day at a project CA research site, where elements of CA research trials were ongoing. This was thought as an assessment of CA elements by visiting farmers.

The AF-team's monitoring design comprised monthly visits by one team member to the three afforestation sites to monitor the establishment and growth of trees and discuss with the farmers specifics of their experience and in-between agronomic practices. This team member also arranged field visits for the collaborating farmers to the projects' afforestation sites to demonstrate the expected benefits in terms of forestry products. Extension materials in the form of a booklet and leaflets were likewise provided to the participating farmers and passed on to other interested individuals. The team leader paid visits to these farmers once every year and occasionally provided advice over the phone to the regularly visiting staff member, who then conveyed this to the farmers.

The SA-team, as outlined earlier, decided to let SANIIRI collect and analyze data of the equipment testing itself. Cautionary explanations were provided that peer reviewed research existed explaining the relationships between technical parameters of soil and accuracy of the estimates. The SANIIRI scientists produced elaborate research reports double checking these.

5.3.4 'Plausible' or 'Implausible' Promises?

From early on in the FTI process, all four teams continuously monitored and regularly evaluated the interaction processes in order to allow for feedback mechanisms and reiterate learning within the teams. The WUA team for example agreed with the WUA representatives and as part of their responsibilities that they would submit a monthly progress report. As these turned out to be solely positive, the team then arranged a WUA performance discussion meeting as well as together with the WUA installed a complaint box for anonymous complaints in front of the WUA office. Additionally project members conducted a perception survey amongst famers in September 2009 and discussed the results in November 2009 with the WUA. Based on this, adjustments to the innovation as well as the innovation process were made. The CA-team monitored less the actual process of interaction but instead the adjustments to the innovation suggested by farmers and also implemented. This was largely done by one project team member, participating in meetings with the farmers or in retrospect interviewing the technical expert. The AF-team visited the farmers on a monthly basis, monitoring the establishment and growth of tress and discussing with the farmers their experiences and explanations for the tree development. These meetings were documented in the style of ethnographic field research notes, later serving as the data basis for a PhD-thesis. Similarly the SA-team documented each meeting with stakeholders in the form of field research notes, but also intensively discussed its experiences with the FTI-facilitator, adjusting its stakeholder selection, ways and styles of interacting and later documenting the stakeholders' reflection on the innovation.

Thus, the monitoring reports of the WUA indicated that the WUA performed well during the first year, especially regarding hardware interventions. The WUA office was refurbished, most farmers had installed hydroposts, the inventory of infrastructure was completed, the pumps were transferred to farmers, and the

Table 5.3.3: Adaptations of Selected Innovations

FTI-Team	Characteristics of Innovation	Adaptations based on Stakeholder Interaction
WUA	Focus on 'soft' measures to enhance ownership of WUA by members - Develop bottom-up WUA using social mobilization by local community mobilizers - Train WUA staff in WUA and irrigation system management - Raise interest of WUA staff through visits to successful WUAs	Ensure basic hardware in addition to 'soft' inputs: - Support for office renovation essential for WUA for indicating its physical presence - low tech means of transport (bicycles) essential to facilitate navigation along hydrological system
CA	- Apply precision levelling followed by minimum tillage, retention of crop residue and appropriate crop rotations for both cotton-wheat and rice wheat systems - Improving soil quality due to organic material	- Zero tillage caused compaction, adapt to minimum and reduced tillage - Crop residue had alternate uses, and thus lower than suggested amounts were used - Use increased seed rate for wheat to avoid initial yield dip - Restrictions on crop rotation due to state order - CA OK for wheat crop not good for second crop in both systems - Farmers scared to practice CA on cotton-wheat systems due to state pressure
AF	- To grow Elaegmus Angustifolia, Ulmus Pumila and Populus Euphratica in abandoned crop lands as contiguous plantations for profitable and environmentally suitable land use	- Abandoned crop lands need to be first declared by the state as unfit for arable crops - Preference as border trees but not as contiguous plantations - Preference for hybrid poplar - Why plant and nurture trees that grow as weeds? - To ensure irrigation water for initial post-plantation phase competes with water use for cotton; ensure water allocation for trees in crop lands
SA	- EM38-based salinity assessment - Fast, maps up to date when produced - Calibration work necessary precondition of usage - Data not as detailed as with conventional method	- Verification of calibration curves on various types of soil seen as essential by the stakeholder

Source: Authors' compilation based on Ul Hassan et al. 2010.

shared canals and drains had been cleaned. In preparation for droughts, a central lake in the WUA (Oberkircher et al. 2011) was filled with water whenever water was abundant. The year 2009 however was all along a water abundant year, in which the WUA received 275 % of its planned water supply between July and September.[4]

The WUA staff believed to have raised the water users' awareness about the WUA's responsibilities and tasks, while some farmers continued to ignore WUA's rules and took water when needed. The efforts for strengthening the WUA in the first 6 months were characterized by a high degree of enthusiasm. Nevertheless, the heavy work load of the WUA staff for installing pumps and solving electricity problems negatively impacted on the time invested into awareness raising measures amongst water users. Furthermore, fee collection from water users remained difficult. Until June 2009, only 15 % of the fees for kitchen garden owners and 15 – 20 % for commercial farmers under state-plan could be collected. The total outstanding debt of the farmers to the WUA was estimated as around 2.5 million Uzbek Soums.[5] Some farmers claimed that they had paid, but possessed no record of payments since part of their farm was previously cultivated by someone else.[6] The WUA identified the farmers' perception of the role of the WUA and of water being a free of charge resource as the main reason for the arising water management problems. According to the farmers nevertheless, it was the WUA's task to provide irrigation water in accordance with the crop demand and timing, and without the users' involvement, which is comparable to the "water master" during Soviet time (Veldwisch 2008). Overall, the perception survey indicated that while an increasing number of commercial (*fermers*) and subsistence (*dehqon*) farmers were aware of the WUA's existence, the aspired feeling of ownership and identification of the WUA as an organization of the water users had not been achieved.

The CA experiments indicated the following findings: a) any changes in the application rates of fertilizers, e.g. non-use, or varied amounts of nitrogen fertilizer, were highly questioned by farmers, irrespective of whether it might yield better results; b) the farmers in their own assessment of the experiments did not distinguish between plowed and unplowed winter wheat plots, whilst the tillage practice and retention of crop residues on cotton plots attracted wide attention; c) farmers expressed their interest in receiving a cost-benefit analysis

---

4 The filling of lakes thus counteracted as it led to a rising of groundwater to the root zone of crops. The farmers had planted an earlier maturing and drought resistant variety of cotton, which badly failed. Twenty of the twenty-one farmers could not meet their state plan and thus got into debt.
5 According to the official conversion rate on September 2, 2010 this converts into Euro 1213.85 (Source: http://coinmill.com/UZS_calculator.html).
6 In November-December 2008, a 'farm optimization' program by the central government merged cotton and wheat farms (10 – 25 ha) into farms of an average size of 80 – 150 ha. This change meant that many of the old farmers who had to pay off WUA-debts were no longer farmers, and thus the WUA had no means of cost recovery. The only way to collect defaulted amounts from such farmers was by launching a time- and finance-consuming lawsuit.

on cotton sown under CA; d) since planting wheat on cotton fields is a common practice, farmers saw no difference between their practice and the researchers' suggestion on amending cotton fields with wheat straw. Consequently, the status of wheat stand was not associated with the level of crop residues, which indicated the need for better arguments to convince farmers to retain residue, as the effect of residues on soil improvement could not be evaluated visually only. Therefore, alternative solutions for crop residue management (e.g. length and amount of residues left from harvesting; fuel wood substitutes) should be offered to farmers in the future (Kienzler et al. 2012), complementing the prevalent use of residue as livestock feed and fuel. Furthermore, winter wheat could be planted as part of CA (like minimum tillage into cotton or surface seeding into rice).

The AF experimentation on all three sites experienced a lower than expected survival rate of the trees. While on the site, where hybrid poplars were planted as field borders by the farmer's preference, the saplings largely died; on the other two sites, Russian olives survived and grew better than poplar and Elagmus. From the perspective of the researchers, the following reasons could be identified: a) late planting, b) using cuttings instead of saplings, c) choice of species (hybrid poplar) due to financial considerations rather than ecological (i.e. tolerance to salinity and water stress). The later two, according to the researchers, could partly be explained with the limited farmers' experience in plantation forestry on saline soils and would require further capacity development in the future and importantly before the actual planting. Furthermore it became clear that the interaction process had not been able to build the hoped for degree of ownership amongst the farmers, meaning that despite a clear division of responsibilities, roles and input provision at the beginning, the farmers largely perceived the experiments as the researchers' trials and consequently relied on the researchers to manage and look after the plantations.

Regarding the validation of EM38, SANIIRI's analysis indicated that EM38 was – with some caution – accepted as a valid tool for rapid salinity assessment over large areas. The device performs well after calibration of the different soil textures (Akramkhanov et al. 2008). As the next step, the team planned to discuss the results with the technical specialist of SANIIRI and carry out a matrix ranking of various methods of soil salinity assessment in comparison with EM38.

5.3.5 Concluding Remarks & Lessons Learnt

The project's approach to the development of agricultural innovations comprised a stepwise transition from multi- to inter- and finally transdisciplinary research. The scientists developed innovations and assessed their scientific promise through a multidisciplinary approach. Once developed, the group of scientists screened and ranked innovations using an interdisciplinary approach in a participatory manner for testing and adaptation under real-life conditions. The inclusion of the stakeholders in joint experimentation made the research process transdisciplinary. All this was accompanied by a step-wise capacity-development of scientists – and selected stakeholders after they joined the process –, which was undertaken through a series of workshops and a continuous facilitation of the processes.

The findings illustrate the degree to which a participatory transdisciplinary process to innovation development and diffusion in the Khorezm region of Uzbekistan poses a challenge while at the same time is a necessity if locally acceptable innovations are to be developed. As such, the 'Follow the Innovation' experience has generated a number of lessons both for the local acceptance of the chosen innovations as well as the innovation development approach.

As far as the innovations are concerned, it became clear that relatively simpler innovations, such as the use of the EM38 equipment as a tool to monitor and map salinity, can be easily verified under any setting within a relatively short period of time, provided the right stakeholder is identified and interested in contributing to the transdisciplinary process. The more complex innovations such as alternative land use concepts (here, afforestation and conservation agriculture) which require adjustment in various domains including legal and administrative settings (e.g. exempting the marginal cropland parcels from the state order) and cropping tradition (minimum till, choice of tree species) are nevertheless resource-intensive and need to involve multiple levels of stakeholders (farmers, agricultural extension services and authorized land-use planners). Moreover, due to long waiting periods needed for i.e. the growing of trees until harvesting marketable products and for conservation agricultural practices to impact soil fertility neither can be assessed for their socio-cultural impact within a three-year time span only. In particular, conservation agriculture is complex in both, its technical base (laser leveling, raised beds, specific planters and cultivators, etc.) and institutional requirements (suitability for specific cropping systems, state order for key crops, competing uses of crop residue as feed and fuel) that require a prolonged time period for a possible local adoption and a consequent assessment of impacts on the involved stakeholders and the environment. The same holds true for the institutional innovation in strengthening WUAs through social mobilization. Designed around a single hypothesis it might prove working in the beginning, but the long-term validity can only be assessed after the innovation has moved from joint experimentation to a "laissez faire" interaction between researchers and stakeholders during which the innovation sustains, develops further and potentially even spreads to other WUAs. Even if fully validated, the CA and EM38 innovations will continue to face the problem that the required equipment is cost-intensive and beyond the reach of the targeted stakeholders. This problem of high- versus low-external-input technologies (Röling 2009: 25ff) poses an immense challenge for wide scale adoption. In the case of the EM38 innovation furthermore a significant reorganization of work processes and reevaluation of more current, but less specific data versus outdated but more detailed data is required on the side of the local stakeholders, which, in a context where the national government is as main decision-maker strongly involved in agriculture, requires a national level decision.

With regard to the FTI process, the main challenges relate to (a) knowledge creation and dissemination in rural Uzbekistan, (b) administrative challenges, (c) scientists' versus farmers' knowledge, (d) team composition and organization, (e) contested transdisciplinary cooperation as discussed previously (Hornidge et al. 2011). We would like to add here a number of practical lessons learnt:

1) Introduce a participatory, transdisciplinary process to innovation research, such as the one illustrated here, as early as possible and earlier than done in the Khorezm project into the project activities. The FTI component in the Khorezm, Uzbekistan project was launched in order to adapt the project innovations to the local setting or identify the conditions for a more likely adoption. The relatively late introduction in terms of the project cycle (six years into the project) nevertheless meant that the innovations had already been developed to a substantial degree without direct involvement of non-scholarly stakeholders and a modifying of the innovations according to the farmers' needs was only possible to a limited degree. The space for choosing an entirely different innovation or agricultural sub-sector to be innovated by addressing aspects, in which the farmer's legal (in terms of the individual's decision-making power) and financial 'window of opportunity' (Röling 2009) to innovate might be bigger or more clearly defined, did not exist anymore. Consequently and despite all four teams incorporating to some degree the ideas and opinions of the stakeholders into the innovations[7], the attempted 'participatory innovation adaptation' developed into a 'participatory innovation validation'. Furthermore and also due to the late introduction, the overall transdisciplinary process was merely given a time span of three years. Yet, as trust forms a crucial basis for the exchange of different types of knowledge, and especially of implicit, tacit knowledge,

---

7 The WUA team adjusted its innovation focusing on 'soft' aspects only to a suggested blend of 'hard' and 'soft' attributes. The CA team accepted the fertilizer and seed rates as well as crop choices suggested by stakeholders, and the AF team incorporated farmers' suggestions on agroforestry practices and species.

while trust building takes time, 5–6 years would have been the possible minimum.
2) The nurturing of a participatory, transdisciplinary process to innovation development requires substantial financial resources and well trained local staff, able to bridge the gap between foreigners and locals, researchers and local stakeholders. Staff continuity is therefore important.
3) The relevance of the transdisciplinary process with regard to the overall project aim and therefore the roles and responsibilities of the team members should be clarified right from the start and consequently appropriate time be allocated. All four outlined FTI processes were regularly disrupted by conflicting scientific and teaching responsibilities, which drew the attention of staff away.
4) Regarding role distribution within the teams, it is important to pay specific attention to agreements on roles and leadership in the teams and, particularly, the tension between technical leadership provided by the scientist mostly involved in the innovation and FTI process leadership. In our experience, the initial decision by the teams to identify the senior scientist either responsible for the development of the innovation or coming from the same discipline as team leader did not always prove useful. Instead the created hierarchy along the lines of technical expertise rather than process knowledge at times proved demotivating for team members more involved in facilitating the stakeholder interaction processes. This led to a separation of process and technical leadership in three out of four teams.
5) Continuous project-internal as well as punctual external facilitation, building the capacities for and adding legitimacy to nurturing the participatory transdisciplinary processes proved extremely useful. When interaction with stakeholders increases, people tend to give less attention to look back critically and identify gaps. The series of FTI training and review events addressed this, encouraging self- and process criticism in team presentations and discussions (Veldhuizen et al. 2010). They furthermore proved useful to receive internal and external reviews of the team process documentation and assist in indicating points for critical reflection. To facilitate critical analysis, team members were helped to analyze their own process documentation in an attempt to prepare scholarly publications based on their experiences. The later also acted as incentive for the researchers.
6) While this reflection based on the process documentation proved useful, it also brought to light, that the whole idea and process of documenting the processes of interaction had been interpreted by each team differently, and the level, focus and detail of the documentation varied greatly. In retrospect a more detailed clarification of the need and purpose of, besides fostering, also documenting the transdisciplinary processes would have been useful.

7) Continous capacity-development in participatory methods and concepts of transdisciplinary innovation research within the interdisciplinary teams of researchers as well as the transdisciplinary teams with stakeholders proved crucial. A great challenge therefore posed the high degree of staff and expertise transitions that drastically influenced the performance of the interdisciplinary teams (WUA, SA and AF).
8) Yet, one of the key process challenges with regard to content, but influenced by the practical process challenges mentioned, was the identification of stakeholders and engaging with them in a systematic way. The SA-team undertook a systematic and in-depth analysis of stakeholders, their mandates and perceived benefits in joint experimentation, which consumed most of the first year of the FTI process. The WUA- and AF-teams also conducted substantial stakeholder analyses but then decided to focus in their actual engagement on, in the WUA-team, one selected WUA and not paying attention to others, and, in the AF-team, three farmers, but keeping an eye out for others. The CA-team decided, without an initial stakeholder analysis, to select stakeholders that had previously interacted with the project. In previous interactions nevertheless the stakeholders' ownership in the nature of interaction had not been of immediate importance, but was more concentrated on the technical nature of the innovation. This experience of past, but differently shaped interaction, in the FTI-process resulted in a lack of ownership on the side of the stakeholders. We therefore learned that significant effort and patience should be allocated for identifying stakeholders and, together with them, fostering and continuously negotiating the overall process.

References

Abdullaev I, Franz J, Oberkircher L, Hoffman I, Nizamedinkhodjaeva N, Ataev J, Tischbein B, Schorcht G, Jumaniyazova Q, Djanibekov N (2008) Work Plan Follow the Innovation Activity: Improving WUA performance through Social Mobilization and Institutional Development. Version 3

Akramkhanov A, Sommer R (2008) Comparison and sensivity of measurement techniques for spatial distribution of soil salinity. Irrigation and Drainage Systems 22: 115–126

Akramkhanov A, Tischbein B, Ibragimov H (2008) FTI- Flexible Irrigation Scheduling and Salinity Assessment-Roadmap.

Bijker W, Law J (Eds) (1997) Shaping Technology / Building Society: Studies in Sociotechnical Change. Cambridge: MIT Press

Douthwaite B (2002) Enabling Innovation: A Practical Guide to Understanding and Fostering Technological Change. London: Zed Books Limited.

Douthwaite B, de Haan NC, Manyong V, Keatinge D (2001) Blending "hard" and "soft"

Science: the "follow-the-technology" Approach to Catalyzing and Evaluating Technology Change' Ecology and Society 5(2)

Duncan A, Barnett A (2005) How Can Research-based Development Interventions be more Effective at Influencing Policy and Practice?, Workshop organized by the Making Market Systems Work Better for the Poor (M4P) programme, October 31st- November 4th, 2005. Hanoi:The Policy Practice

Egamberdiev O, Ibraghimov N, Lamers JPA (2008) Road and Maneuver Map: The introduction of Conservation Agriculture practices in the Khorezm Region

Hall A (2007) Challenges to Strengthening Agriculture Innovation Systems: Where Do We Go From Here? Maastricht, The Netherlands, United Nations University; UNU_MERIT

Hornidge A-K, Hassan Ul MM, Mollinga PP (2011) Transdisciplinary Innovation Research in Uzbekistan – 1 year of 'Following The Innovation', Development in Practice, 21: 6, pp. 825–838

Hornidge A-K, Hassan Ul MM, Mollinga PP (2011) Transdisciplinary Innovation Research in Uzbekistan – 1 year of 'Following The Innovation', Development in Practice, 21: 6, pp. 825–838

Hornidge A-K, Hassan Ul MM, Mollinga PP (2009) 'Follow the Innovation' – A Joint Experimentation & Learning Approach to Transdisciplinary Innovation Research", ZEF Working Paper Series. Vol. 39. Bonn: Zentrum für Entwicklungsforschung

Khamzina A, Lamers JPA, Vlek PLG (2012) Conversion of degraded cropland to tree plantations for ecosystem and livelihood benefits. In: Martius C, Rudenko I, Lamers JPA, Vlek PLG (Eds.) Cotton, water, salts and soums – economic and ecological restructuring in Khorezm, Uzbekistan. Springer, Dordrecht pp. 235–248

Khamzina A, Lamers JPA, Vlek PLG (2008) Tree establishment under deficit irrigation on degraded agricultural land in the lower Amu Darya River region, Aral Sea Basin. For. Ecol. Manage., 255(1): 168–178

Khamzina A, Lamers JPA, Martius C, Worbes M, Vlek PLG (2006) Potential of nine multipurpose tree species to reduce saline ground water tables in the lower Amu Darya River region of Uzbekistan. Agrofor. Syst., 68: 151–165

Khamzina A, Lamers JPA, Worbes M, Botman E, Vlek PLG (2006) Assessing the potential of trees for afforestation of degraded landscapes in the Aral Sea Basin of Uzbekistan. Agrofor. Syst., 66: 129–141

Kienzler K, Lamers JPA, McDonald A, Mirzabaev A, Ibragimov N, Egamberdiev O, Ruzibaev E, Akramkhanov A (2012) Conservation agriculture in Central Asia – What do we know and where do we go from here? Field Crops Research 132, 95–105

Lamers JPA, Khamzina A, Botmann E, Kan I, Bobojanov I (2008). Work Package 420 / FTI component Afforestation. Work Plan

Martius C, Rudenko I, Lamers, JPA, Vlek PLG (Eds.) (2012) Cotton, Water, Salts and Soums – Economic and Ecological Restructuring in Khorezm, Uzbekistan. Springer Dordrecht Heidelberg London New York, 2012, 419 pp

Mollinga P, Martius C, Lamers JPA (2006) 'Work Package 710, Implementing, Improving and Adapting with target groups: "Follow the Innovation" (FTI)', in Martius C, Lamers JPA, Khamzina A, Mollinga P, Müller M, Ruecker G, Sommer R, Tischbein B, Conrad C, Vlek PLG, Economic and Ecological Restructuring of Land and Water Use in The Region Khorezm (Uzbekistan). Project Phase II: Change-Oriented Research for Sustainable Innovation in Land and Water Use (2007–2010). Bonn: Center for Development Research

Nonaka I, Takeuchi H (1995) The Knowledge-Creating Company: How Japanese Companies Create the Dynamics of Innovation. New York: Oxford University Press

Oberkircher L, Shanafield M, Ismailova B, Saito L (2012) Ecosystem and social construction: an interdisciplinary case study of the Shurkul lake landscape in Khorezm, Uzbekistan. Ecology and Society 16 (4): 20 http://dx.doi.org/10.5751/ES-04511-160420

Oudshoorn N, Pinch T (Eds.) (2003) How Users Matter: The Co-Construction of Users and Technology. Cambridge: MIT Press

Rath A, Barnett A (2006) Innovation Systems, Concepts, Approaches and Lessons from RNRRS, RNRSS Synthesis Study Number 10.The Policy Practice

Röling N (2009) Conceptual and Methodological Developments in Innovation, in Sanginha PC, Waters-Bayer A, Kaaria S, Njuki J, Wettasinha C (Eds.) Innovation Africa. Enriching Farmers' Livelihoods. London: Earthscan

Turaeva-Höhne R, Hornidge A-K (2012) From Knowledge Ecology to Innovation Systems: Innovations in the Sphere of Agriculture in Uzbekistan, Innovation: Management, Policy & Practice, 14: 4, pp. 1819-1843

Turaeva-Höhne R (2012) Innovation policies in Uzbekistan – Path taken by ZEFa project on innovations in the sphere of agriculture ZEF Working Paper Series No. 90, Bonn: Zentrum für Entwicklungsforschung

Ul Hassan MM (2008) Research Concept and Progress so far: WP 710 "Follow the Innovation". A presentation made to the mid-term evaluation team in September, 2008 at ZEF/UNESCO Office, Urgench, Uzbekistan

Ul Hassan M and Hornidge AK (2010) "'Follow the Innovation' – The second year of a joint experimentation and learning approach to transdisciplinary research in Uzbekistan," ZEF Working Paper Series. Vol. 63. Bonn: Zentrum für Entwicklungsforschung

Ul Hassan M, Hornidge A-K, van Veldhuizen L, Akramkhanov A, Rudenko I, Djanibekov N (2011) "Follow the Innovation: Participatory Testing and Adaptation of Agricultural Innovations in Uzbekistan – Guidelines for Researchers and Practitioners", Bonn: Center for Development Research, in Collaboration with ETC Agriculture, the Netherlands

Veldhuizen L (2010) Taking Stock and Looking for Impact: Report of the ZEF Follow -The-Innovation Training Workshop IV. 2-5 November, Urgench, Uzbekistan. Leusden: ETC Ecoculture

Veldwisch GJA 2008 Cotton, Rice & Water. The Transformation of Agrarian Relations, Irrigation Technology and Water Distribution in Khorezm, Uzbekistan, Dissertation. University of Bonn. Germany

Wall C (2008) Argorods of Western Uzbekistan. Knowledge Control and Agriculture in Khorezm. Münster, Lit Verlag

Wehrheim P, Schoeller-Schletter A, Martius C (Eds.) (2008) Continuity and change: land and water use reforms in rural Uzbekistan Socio-economic and legal analyses from the region Khorezm.(IAMO) Studies on the agricultural and food sector in Central and Eastern Europe. Halle/Saale, Germany, Leibniz-Institut für Agrarentwicklung in Mittel- und Osteuropa (IAMO), pp 203

Section 6: Conclusions and Options for Action

John P. A. Lamers, Paul L. G. Vlek, Asia Khamzina,
Bernhard Tischbein, Inna Rudenko

Conclusions, recommendations and outlook

6.1 Background

Since the 1950s, the goal to rapidly increase cotton production has been the major driving force in the expansion of the irrigated area in the five socialistic republics of the Soviet Union in Central Asia: Kazakhstan (16.6 million inhabitants as of 2012), Kyrgyzstan (5.5 million), Tajikistan (7.6 million), Turkmenistan (5.1 million), and Uzbekistan (29.5 million). The cultivated area grew consequently from 1.7 million ha (Mha) in 1920 to about 8 million ha in 1990 and to about 11 million ha by 2005. Although this remained only a small fraction of the total area in Central Asia, which spans over 4 003 400 km^2, it made Central Asia one of the largest irrigated zones worldwide. Consequently, irrigated agriculture became the centerpiece of the livelihoods of the ca. 70 % rural people of the total 63 million inhabitants. Uzbekistan alone became the sixth largest raw cotton producer worldwide and the second largest exporter of this "white gold". The three-fold population increase in Central Asia in the last 5–6 decades (Shiklomanov and Rodda 2003) went hand in hand with the expansion of the irrigated croplands, which in turn tripled water consumption to about 96.3 km^3, with over 90 % of the total water consumption used for irrigated agriculture (Orlovsky et al. 2000). The unsustainable land management practices led to waterlogging and soil salinization causing widespread environmental degradation, especially of the land resources.

After the establishment of the Socialist Republic of Uzbekistan in 1924, the land-tenure systems from the historic Khanates were replaced by a land-tenure system based on political control, with the aim of cotton production and with the state as the sovereign of all water and land. With the development of the centrally organized and managed agricultural production units (*kolkhoz* and *sovkhoz*) during the Soviet era, the focus of agricultural land use became unilateral, and this has not changed much since. The research approach of the project partnership[1] aimed, therefore, to improve and test potential sustainable development options for land and water use considering technologies, institutions and policies. Furthermore, water and land use were viewed as integrated elements. For example, for coping with soil salinization, salt leaching requires prior land leveling. This means that various management actions directed towards one resource can have an impact on another and vice versa. This type of interaction is seen not only in crop cultivation practices but also in institutional measures such as the state order and its impacts on water management. Therefore, as a first step of the analyses, the characteristics of the production systems, including rural households (*dehqons*) and private farmers (*fermers*), were analyzed (**Chapter 2.1; Chapter 1.2**). This was followed by an assessment of the present management practices of land and water use. On the one hand, the reasons for the low water use efficiency were analyzed and on the other hand, experiments with innovations to improve the water use efficiency by testing alternative field level and system level irrigation practices were reported on (Table 6.1). However, in order to become effective, such technical options must be flanked by an enabling institutional and policy environment (Table 6.2). This integrated assessment furthermore not only addressed the financial feasibility of changing production technologies but also analyzed the role of society, policies and institutions in increasing the efficiency of land and water use. After all, both the farming population and society at large must have an interest in implementing innovations. When recommending options for technical, institutional and policy change, it must be clear that these interventions have to be agronomical and financially superior to traditional practices, fit the socio-economic environment of the farming systems, and be durable. This requires not a single, standardized technology, innovation or solution, but multiple options that are flexible enough to be adapted by farmers according to their means and the aims they pursue. Therefore, the perception of farmers and land managers was considered in the development of selected innovations.

6.2 Options to increase water use efficiency

Soil degradation in the study region is predominantly caused by soil salinization and water logging, triggered however by a wide range of factors such as poorly functioning drains and irrigation structures lacking drainage, a continuous salty recharge of the groundwater by percolation, seepage losses to the groundwater during conveyance and distribution due to unlined channels, excessive irriga-

---

[1] Center for Development Research (ZEF), Germany, in collaboration with the Science Sector of UNESCO and the State University of Urgench (UrDU) in Khorezm.

tion with low efficiency at field level, excessive water use during pre-season leaching, and the increasing use of saline water sources for irrigation (groundwater and drainage effluent). Decreasing soil salinity and increasing water use efficiency could be achieved through a combination of technical interventions (Table 6.1) that consist of (i) improving the operation of irrigation (from field to system level), (ii) abolishing (the most severe) bottlenecks of the system by rehabilitation, and (iii) re-designing and introducing modern irrigation techniques. Deriving such options, also for concurrent water saving, is based on analyzing the current deficits of water management and understanding the reasons for these deficits. These deficits are tackled by the above-mentioned options in a targeted way. To implement the technical options and make them work in the long run necessitate supportive economic tools and an enabling policy and institutional environment (Table 6.2).

Table 6.1: Technical options for action to improve irrigation water use efficiency at different levels of the irrigation and drainage systems

Options for action	Anticipated effect on water use	Section
Field level		
Adjusting irrigation time and amount based on flexible irrigation scheduling at field level (using tools for modeling water fluxes & balances).	Optimizing fulfillment of time-depending and site-specific crop water demand (in case of sufficient supply); avoiding of water stress (in case of insufficient supply); minimizing impact of water stress on yield (controlled deficit irrigation).	3.1
Determining amount (and timing) of leaching (using tools for modeling water fluxes and salt dynamics/ balances in combination with advanced field monitoring (EM 38)).	Site-specific leaching amount; improved leaching effectiveness.	3.1; 3.3
Optimizing application discharge (considering irrigation method, soil characteristics, field geometry, irrigation amount).	Improving application uniformity and efficiency.	3.1
Advanced handling of furrow irrigation (surge flow; alternate furrow). Intermittent rice irrigation strategies.	Improving application uniformity and efficiency. Raising water productivity of rice irrigation.	3.1, 4.1 2.3
Double-side irrigation.	Improving application uniformity and efficiency in case of zero-slope lands.	3.1
Laser-guided leveling.	Improving application uniformity, efficiency and leaching effectiveness.	3.1, 4.1

(Continued)		
Introduction of equipment for discharge dosage.	Achieving appropriate irrigation amount and application efficiency.	3.1 3.1, 4.1
Introduction of modern irrigation techniques.	Improving irrigation efficiency and water productivity by more targeted irrigation.	3.1
Decentralized storage (lakes, aquifer, artificial storage).	Improving temporal matching of supply made available by the irrigation system and demand.	
System level		
Linked irrigation scheduling-groundwater model.	Optimizing water distribution on WUA/WCA*-level. Conjunctive use of surface and groundwater. Assessing the impact of improved efficiency on groundwater recharge and drainage.	3.1, 3.2
Assessing irrigation performance.	Detecting problematic areas, reasons for problems and improving large-scale allocation.	3.1
Groundwater monitoring.	Assessing drainage performance. Options (regions) for conjunctive use. Improving input for balancing approaches.	3.2

*WUA or Water Users Associations were renamed in Water Consumers Associations (WCA) in 2009

The research findings underline that the elaboration of a functional drainage system in the study region is a first step to address overall soil degradation, and must aim to manage the groundwater levels during the irrigation season. However, this is not only a matter of field drainage as is often mentioned, but also of an ill-functioning outlet system that requires new hydraulic structures to control the water level and discharge from the drainage system (Akramkhanov et al. 2010). Hence, improvements of the drainage network should focus on the main and local outlets first before more costly interventions such as, for example, narrowing drainage ditch spacing and a more widespread introduction of tile drainage. Although such interventions may not be feasible for administrative or financial reasons, approaches must concurrently center on several components and on different levels of the irrigation system, i.e., field/farm and conveyance and distribution systems (Table 6.1). Only then is there scope for lowering the gross irrigation requirements and water withdrawals and for increasing the ability to adapt to the supply limitation and variability.

Field-level actions include (i) lowering irrigation water losses through optimized irrigation scheduling considering time-dependent and site-specific requirements; (ii) introducing precise land leveling (e.g., laser-guided) and using equipment to measure and control discharge; (iii) advancing current water application methods (e.g., double-side surge-flow); and (iv) in those cases where the impact of these measures is insufficient, the introduction of water-saving irrigation systems (e. g., drip and micro-sprinkler irrigation). But for this option, a prerequisite would be the cultivation of crops of sufficient economic value to reach cost recovery and on the most suitable soils while tackling the most severe bottlenecks in the drainage system. In the absence of such measures, capillary rise and salt accumulation could still occur, leading to an increased demand for leaching water. Those areas that cannot be productive despite these measures should probably be taken out of cultivation and be dedicated to alternative uses such as afforestation (Khamzina et al. 2012, **Chapter 3.5**).

Table 6.2: Institutional and policy-oriented options for action to support and enable the use of improved technologies for increasing land and water use efficiency.

Options for action	Anticipated effect on land and water use	Section
Institutional level.		
Flexible water pricing (taking into account irrigation months, crops, location) and progressive water tariffs (higher fees with increased water use).	Recognition of the (monetary) value of water leading to water saving and increased water use efficiency; conducive for adoption of innovations.	4.2
Stability of irrigation water supply.	Higher water use efficiency; reduction of water waste; lowering of groundwater level and in turn secondary soil salinization.	4.2
Farmer-supporting services (agricultural implements, closer links with research and other institutions).	Better informed farmers and land users; increased adoption of innovations; increased efficiencies of land and water; water saving.	4.3, 5.2, 5.3
Policy level		
Shift water use from the water-demanding agricultural sector to the less water consuming industrial sector (cotton processing industry).	Reducing water use and coping with water scarcity with lowest possible detriment to the regional economy.	2.4, 4.1
Easing cotton policy restrictions.	Increased resource use efficiencies; increased investments and adoption of innovations; modernization of management; reduced rate of natural resource degradation.	4.1, 4.2, 4.3
Land tenure security.	Increased investments and adoption of innovations; modernization of management; reduced rate of resource degradation.	4.2
Increasing farmer-markets links	Higher yields; cheaper and higher quality products (fruits and vegetables); trade and favorable balance of trade.	4.5

At **system level**, water losses during conveyance and distribution, which are highest in the low-hierachy canals, can be counterbalanced through (i) the application of water distribution and allocation models as a basis for appropriately coordinating water allocation with key stakeholders (farmers and water managers), (ii) improving the coordination between irrigation activities in the system and at the interface between field and system/network level, and (iii) rehabilitating the hydraulic structures for discharge and water level control (to implement model-based schedules for easing the present operational losses).

To reduce water losses through percolation and seepage in the system, channel lining is most effective, but also expensive and resource demanding. In case resources are in short supply, prioritizing canals and canal reaches to be lined could be an intermediate step. Since percolation and seepage are predominantly driven by (a) the water level difference between canal and the groundwater (hydraulic gradient) and (b) the material (soil texture) of the canal bottom and embankment, these parameters are consequently the major criteria for setting such priorities. The options to reduce percolation and seepage are to be formulated for the three most typical conditions explained below. Improving the maintenance is obviously an option for virtually all irrigation canals and drainage ditches (collectors), as this would in any case improve the performance (discharge in relation to water level) over time.

- Type 1: This refers to canals of high hierarchy with a canal water level lower than the groundwater throughout (or nearly throughout) the irrigation season. Since the water level is mainly driven by the hydraulic gradient, groundwater moves towards the canal and (nearly) no losses occur through percolation and seepage, hence no additional measures other than maintenance of the drains are prioritized.
- Type 2: To enable irrigation without pumping, canals of low-hierarchy often have been elevated in the study region, creating in turn a hydraulic gradient towards the groundwater. This drives percolation and seepage losses throughout all irrigation events. Therefore, elevated canals consisting of material with a high hydraulic conductivity should be prioritized during lining, because in this way the highest water-saving impact can be achieved. Furthermore, those canals with the highest discharge capacity could be rehabilitated and lined, since in such cases the highest amount of water can be saved, although this again demands high investment costs.
- Type 3 refers to canals of mid-hierarchy with a hydraulic gradient that changes over time, which in turn creates periods of losses and gains. An effective maintenance of these irrigation canals is to be prioritized, because this could keep the hydraulic capacity at a high level, which reduces percolation and seepage losses.

In locations and sites of high canal elevation (low-canal hierarchy), seepage and percolation occur all year round. This necessitates a lowering or lining of elevated channels, which is resource demanding. Furthermore, an effective maintenance of irrigation canals keeps hydraulic capacity at a high level and as a consequence reduces percolation and seepage in relation to discharge. In the canals of mid-hierarchy, the situation changes over time, and periods of infiltration are followed by periods of exfiltration, which necessitates specific measures depending on the groundwater and canal water level.

The findings show that a promotion of technical options at the field and system levels (Table 6.1) in Uzbekistan not only demands a fundamental shift in the present farming system and its management but also necessitates institutional and policy support (Table 6.2). For example, the introduction of a direct pricing for irrigation water, which is subject of continuous debate, is very likely to negatively impact the incomes of agricultural producers, but has a high potential to reduce water use and increase water use efficiencies. Several options can be considered to cushion the introduction of a water price such as (i) the gradual promotion of water pricing to allow water users to adjust to higher production costs, (ii) water pricing differentiated according to irrigation months, crops or location within the irrigation network, and (iii) investing the collected payments for maintaining the irrigation and drainage systems and supporting the adoption of new production technologies, crops and varieties. Due to the strong interrelationship of land and water use, water pricing based on a combination of soil salinity dynamics and water use could be considered. One incentive-disincentive scheme addressing soil salinity and water use efficiency simultaneously suggests that farmers/land users who use large amounts of water, which result in higher soil salinity, would be subject to a heavier tax (Akramkhanov et al. 2010). In contrast, farmers who use low amounts of water and generate good land conditions would earn a tax reduction or even a financial bonus. Hence, farmers who choose to invest in their land or increase water use efficiency by using options that reduce soil salinity would be subject to monitoring before and after these measures at determined intervals and at a given time of the year to determine their eligibility for the benefits.

Since independence, Uzbekistan has preferred pursuing capital-intensive technologies rather than a capacity-building path aiming at increasing knowledge and awareness of farmers and land users. The transfer of knowledge has in practice thus not been pursued at the same intensity as the promotion of technical innovations to boost agricultural production and, concurrently, farmers have implemented only few innovations. Whereas worldwide agricultural service organizations are used to bridge the gap between the generators and users of agricultural knowledge, this has hardly occurred in the various stages of farm restructuring in Uzbekistan. The services and institutions (e.g., phytosanitary and soil laboratories, machinery parks, etc.) compulsory to support the declared restructuring to modern agriculture underwent profound changes following independence. In particular, the previous high level of subsidies was abandoned, and farmers now have to pay the full price for services and agricultural inputs, although only a few agro-service organizations are able to meet the demands of farmers and fill their knowledge gap. The service providers are unpopular, also due to their top-down attitude and their mandates to represent the state interests in fulfilling the state targets for cotton and wheat. Overall, the findings suggest developing and introducing an overarching agricultural advisory concept rather than further pursuing the development of individual service providers. These extension services, yet to be created, should also organize intensive training and education of farmers aside from informing farmers about the availability of necessary agricultural equipment such as seeders and planters. The extension staff may also benefit from adopting a "salad bar" type of approach: Let farmers choose the topics they want to be taught on and let the new organizations and staff serve them, and not the other way around.

Farmers' land-lease rights are non-transferable usufruct rights, indicating that users cannot sell, mortgage, or exchange land, whilst *dehqons* are granted land for a lifetime, and the land rights are inheritable. It is commonly argued that low productivity is to be expected from land users who do not own the land they cultivate. The reasoning is that they lack the motive of self-interest, and in turn are reluctant to invest, for instance, in soil quality improvements or soil conservation practices, or in long-term investments, e.g., in water saving techniques. Although financial assessments indicated multiple benefits, secure land tenure did not lead to increased investments in land quality in other places of the world (e.g., Toulmin 1993; Napier 1992, 1995). This controversial aspect still remains unclear in Uzbekistan.

6.3 Options to increase land-use efficiency

Through the intensification of irrigated agriculture in the region over many decades, the top soil layers have been homogenized whereas the underlying layers have remained multi-layered (Akramkhanov et al. 2012). Irrigation thus resulted in the development of anthropogenic soils overlaying more natural formations. These practices have led to widespread soil salinity especially in the top 60 cm of the soil. All topsoils in the region suffer from soil salinization to a lesser or greater degree, which necessitates leaching (up to three times in winter and early spring) to flush the salts from the soil. If this is not done, soil salinity will impair crop performance. An effective, near-time, comprehensive and cheap assessment of soil salinity detection is hence of help to increase effectiveness by targeting the leaching in terms of timing, amount and location closer to the spatial-temporal leaching requirements. A rapid and reliable method was therefore developed as an alternative to the cumbersome procedure of soil sampling and laboratory analyses to determine soil salinity levels. This alternative method can be easily adopted by local organizations engaged in soil salinity monitoring or mapping or in soil reclamation.

The combined findings from GIS and ground truthing classified about a third of the croplands within the irrigation areas of Khorezm as degraded (**Chapter 3.6**). These areas, scattered throughout the region, are prone to further degradation and should therefore be considered for rehabilitation. This should be guided by the principle of efficiency. For instance by setting aside marginal croplands for alternative uses that are both ecologically and economically beneficial, and by using the spared resources to the benefit of the productive croplands. The initial findings of this concept reveal benefits from mixed-species tree plantations on the marginal, salt-affected croplands (Khamzina et al., 2012). These planted forests can be not only used for fuelwood, fruit, fodder and timber production but also for improving soil fertility through the input of organic matter. This would also contribute to carbon sequestration, thus reducing global warming. Particularly the N2-fixing tree species *Elaeagnus angustifolia* L. and *Robinia pseudoacacia* L. were effective in replenishing soil nutrient stocks. Furthermore, the localized addition of low amounts of phosphorus (P) fertilizer further enhanced N-fixing and growth potential (**Chapter 3.5**). Although in-depth financial analyses on the use of such P applications are still due, the first estimates illustrate that a P application is a low-cost management option to increase the land value. Research on the management of tree hedgerow systems in the landscape, previously established to combat wind erosion, showed that it was highly necessary to improve design and maintenance and harvesting techniques to foster the windbreak structure and function (**Chapter 3.4**). Otherwise there is a growing risk that the tree windbreaks will be used merely as a source of fuel to satisfy domestic energy demands.

6.4 Outlook for restructuring

Following independence, various reforms have been implemented to strengthen the productivity of irrigated agriculture in Uzbekistan. A certain level of responsibility was transferred to the land and water users, whilst concurrently a large number of small farms were established. To counterbalance the adverse effects of a large amount of small-scale production units operating within an infrastructural set-up designed during the Soviet era for a small number of large-scale production units, decentralized institutions had to be created. The WUA/

WCA at the lower hierarchical level, for instance, were mandated to fill the gaps at the level of the former state and collective farms and to increase irrigation performance and sustainability. But people and organizations were insufficiently prepared and had neither experience nor adequate funding to deal with increasing water allocation and distribution conflicts themselves, resulting in decreasing water use efficiency and low water fee collection rates from WUA members.

Although the potential of increasing irrigation performance and sustainability had previously been assessed to be high for Central Asia, hardly any visible results have been seen yet. The greater responsibility given to the water users and the introduction of participatory irrigation management approaches have not led to increased productivity. The increase in funds for operation and maintenance of the irrigation and drainage systems has also had little effect. Since the reforms showed only limited impact on the ground, in 2008 a land consolidation process was initiated (**Chapter 2.1**). However, the present farming population cannot abandon agriculture altogether and/or shift to less labor-intensive industrial or factory work, because such alternatives simply do not exist. Since it is highly likely that agriculture will remain the main livelihood option of the rural population for some time to come, more needs to be done for the livelihood security of the rural households that depend on irrigated agriculture. Various options can be offered in this regard by research findings. A limited number of these findings have already passed government approval, but many more have been prepared and can be proposed for implementation to increase resource use efficiency.

During 2001–2011, the Government of Uzbekistan (GoU) frequently underlined that the right measures for reaching sustainable resource use cannot be the same for all countries of the former Soviet Union (Spoor and Visser 2001). In contrast to other countries, the GoU keeps a strong grip on the national and regional economy to continue its chosen, gradual path of change and introducing economic reforms. This strategy has made the Uzbek economy more resilient to the recent global financial turmoil than in many other countries. The GoU has kept its conservative view on crop production, which leaves little room for changes in the agricultural production sector (**Chapters 2.1, 4.1, 4.2, 4.3**). As long as the GoU maintains its priorities, only gradual changes can be expected. The state order practice will remain, although the farming population in Uzbekistan has demonstrated its ability to increase resource use efficiency (**Chapter 4.1**) and income once exempted from the state orders for cotton-wheat production. Furthermore, the present cautious nature of reforms and their introduction continue to generate new distortions, especially among the farming population. For instance, most cropland is cultivated by farmers who can sell only a few commodities to domestic markets (**Chapter 4.5**), since they still have to answer to state bodies that impose quotas on crops such as cotton and wheat yet fail to supply basic production factors such as seeds, fertilizers or pesticides, which remain deficient (**Chapters 4.3, 5.2**). The sustainable irrigation practices remain hampered not only by national budgetary constraints, insufficient capability of the institutions responsible for the water supply and distribution and management but also by the decrease in water resources due to climate change as well as population growth upstream.

The interdisciplinary research on water and land use revealed that the present poor performance of irrigation water management is not only a technical matter (**Chapter 3.1**), but is rooted also in economic and institutional systems not interested in raising the efficiency of irrigation water management (**Chapter 2.4**). Only an integrated approach to improve the technical and socio-economic environment will improve the functionality of the irrigation and drainage networks, which were designed to suit a small number of large farms and not the anticipated large number of small farms established through the first reforms. The findings illustrate the opportunities for cutting irrigation water demand through smart delivering systems aiming at curbing application wastage, or freeing irrigation water through efficiency rather than equipment and technologies alone. A further approach provides a reliable prognosis of future supply and demand of resources (water as well as land for the production of food, feed, fuel and fiber), and ways to reach the overall goal of curbing water demand and produce more with less. Findings (**Chapter 2.4**) point also at the shift in water use from the high water-demanding agricultural sector (e.g., cotton production) to the less water consuming industrial sector (e.g., cotton processing). Such shifts have a large potential to increase overall water use efficiency, although such options mainly become effective in the mid and long term.

Any approach to increase irrigation efficiency and water productivity has to realize that current deficiencies – low irrigation efficiency, water stress and low water productivity – are not only caused by water quantities but are due also to the low temporal reliability of the water supply to the water users. This is evidenced especially in periods with a supply exceeding the demand and in periods with insufficient or missing supply during peak-demand times. As long as the temporal stability of supply is unreliable, water users demonstrate a reduced interest to invest in water-saving technologies, since these mainly tackle the problem of water quantities. Subsequently, raising the temporal reliability of supply remains a prerequisite when aiming at water savings at the field level, since this would (i) reduce the tendency of farmers to over-irrigate (at the moment driven by the insecurity of water delivery), (ii) ease the coordination of irrigation with the use of other agricultural inputs, and (iii) support the introduction of controlled deficit irrigation (by lowering the risk caused by temporal unreliability of supply) in periods of non-avoidable water stress. Especially the latter approach is likely to dominate in the future in the tail-end location of Khorezm, which is vulnerable to variability of available water supply.

Climate change and sharpening competition for water in the upstream parts of the basin will increase the variability of water availability even more, creating in turn a greater need for improved water management in downstream regions to cope with this variability, e.g., by improved forecasting of water availability (Schieder and Ximing 2008). Such forecasts can be useful for framing scenarios to deal with limited supply, using simulation models based on different water amounts entering Khorezm (or areas fed by main canals). They could provide flexible irrigation scheduling with high spatio-temporal resolution including controlled deficit irrigation diversification and adaptation of cropping patterns. Obviously, the higher temporal reliability needs a re-arrangement of water institutions, using decentralized storage options (lakes, groundwater, artificial storage) and/or providing (technical) alternatives.

In Uzbekistan, low irrigation water use efficiency, secondary soil salinization, and water logging as well as high fuel consumption (e.g., for pumping irrigation water) and high greenhouse gas emissions (Scheer et al. 2008) are typical for irrigated agriculture. With the global warming and increasing oil prices there is a need to lower fuel and general energy consumption and to reduce greenhouse gas emissions. Here innovative approaches are urgently needed. Furthermore, administrators still underestimate the benefits of critical services such as land leveling and no-till/raised bed planting as potential employment opportunities for the jobless, rural youth. These also present opportunities for the small-scale manufacturing and transport-related sectors.

Agricultural research previously played a significant role in the Soviet Union where much knowledge was generated and passed to users. Yet, nowadays, agricultural research findings are very slowly adopted by practitioners (**Chapter 5.1, 5.2**). This can be traced to poor communication but also to a misfit of technologies regarding farmers' needs. Farmers are familiar with the perils and inadequacies of Uzbekistan's public agricultural research system, and are aware that these need not only more resources, but also a change in paradigm, and closer links between farmers and researchers. At present, the academic research organizations are involved also with transfer of knowledge, since agricultural extension and advisory services do not exist although some elements of agricultural extension are provided by different agro-service organizations. Associations of farmers represent these services to some extent at the regional and district levels, and are mandated to render juridical, marketing, transport, agrotechnical and other services to farmers. However, the notions of principles of advisory work have not been introduced or understood, nor is the present extension staff capable of delivering the due services. Additionally, farmers cannot transfer their concerns, as these remain hardly heard and are only channeled further through the farmer associations. Extension and farmer education programs are used in many countries as policy instruments for improving agricultural productivity and/or protecting the environment. Such agricultural extension organizations could be tasked to voice the knowledge demands by farmers and farming communities in addition to bridging the knowledge gap of the farmers.

The present land and water use practices are prioritized to assure high cotton and wheat yields. But this approach does not deal with two critical issues: reducing the general water insecurity and improving the degraded soils. These challenges require a long-term commitment necessitating that actions are taken now, especially given the potentially high costs of inaction. A shift from the present land use and irrigation practices through a restructuring of land and water use is possible, feasible and compulsory.

6.5 References

Akramkhanov A, Ibrakhimov M, Lamers JPA (2010) Managing soil salinity in the lower reaches of the Amudarya delta: How to break the vicious circle? Case Study #8–7, In: Per Pinstrup-Andersen and Fuzhi Cheng (Eds.) Food Policy for Developing Countries: Case Studies. 13 pp

Akramkhanov A, Kuziev R, Sommer R, Martius C, Forkutsa O, Massucati L (2012) Soils and soil ecology in Khorezm. In: Martius C, Rudenko I, Lamers JPA, Vlek PLG (Eds.) Cotton, water, salts and soums – economic and ecological restructuring in Khorezm, Uzbekistan. Springer, Dordrecht, pp. 37–58

Djanibekov N, Van Assche K, Bobojonov I, Lamers JPA (2012) Farm Restructuring and Land Consolidation in Uzbekistan: New Farms with Old Barriers. Europe-Asia Studies, 64 (6): 1101–1126

Khamzina A, Lamers JPA, Vlek PLG (2012) Conversion of degraded cropland to tree plantations for ecosystem and livelihood benefits. In: Martius C, Rudenko I, Lamers JPA, Vlek PLG (Eds.) Cotton, Water, Salts and Soums – Economic and Ecological Restructuring in Khorezm, Uzbekistan. Springer, Dordrecht, pp 235–248

Napier TL (1992) Property rights and adoption of soil and water conservation practices. p. 193–202 In: Arsyad SA, Istiqlal AQ Sheng T, Moldenhauer W (Eds.) Conservation Policies for Sustainable Hillslope Farming.

Napier TL (1995) Socio-Economic factors affecting adoption of soil and water conservation practices in lesser-scale societies. In: Buerkert B, Allison B,Oppen M v (Eds.) Wind Erosion in West Africa: The problem and its Control. December 5–7, 1994, Stuttgart-Hohenheim, Germany.

Orlovsky N, Glanz M, Orlovsky L (2000) Irrigation and Land degradation in the Aral Sea Basin. Pages 115–125 in Breckle SW, Vesle M, Wuecherer W (Eds.) Sustainable Land Use in Deserts. Springer Verlag Heidelberg, Germany

Scheer C, Wassmann R, Kienzler K, Ibraghimov N, Lamers JPA, Martius C (2008) Methane and nitrous oxide fluxes in annual and perennial land-use systems of the irrigated areas in the Aral Sea Basin. Global Change Biology 14, 1–15

Schieder T, Ximing C (2008) Analysis of water use and allocation for the Khorezm region in Uzbekistan using an intergrated economic-hydrological model. In: Wehrheim P, Schoeller-Schletter A, Martius C (Eds.) Continuity and change: land and water use reforms in rural Uzbekistan. Halle/Saale, Germany, Leibniz-Institut für Agrarentwicklung in Mittel- und Osteuropa, pp. 105–127

Shiklomanov IA, Rodda JC (Eds.) (2003) World Water Resources at the Beginning of the 21st Century. Cambridge University Press, New York, 435 pp

Spoor M, Visser O (2001) The State of Agrarian Reform in the Former Soviet Union. Europe-Asia Studies (Former Soviet Studies) 53: 885–901

Toulmin C (1993) Gestion de Terroir. Principes, premiers enseignements et consequences opérationelles. Document de travail préparé pour le bureau des Nations-Unies pour la Région Soudano-Sahélienne (BNUS), Programme Zones Arides, Institute International pour l'Environnement et le Développement, Londres

List of abbreviations and acronyms

ADB	Amu Darya site Beruni
ADF	Alternate dry furrow
AGSO	Agroservice organizations
AE	Adult Equivalent
AMTP	Alternative machinery and tractor park
AquaCrop	FAO crop-model to simulate yield response to water
ASB	Aral Sea Basin
ASL	Above sea level
AV	A-value
AWU	Agricultural water use
BMBF	Federal Ministry of Education and Research (Germany)
BNF	Biological N_2 fixation
CA	Conservation agriculture
CAC	Central Asian countries
CDM	Clean Development Mechanism
CP	Crude protein
CR	Conveyance ratio
CT	Conventional tillage
CVC	Cotton value chain
DAS	Days after sowing
DF	Depleted fraction
DM	Dry matter
DPR	Delivery performance ratio
DR	Drainage ratio
DSR	Direct seeded rice
ETa	Evapotranspiration
FAR	Field application ratio
FTI	'Follow the Innovation'
FTT	'Follow the Technology'

GIS	Geographical Information System
GPS	Global Positioning System
GW	Groundwater
GWP	Global Warming Potential
HI	Harvest index
HRU	Hydrological response unit
IWRM	Integrated water resources management
IWU	Industrial water use
LU	Livestock units
MAWR	Ministry of Agriculture and Water Resources of Uzbekistan
ME	Metabolizable energy
MTP	Machinery and tractor park
Ndfa	Nitrogen derived from atmosphere
NDM	Nitrogen difference method
NE	Nitrogen enrichment
NP	N (nitrogen)-productivity
OblStat	Regional Statistical Department
Obl(Sel)Vodkhoz	Regional Department of Agriculture and Water Resources Management
Oliy Majlis	Parliament of Uzbekistan
RET	Relative evapotranspiration
RGR	Relative growth rates
RMSE	Root mean square error
SE	Standard error of the mean
SHDI	Shannon diversity index
SOC	Soil organic carbon
TVW	Total virtual water
UZS	Uzbek Soum (currency)
Uzstandart	State agency of standardization
VCA	Value chain analysis
WAD	Wet and dry irrigation
WCA	Water Consumers Association
WFA	Water footprint analysis
WUA	Water Users Association
ZEF	Center for Development Research, University of Bonn
ZT	Zero tillage

Glossary of Latin, Russian and Uzbek words

bazarkom	Administrator of a local bazaar (Uzbek)
boni and mali	"good" and "bad" (Latin)
bonitet	Soil quality indicator; an aggregate of several parameters ranging from field characteristics (morphology, etc.) to results of laboratory analyses for various soil properties (fertility, chemistry, etc.) (Russian)
dehqon	Rural (peasant) household (Uzbek)
fermer	A private farmer (Uzbek)
hokim	Local governor, indicating the chairman of the local council and head of local administration at provincial, district levels and city levels (Uzbek)
hokimiyat	State administrative organisation at provincial, district and city levels (Uzbek)
in situ	Refers to "on-site" data (Latin)
kolkhoz	Large-sized collective farm (Russian abbreviation for 'kollektivnoye khozyaistvo')
Nisholda	Traditional sweet (Uzbek)
oblast	Administrative territorial divisions akin to provinces (Russian)
plov	National dish cooked from rice (Uzbek)
rais	Head of the former collective farm, and now of the MTP (Uzbek)
Reestr	State database (Russian and Uzbek)
rayon	Administrative division, indicating a district (Russian)
shirkat	Agricultural cooperative (Uzbek)
sovkhoz	Large-sized state farm (Russian abbreviation of 'sovietskoye khozyaistvo')
Soums	Same as UZS (Uzbek currency)
tamorqa	A household plot, located in the irrigated area (Uzbek)

List of authors/co-authors and postal addresses

ANNA-KATHARINA HORNIDGE
Center for Development Research (ZEF), University of Bonn, Walter-Flex-Str. 3, 53113 Bonn, Germany, hornidge@uni-bonn.de

AHMAD M. MANSCHADI
University of Natural Resources and Life Sciences, Vienna – Department of Crop Sciences, Konrad Lorez Str. 24, 3430 Tulln, Austria, manschadi@boku.ac.at

AKMAL AKRAMKHANOV
ICARDA-CAC, Osiyo street 6, 106, Tashkent 100000, Uzbekistan, api001@yahoo.com

ALEXANDER TUPITSA
Center for Development Research (ZEF), Walter-Flex-Str. 3, 53113 Bonn, Germany, zefalex@yahoo.com

ALEXANDRA CONLIFFE
University of Oxford, c/o 25 Cole Street, apt. 1102, UK, alexandra.conliffe@ouce.ox.ac.uk

ANASTASIYA SHTALTOVNA
3750 Rue du Frere Andre app. 6, H3v1b3, Montreal, Quebec, Canada

ANIK BHADURI
Global Water System Project (GWSP), International Project Office, Walter-Flex-Strasse 3, 53113 Bonn, Germany, abhaduri@uni-bonn.de

ASIA KHAMZINA
Center for Development Research (ZEF), Walter-Flex-Str. 3, 53113 Bonn, Germany, asia.khamzina@uni-bonn.de

AZIZ KARIMOV
International Livestock Research Institute, Vietnam country office, 17 A Nguyen Khang street, Trung Hoa ward, Cau Giay district, Hanoi, Vietnam, Karimov@wider.unu.edu

BERNHARD TISCHBEIN
Center for Development Research (ZEF), Walter-Flex-Str. 3, 53113 Bonn, Germany, tischbein@uni-bonn.de

BENJAMIN SCHRAVEN
Deutsche Institut für Entwicklungspolitik (DIE), Tulpenfield 6, 53113, Bonn, Germany

CHRISTOPHER CONRAD
Remote Sensing Unit, University of Würzburg, Am Hubland, 97074 Würzburg, Germany, christopher.conrad@uni-wuerzburg.de

CHRISTOPHER MARTIUS
Center for International Forestry Research (CIFOR), Jalan CIFOR, Situ Gede, Bogor Barat 16115, Indonesia, c.martius@cgiar.org

DIANA B. SHERMETOVA
Urgench State University, 14, Khamid Olimjan Street, Khorezm, 220100 Uzbekistan

DILFUZA DJUMAEVA
Center for Development Research (ZEF), Walter-Flex-Str. 3, 53113 Bonn, Germany, djumaeva_dela@yahoo.com

DILOROM FAYZIEVA
Institute of Irrigation and Water Problems, 11, Karasy-4, Tashkent, 100187 Uzbekistan, dfayzieva@gmail.com

ELENA N. GINATULLINA
Institute of Irrigation and Water Problems, 11, Karasy-4, Tashkent, 100187 Uzbekistan, e-ginatullina@yandex.ru

ERKIN RUZIBAEV
Urgench State University and NGO KRASS, Hamid Olimjon Str. 14, 220100 Urgench, Uzbekistan

EVGENIY BOTMAN
Republican Scientific Production Center for Decorative Gardening and Forestry, Darkhan 111104, Zangiota district, Tashkent province, Uzbekistan, darhanbek@yandex.ru

FAZLULLAH AKHTAR
Center for Development Research (ZEF), Walter-Flex-Str. 3, 53113 Bonn, Germany, fakhtar@uni-bonn.de

GUNTHER SCHORCHT
green spin UG (haftungsbeschränkt), Josef-Martin-Weg 54/2 97074 Würzburg, Germany, gunther.schorcht@uni-wuerzburg.de; schorcht@greenspin.com

IHTIYOR BOBOJONOV
Leibniz Institute of Agricultural Development in Central and Eastern Europe (IAMO), Theodor-Lieser-Str.2, 06120, Halle (Saale), Germany, ihtiyorb@yahoo.com

INNA RUDENKO
Urgench State University and NGO KRASS, Hamid Olimjon Str. 14, 220100 Urgench, Uzbekistan, irudenko@mail.ru

JOHN P. A. LAMERS
Center for Development Research (ZEF), Walter-Flex-Str. 3, 53113 Bonn, Germany, j.lamers@uni-bonn.de

KRISHNA P. DEVKOTA
Crop and Environmental Science Division (CESD), DAPO Box 7777, Metro, Manila, Philippines, K.Devkota@cgiar.org

KRISTOF VAN ASSCHE
Planning, Governance & Development, Faculty of Extension, University of Alberta, Jasper Ave, T6G 2R3, Edmonton, Alberta, Canada; Wageningen University, Droevendaalsesteeg 4, 6708 PB Wageningen, The Netherlands; Center for Development Research (ZEF), University of Bonn, Walter-Flex-Str. 3, 53113, Bonn, Germany, vanassch@ualberta.ca

KUDRAT NURMETOV
Department of Economics, Faculty of Economics and Management, Slovak University of Agriculture in Nitra, Tr. A. Hlinku 2, 949 76 Nitra, Slovakia, kudrat_n@mail.ru

LAUREL SAITO
Department of Natural Resources and Environmental Science and Graduate Program of Hydrologic Sciences, Mail Stop 186, 1664 N. Virginia Street, Reno, NV 89557, USA, lsaito@cabnr.unr.edu

LAURENS VAN VELDHUIZEN
AgriCulture, ETC Foundation, PO Box 64, 3830 AB Leusden, the Netherlands, l.van.veldhuizen@etcnl.nl

LISA ATWELL
University of Nevada Reno, Department of Natural Resources and Environmental Science, Mail Stop 186, 1664 N. Virginia Street, Reno, NV 89557, USA

MAKSUD BEKCHANOV
Center for Development Research (ZEF), Walter-Flex-Str. 3, 53113 Bonn, Germany, maksud@uni-bonn.de; International Water Management Institute (IWMI), 127 Sunil Mawatha, Pelawatte, Battaramulla, Sri Lanka, m.bekchanov@cgiar.org

MARGARET SHANAFIELD
National Centre for Groundwater Research and Training, IS&T Building, via Carpark 15 Ring Road, Bedford Park SA 5042 Australia

MARTIN WORBES
Georg-August University of Goettingen, Grisebach Str. 6, 37130 Goettingen, Germany, mworbes@gwdg.de

MEHMOOD UL-HASSAN
World Agroforestry Center, Room 121, ICRAF House, UN Avenue, Gigiri 30677 00100, Nairobi, Kenya, m.hassan@cgiar.org

MIGUEL NIÑO-ZARAZÚA
United Nations University's World Institute for Development Economics Research, Katajanokanlaituri 6 B, FI-00160 Helsinki Finland, Miguel@wider.unu.edu

MINA K. DEVKOTA
CIMMYT International South Asia Regional Office, Singh Darbur Plaza Marg, Kathmandu, Nepal, M.Devkota@cgiar.org

MIRZAKHAYOT IBRAKHIMOV
Khorezm Rural Advisory Support Service (KRASS), Hamid Olimjon Str., 14, 220100 Urgench, Uzbekistan, hayot_i@yahoo.com

NODIR DJANIBEKOV
Leibniz Institute of Agricultural Development in Central and Eastern Europe (IAMO), Theodor-Lieser-Str.2, 06120, Halle (Saale), Germany, nodir79@gmail.com

OYBEK EGAMBERDIEV
Urgench State University and NGO KRASS, Hamid Olimjon Str. 14, 220100 Urgench, Uzbekistan, oybek_72@yahoo.com

PAUL L.G. VLEK
Center for Development Research (ZEF), Walter-Flex-Str. 3, 53113 Bonn, Germany, p.vlek@uni-bonn.de

PETER P. MOLLINGA
School of Oriental and African Studies, University of London, Thornhaugh Street, Russell Square, London WC1H 0XG United Kingdom, pm35@soas.ac.uk

PULATBAY KAMALOV
Urgench State University, Khamid Olimjan str., 14, Urgench 220100, Uzbekistan, kamolpulat@mail.ru

RAJ K. GUPTA
Research Station Developments, Borlaug Institute for South Asia, C/o CIMMYT India Office, NASC Complex, Todapur Road, Pusa, New Delhi 110012, India, Rajbisa2013@gmail.com

RUZUMBAY ESHCHANOV
Urgench State University, Hamid Olimjon Str. 14, 220100 Urgench, Uzbekistan, ruzimboy@mail.ru

SANJAR DAVLETOV
Urgench State University, Hamid Olimjon Str., 14, 220100 Urgench, Uzbekistan, sanjar-22@mail.ru

SEBASTIAN FRITSCH
green spin UG (haftungsbeschränkt), Josef-Martin-Weg 54/2 97074 Würzburg, Germany, fritsch@greenspin.com

SHIRIN BABAJANOVA
Urgench State University, Hamid Olimjon Str. 14, 220100 Urgench, Uzbekistan, shirinka_74@mail.ru

SUDEEP CHANDRA
University of Nevada Reno, Department of Natural Resources and Environmental Science, Mail Stop 186, 1664 N. Virginia Street, Reno, NV 89557, USA

TERESA DÜRBECK
Natural Resource Assessment project, Dürbeck Consulting, Waldfriedhofstr 58, D-81377 München, Germany, teresa@duerbeck.de

USMAN KHALID AWAN
International Water Management Institute (IWMI), 12 KM, Multan Road, Thoker Niaz Beg, Lahore 53700, Pakistan, U.K.Awan@cgiar.org

UTKUR DJANIBEKOV
Center for Development Research (ZEF), Walter-Flex-Str. 3, 53113 Bonn, Germany, utkurdjanibekov@yahoo.com, Production Economics Group, Institute for Food and Resource Economics, University of Bonn, Meckenheimer Allee 174, 53115 Bonn, Germany,u.djanibekov@ilr.uni-bonn.de

V.S. SARAVANAN
Center for Development Research (ZEF), Walter-Flex-Str. 3, 53113 Bonn, Germany, s.saravanan@uni-bonn.de

YADIRA MORI CLEMENT
Agriculture and Food Economics (AFECO) programme, University of Bonn, Nussallee 21, 53115 Bonn, Germany, yadira.mori.clement@gmail.com